T0281247

INTRODUCTION TO COMPUTATIONAL FLUID DYNAMICS

Introduction to Computational Fluid Dynamics introduces all the primary components for learning and practicing computational fluid dynamics (CFD). The book is written for final year undergraduates and/or graduate students in mechanical, chemical, and aeronautical engineering who have undergone basic courses in thermodynamics, fluid mechanics, and heat and mass transfer. Chapters cover discretisation of equations for transport of mass, momentum, and energy on Cartesian, structured curvilinear, and unstructured meshes; solution of discretised equations, numerical grid generation, and convergence enhancement. The book follows a consistent philosophy of control-volume formulation of the fundamental laws of fluid motion and energy transfer and introduces a novel notion of "smoothing pressure correction" for solution of flow equations on collocated grids within the framework of the well-known SIMPLE algorithm.

There are over 50 solved problems in the text and over 130 end-of-chapter problems. Practicing industry professionals will also find this book useful for continuing education and refresher courses.

Professor Anil W. Date obtained his bachelor's degree in mechanical engineering from Bombay University; his master's degree in thermo-fluids from UMIST Manchester, UK; and his doctorate in heat transfer from Imperial College of Science and Technology, London. He has been a member of the Thermo-Fluids-Engineering group of the Mechanical Engineering Department at IIT Bombay since 1973. Over the past thirty years, he has taught courses at both undergraduate and postgraduate level in thermodynamics, energy conversion, heat and mass transfer, and combustion. He has been engaged in research and consulting in thermo-fluids engineering and is an active reviewer of research proposals and papers for various national and international bodies and journals. He has been Editor for India of the *Journal of Enhanced Heat Transfer* and has contributed research papers to several international journals in the field. He has been a visiting scientist at Cornell University and a visiting professor at the University of Karlsruhe, Germany. He has delivered seminar lectures at universities in Australia, Hong Kong, Sweden, Germany, UK, USA, and India. Professor Date derives great satisfaction from applying thermo-fluid science to rural-technology problems in India and has taught courses in science, technology, and society and in appropriate technology at IIT Bombay. Professor Date is a Fellow of the Indian National Academy of Engineering (FNAE).

Introduction to Computational Fluid Dynamics

ANIL W. DATE

Indian Institute of Technology, Bombay

CAMBRIDGE
UNIVERSITY PRESS

CAMBRIDGE UNIVERSITY PRESS
Cambridge, New York, Melbourne, Madrid, Cape Town,
Singapore, São Paulo, Delhi, Tokyo, Mexico City

Cambridge University Press
32 Avenue of the Americas, New York, NY 10013-2473, USA

www.cambridge.org
Information on this title: www.cambridge.org/9780521140058

First published 2005
Reprinted 2008
First paperback edition 2009

A catalog record for this publication is available from the British Library

Library of Congress Cataloging in Publication data

Date, Anil Waman.
Introduction to computational fluid dynamics / Anil W. Date.
  p. cm.
Includes bibliographical references and index.
ISBN 13: 978-0-521-85326-2 (hardback)
ISBN 10: 0-521-85326-5 (hardback)
1. Fliud dynamics – Mathematics – Textbooks. 2. Numerical analysis – Textbooks. 1. Title.
TA357.D364   2005
620.1'064'015186 – dc22   2005012195

ISBN 978-0-521-85326-2 Hardback
ISBN 978-0-521-14005-8 Paperback

Dedicated to the memory of

Aai, Kaka, and Walmik

Contents

Contents

Nomenclature

Only major symbols are given in the following lists.

$AE, AW, AN,$	
AS, AP, Sp, A_k	Coefficients in Discretised Equations
B	Body Force (N/kg) or Spalding Number
C_p	Constant-Pressure Specific Heat (J/kg-K)
C_v	Constant-Volume Specific Heat (J/kg-K)
D	Mass Diffusivity (m^2/s)
e	Turbulent Kinetic Energy or Internal Energy (J/kg)
f	Fanning Friction Factor Based on Hydraulic Diameter
Gr	Grashof Number
h	Enthalpy (J/kg) or Heat Transfer Coefficient (W/m^2-K)
k	Thermal Conductivity (W/m-K)
M	Molecular Weight or Mach Number
Nu	Nusselt Number
P	Peclet Number
P_c	Cell Peclet Number
Pr	Prandtl Number
p	Pressure (N/m^2)
q	Heat Flux (W/m^2)
q''', Q'''	Internal Heat Generation Rates (W/m^3)
R	Residual or Gas Constant (J/kg-mol-K)
Re	Reynolds Number
S, Su	Source Term
Sc	Schmidt Number
St	Stanton or Stefan Number
T	Temperature (°C or K)
t	Time (s)
u, v, w	x-, y-, z-Direction Velocities (m/s)

u_i	Velocity in x_i, $i = 1, 2, 3$ Direction
V	Volume (m^3)

Greek Symbols

α	Under relaxation Factor or Thermal Diffusivity (m^2/s)
β	Under relaxation Factor for Pressure or Coefficient of Volume Expansion (K^{-1})
δ	Boundary Layer Thickness (m)
Δ	Incremental Value
ϵ	Turbulent Energy Dissipation Rate (m^2/s^3)
Ψ	Stream Function or Weighting Factor
Φ	General Variable or Dimensionless Enthalpy
Γ	General Exchange Coefficient $= \mu, \rho D,$ or k/Cp
κ	Constant in the Logarithmic Law of the Wall
μ	Dynamic viscosity ($N\text{-}s/m^2$)
ν	Kinematic Viscosity (m^2/s)
ω	Species Mass Fraction or Dimensionless Coordinate
ρ	Density (kg/m^3)
λ	Second Viscosity Coefficient or Latent Heat (J/kg)
λ_1	Multiplier of $p - \overline{p}$
σ	Normal Stress (N/m^2)
θ	Dimensionless Temperature
τ	Shear Stress (N/m^2) or Dimensionless Time

Subscripts

P, N, S, E, W	Refers to Grid Nodes
n, s, e, w	Refers to Cell Faces
eff	Refers to Effective Value
f	Refers to Cell Face
l	Liquid or Liquidus
m	Refers to Mass Conservation, Mixture, or Melting Point
s	Solid or Solidus
sm	Refers to Smoothing
sup	Superheated
T	Transferred Substance State
x_i	Refers to x_i, $i = 1, 2, 3$ directions

Superscripts

l	Iteration Counter
o	Old Time
u, v	Refers to Momentum Equations

| — | Multidimensional Average |
| ' | Correction |

Acronyms

1D	One-Dimensional
2D	Two-Dimensional
3D	Three-Dimensional
ADI	Alternating Direction Implicit
CDS	Central Difference Scheme
CFD	Computational Fluid Dynamics
CG	Conjugate Gradient Method
CONDIF	Controlled Numerical Diffusion with Internal Feedback
DNS	Direct Numerical Simulation
GMRES	Generalised Minimal Residual Method
GS	Gauss–Seidel Method
HDS	Hybrid Difference Scheme
HRE	High Reynolds Number Model
IOCV	Integration over a Control Volume Method
LHS	Left-Hand Side
LRE	Low Reynolds Number Model
LU	Lower-Upper Decomposition
ODE	Ordinary Differential Equation
PDE	Partial Differential Equation
POWER	Power-Law Scheme
RHS	Right-Hand Side
SIMPLE	Semi-Implicit Method for Pressure Linked Equations
TDMA	Tridiagonal Matrix Algorithm
TSE	Taylor Series Expansion Method
TVD	Total Variation Diminishing
UDS	Upwind Difference Scheme

Preface

During the last three decades, computational fluid dynamics (CFD) has emerged as an important element in professional engineering practice, cutting across several branches of engineering disciplines. This may be viewed as a logical outcome of the recognition in the 1950s that undergraduate curricula in engineering must increasingly be based on *engineering science*. Thus, in mechanical engineering curricula, for example, the subjects of fluid mechanics, thermodynamics, and heat transfer assumed prominence.

I began my teaching career in the early 1970s, having just completed a Ph.D. degree that involved solution of partial differential equations governing fluid motion and energy transfer in a particular situation (an activity not called CFD back then!). After a few years of teaching undergraduate courses on heat transfer and postgraduate courses on convective heat and mass transfer, I increasingly shared the feeling with the students that, although the excellent textbooks in these subjects emphasised application of fundamental laws of motion and energy, the problem-solving part required largely varied mathematical tricks that changed from one situation to another. I felt that teachers and students needed a chance to study relatively more *real* situations and an opportunity to concentrate on the physics of the subject. In my reckoning, the subject of CFD embodies precisely this scope and more.

The introduction of a five-year dual degree (B. Tech. and M. Tech.) program at IIT Bombay in 1996 provided an opportunity to bring new elements into the curriculum. I took this opportunity to introduce a course on computational fluid dynamics and heat transfer (CFDHT) in our department as a compulsory course in the fourth year for students of the thermal and fluids engineering stream. The course, with an associated CFDHT laboratory, has emphasised *relearning* fluid mechanics and heat and mass transfer through obtaining *numerical* solutions. This, of course, contrasts with the *analytical* solutions learnt in earlier years of the program. Through teaching of this CFDHT course, I discovered that this relearning required attitudinal change on the part of the student. Thus, for example, the idea that *all* 1D conduction problems (steady or unsteady, in Cartesian, cylindrical, or spherical coordinates, with constant or variable properties, with or without area change, with or without internal heat generation, and with linear or nonlinear boundary conditions) in a typical undergraduate textbook can be solved by a *single* computer program based on a *single* method is found by the students to be new. Similarly, the idea that a numerical instability in an unsteady conduction problem essentially represents violation of the second law of thermodynamics is found to be new because no book on *numerical analysis* treats it as such. Nothing encourages a teacher to write a book more than the discomfort expressed by the students. At the same time, it must be mentioned that when a student succeeds in writing a generalised computer program for 1D conduction in the laboratory part of the course through struggles of *where and how do I begin*, of debugging, of comparing numerical results with analytical results, of studying effects of parametric variations, and of plotting of results, the computational activity is found to be both enlightening and entertaining.

I specifically mention these observations because, although there are a number of books bearing the words *Computational Fluid Dynamics* in their titles, most emphasise numerical analysis (a branch of applied mathematics). Also, most books, it would appear, are written for researchers and cover a rather extended ground but are usually devoid of exercises for student learning. In my reckoning, the most notable exception to such a state of affairs is the pioneering book *Numerical Heat Transfer and Fluid Flow* written by Professor Suhas V. Patankar. The book emphasises *control-volume* discretisation (the main early step to obtaining numerical solutions) based on physical principles and strives to help the reader to *write* his or her own computer programs.

It is my pleasure and duty to acknowledge that writing of this book has been influenced by the works of two individuals: Professor D. B. Spalding (FRS, formerly at Imperial College of Science and Technology, London), who unified the fields of heat, mass, and momentum transfer, and Professor S. V. Patankar (formerly at University of Minnesota, USA), who, through his book, has made CFD so lucid and SIMPLE.[1] If the readers of this book find that I have mimicked writings of these two pioneers from which several individuals (teachers, academic researchers, and consultants) and organisations have benefited, I would welcome the compliment.

I have titled this book as *Introduction to Computational Fluid Dynamics* for two reasons. Firstly, the book is intended to serve as a textbook for a student uninitiated in CFD but who has had exposure to the three courses mentioned in the first paragraph of this preface at undergraduate and postgraduate levels. In this respect, the book will also be found useful by teachers and practicing engineers who are increasingly attracted to take refresher courses in CFD. Secondly, CFD, since its inception, has remained an ever expanding field, expanding in its fundamental scope as well as in ever new application areas. Thus, turbulent flows, which are treated in this book through *modelling*, are already being investigated through direct numerical simulation (DNS). Similarly, more appropriate constitutive relations for multiphase

---

[1] The reader will appreciate the significance of capital letters in the text.

flow or for a reacting flow are being explored through CFD. Newer application areas such as heat and mass transfer in biocells are also beginning to be explored through CFD. Such areas are likely to remain more at the research level than to be part of regular practice and, therefore, a student, over the next few years at least, may encounter them in research at a Ph.D. level. It is my belief that the approach adopted in this book will provide adequate grounding for such pursuits.

Although this is an introductory book, there are some departures and basic novelties to which it is important to draw the reader's attention. The first of these concerns the manner in which the fundamental equations of motion (the Navier–Stokes equations) are written. Whereas most textbooks derive or write these equations for a continuum fluid, it is shown in the first chapter of this book that since numerical solutions are obtained in discretised space, the equations must be written in such a way that they are applicable to both the continuum as well as the discretised space. Attention is also drawn to use of special symbols that the reader may find not in common with other books on CFD. Thus, a *mass-conserving* pressure correction is given the symbol p'_m to contrast with the two other pressure corrections, namely, the *total* pressure correction p' and the *smoothing* pressure correction p'_{sm}. Similarly, the velocities appearing at the control-volume faces are given the symbol $u_{f,i}$ to contrast with those that appear at the nodal locations, which are referred to as u_i. Again, in a continuum, the two velocity fields must coincide but, in a discretised space, distinction between them preserves clarity of the physics involved. Novelty will also be found in the discussion of physical principles behind seemingly mathematical activity governing the topics of numerical grid generation and convergence enhancement. It is not my claim that the entire material of the book can be covered in a single course on CFD. It is for this reason that 1D formulations are emphasised through dedicated chapters. These formulations convey most of the essential ingredients required in CFD practice.

The ambience of academic freedom, the variety of facilities and the friendly atmosphere on the campus of IIT Bombay has contributed in no small measure to this solo effort at book writing. I am grateful to my colleagues for their cooperation in many matters. I am particularly grateful for having had the association of a senior colleague like Professor S. P. Sukhatme (FNA, FNAE, former Director, IIT Bombay). It has been a learning experience for me to observe him carry out a variety of roles (including as writer of two well-received textbooks on heat transfer and solar energy) in our institute with meticulous care. Hopefully, some rub-off is evident in this book. I have also gained considerably from my Ph.D. and M.Tech. students who through their dissertations have helped validate the computer programs I wrote.

I would like to express my special gratitude to Mr. Peter Gordon, Senior Editor (Aeronautical, Biomedical, Chemical, and Mechanical Engineering), Cambridge University Press, New York, for his considerable advice and guidance during preparation of the manuscript for this book.

Finally, I would like to record my appreciation of my wife Suranga, son Kartikeya, and daughter Pankaja (Pinky) for bearing my absence on several weekends while writing this book.

Mumbai
June 2004   Anil W. Date

1 Introduction

1.1 CFD Activity

Computational fluid dynamics (CFD) is concerned with numerical solution of differential equations governing transport of mass, momentum, and energy in moving fluids. CFD activity emerged and gained prominence with availability of computers in the early 1960s. Today, CFD finds extensive usage in basic and applied research, in design of engineering equipment, and in calculation of environmental and geophysical phenomena. Since the early 1970s, commercial software packages (or computer codes) became available, making CFD an important component of engineering practise in industrial, defence, and environmental organizations.

For a long time, design (as it relates to sizing, economic operation, and safety) of engineering equipment such as heat exchangers, furnaces, cooling towers, internal combustion engines, gas turbine engines, hydraulic pumps and turbines, aircraft bodies, sea-going vessels, and rockets depended on painstakingly generated empirical information. The same was the case with numerous industrial processes such as casting, welding, alloying, mixing, drying, air-conditioning, spraying, environmental discharging of pollutants, and so on. The empirical information is typically displayed in the form of *correlations* or tables and nomograms among the main influencing variables. Such information is extensively availed by designers and consultants from handbooks [55].

The main difficulty with empirical information is that it is applicable only to the limited range of scales of fluid velocity, temperature, time, or length for which it is generated. Thus, to take advantage of economies of scale, for example, when engineers were called upon to design a higher capacity power plant, boiler furnaces, condensers, and turbines of ever higher dimensions had to be designed for which new empirical information had to be generated all over again. The generation of this new information was by no means an easy task. This was because the information applicable to bigger scales had to be, after all, generated via laboratory-scale models. This required establishment of *scaling laws* to ensure geometric, kinematic, and dynamic similarities between models and the full-scale equipment. This activity required considerable experience as well as ingenuity, for it is not an easy matter to simultaneously maintain the three aforementioned similarities. The activity had to, therefore, be supported by flow-visualization studies and by simple (typically, one-dimensional) analytical solutions to equations governing the phenomenon under consideration. Ultimately, experience permitted judicious compromises. Being very expensive to generate, such information is often of a proprietary kind. In more recent times, of course, scaling difficulties are encountered in the opposite direction. This is because electronic equipment is considerably miniaturised and, in materials processing, for example, the more relevant phenomena occur at microscales (even molecular or atomic scales where the continuum assumption breaks down). Similarly, small-scale processes occur in biocells.

Clearly, designers need a design tool that is *scale neutral*. The tool must be scientific and must also be economical to use. An individual designer can rarely, if at all, acquire or assimilate this scale neutrality. Fortunately, the fundamental laws of mass, momentum, and energy, in fact, do embody such scale-neutral information. The key is to solve the differential equations describing these laws and then to interpret the solutions for practical design.

The potential of fundamental laws (in association with some further empirical laws) for generating widely applicable and scale-neutral information has been known almost ever since they were invented nearly 200 years ago. The realisation of this potential (meaning the ability to solve the relevant differential equations), however, has been made possible only with the availability of computers. The past five decades have witnessed almost exponential growth in the speed with which arithmetic operations can be performed on a computer.

By way of reminder, we note that the three laws governing transport are the following:

1. the law of conservation of mass (transport of mass),
2. Newton's second law of motion (transport of momentum), and
3. the first law of thermodynamics. (transport of energy).

1.2 Transport Equations

The aforementioned laws are applied to an infinitesimally small *control volume* located in a moving fluid. This application results in partial differential equations (PDEs) of mass, momentum and energy transfer. The derivation of PDEs is given in Appendix A.[1] Here, it will suffice to mention that the law of conservation of mass is written for a single-component fluid or for a mixture of several species. When applied to a single species of the mixture, the law yields the equation of *mass transfer* when an empirical law, namely, Fick's law of mass diffusion ($m_i'' = - \rho \, D \, \partial \omega / \partial x_i$),

---

[1] The reader is strongly advised to read Appendix A to grasp the main ideas and the process of derivations.

is invoked. Newton's second law of motion, combined with Stokes's stress laws, yields three momentum equations for velocity in directions x_j $(j = 1, 2, 3)$. Similarly, the first law of thermodynamics in conjunction with Fourier's law of heat conduction $(q_{i,\text{cond}} = -K\,\partial T/\partial x_i)$ yields the so-called energy equation for the transport of temperature T or enthalpy h. Using tensor notation, we can state these laws as follows:

Conservation of Mass for the Mixture

$$\frac{\partial \rho_m}{\partial t} + \frac{\partial (\rho_m u_j)}{\partial x_j} = 0, \tag{1.1}$$

Equation of Mass Transfer for Species k

$$\frac{\partial (\rho_m \omega_k)}{\partial t} + \frac{\partial (\rho_m u_j \omega_k)}{\partial x_j} = \frac{\partial}{\partial x_j}\left[\rho_m D_{\text{eff}} \frac{\partial \omega_k}{\partial x_j}\right] + R_k, \tag{1.2}$$

Momentum Equations u_i $(i = 1, 2, 3)$

$$\frac{\partial (\rho_m u_i)}{\partial t} + \frac{\partial (\rho_m u_j u_i)}{\partial x_j} = \frac{\partial}{\partial x_j}\left[\mu_{\text{eff}} \frac{\partial u_i}{\partial x_j}\right] - \frac{\partial p}{\partial x_i} + \rho_m B_i + S_{u_i}, \tag{1.3}$$

Energy Equation – Enthalpy Form

$$\frac{\partial (\rho_m h)}{\partial t} + \frac{\partial (\rho_m u_j h)}{\partial x_j} = \frac{\partial}{\partial x_j}\left[\frac{k_{\text{eff}}}{C_{pm}} \frac{\partial h}{\partial x_j}\right] + Q''', \tag{1.4}$$

where enthalpy $h = C_{pm}(T - T_{\text{ref}})$, and

Energy Equation – Temperature Form

$$\frac{\partial (\rho_m T)}{\partial t} + \frac{\partial (\rho_m u_j T)}{\partial x_j} = \frac{\partial}{\partial x_j}\left[\frac{k_{\text{eff}}}{C_{pm}} \frac{\partial T}{\partial x_j}\right] + \frac{Q'''}{C_{pm}}. \tag{1.5}$$

In these equations, the suffix m refers to the fluid mixture. For a single-component fluid, the suffix may be dropped and the equation of mass transfer becomes irrelevant. Similarly, the suffix *eff* indicates *effective* values of mass diffusivity D, viscosity μ, and thermal conductivity k. In laminar flows, the values of these *transport* properties are taken from property tables for the fluid under consideration. In turbulent flows, however, the transport properties assume values much in excess of the values ascribed to the fluid; moreover, the effective transport properties turn out to be properties *of the flow* [39], rather than those of the fluid.

From the point of view of further discussion of numerical methods, it is indeed a happy coincidence that the set of equations [(1.1)–(1.5)] can be cast as a single equation for a general variable Φ. Thus,

$$\frac{\partial (\rho_m \Phi)}{\partial t} + \frac{\partial (\rho_m u_j \Phi)}{\partial x_j} = \frac{\partial}{\partial x_j}\left[\Gamma_{\text{eff}} \frac{\partial \Phi}{\partial x_j}\right] + S_\Phi. \tag{1.6}$$

Table 1.1: Generalised representation of transport equations.

Equation	Φ	Γ_{eff} (exch. coef.)	S_Φ (net source)
1.1	1	0	0
1.2	ω_k	$\rho_{\text{m}} D_{\text{eff}}$	R_k
1.3	u_i	μ_{eff}	$-\partial p / \partial x_i + \rho_{\text{m}} B_i + S_{u_i}$
1.4	h	k_{eff} / C_{pm}	Q'''
1.5	T	k_{eff} / C_{pm}	Q''' / C_{pm}

The meanings of Γ_{eff} and S_Φ for each Φ are listed in Table 1.1. Equation 1.6 is called the *transport equation* for property Φ. The rate of change (or time derivative) term is to be invoked only when a transient phenomenon is under consideration. The term $\rho_{\text{m}} \Phi$ denotes the amount of *extensive* property available in a unit volume. The convection (second) term accounts for transport of Φ due to bulk motion. This first-order derivative term is relatively uncomplicated but assumes considerable significance when stable and convergent numerical solutions are to be economically obtained. This matter will become clear in Chapter 3. Both the transient and the convection terms require no further modelling or empirical information.

The greatest impediment to obtaining physically accurate solutions is offered by the diffusion and the net source (S) terms because both these terms require empirical information. In laminar flows, the diffusion term represented by the second-order derivative offers no difficulty because Γ, being a fluid property, can be accurately determined (via experiments) in *isolation* of the flow under consideration. In turbulent (or transitional) flows, however, determination of Γ_{eff} requires considerable empirical support. This is labelled as *turbulence modelling*. This extremely complex phenomenon has attracted attention for over 150 years. Although turbulence models of adequate generality (at least, for specific classes of flows) have been proposed, they by no means satisfy the expectations of an equipment designer. These models determine Γ_{eff} from simple algebraic empirical laws. Sometimes, Γ_{eff} is also determined from other scalar quantities (such as turbulent kinetic energy and/or its dissipation rate) for which differential equations are constituted. Fortunately, these equations often have the form of Equation 1.6.

The term *net source* implies an algebraic sum of sources and sinks of Φ. Thus, in a chemically reacting flow (combustion, for example), a given species k may be generated via some chemical reactions and destroyed (or consumed) via some others and R_k will comprise both positive and negative contributions. Also, some chemical reactions may be exothermic, whereas others may be endothermic, making positive and negative contributions to Q'''. Similarly, the term B_i in the momentum equations may represent a buoyancy force, a centrifugal and/or Coriolis force, an electromagnetic force, etc. Sometimes, B_i may also represent resistance forces. Thus, in a mixture of gas and solid particles (as in pulverised fuel combustion), B_i will represent the drag offered by the particles on air, or, in a fluid flow through a densely filled medium (a porous body or a shell-and-tube geometry), the resistance will be a function of the porosity of the medium. Such empirical resistance laws are often determined from experiments. The S_{u_i} terms represent viscous terms arising from Stokes's stress laws that are not accounted for in the $\frac{\partial}{\partial x_j}[\mu_{\text{eff}} \frac{\partial u_i}{\partial x_j}]$ term in Equation 1.3.

1.3 Numerical Versus Analytical Solutions

Analytical solutions to our transport equations are rarely possible for the following reasons:

1. The equations are three-dimensional.
2. The equations are strongly coupled and nonlinear.
3. In practical engineering problems, the solution domains are almost always complex.

The equations, however, can be made amenable to analytical solutions when simplified through assumptions. In a typical undergraduate program, students develop extensive familiarity with such analytical solutions that can be represented in *closed form*. Thus, in a fluid mechanics course, for example, when fully developed laminar flow in a pipe is considered, a student is readily able to *integrate* the simplified (one-dimensional) momentum equation to obtain a closed-form solution for the streamwise velocity u as a function of radius r. The assumptions made are as follows: The flow is steady and laminar, it is fully developed, it is axisymmetric, and fluid properties are uniform. The solution is then interpreted to yield the scale-neutral result $f \times Re = 16$. The friction factor f is a practically useful quantity that enables calculation of pumping power required to force fluid through a pipe. Similarly, in a heat transfer course, a student learns to calculate reduction of heat transfer rate when insulation of a given thickness is applied to a pipe. In this case, the energy equation is simplified and the assumptions are as follows: Heat transfer is radial and axi symmetric, steady state prevails, and the insulation conductivity may be constant and there is no generation or dissipation of energy within the insulation.

In both these examples, the equations are one dimensional. They are, therefore, *ordinary differential equations* (ODEs), although the original transport equations were PDEs. In many situations, in spite of the assumptions, the governing equations cannot be rendered one dimensional. Thus, the equations of a steady, two-dimensional velocity boundary layer or that of one-dimensional *unsteady* heat conduction are partial differential equations. It is important to recognise, however, that there are no direct solutions to partial differential equations. To obtain solutions, the PDEs are always first converted to ODEs (usually more in number than the original PDEs) and the latter are solved by integration. Thus, in an unsteady conduction problem, the ODEs are formed by the method of separation of variables, whereas, for the two-dimensional velocity boundary layer, the ODE is formed by invoking a *similarity* variable. In such circumstances, often the solution is in the form of a series. We assume, of course, that the reader is familiar with the restrictive circumstances (often of significant practical consequence) under which such analytical solutions are constructed.

Analytical solutions obtained in the manner described here are termed *exact* solutions. They are applicable to every point of the time and/or space domain. The solutions are also called continuous solutions. All the aforementioned solutions are well covered in an undergraduate curriculum and in textbooks (see, for example, [34, 80, 88]).

Unlike analytical solutions, numerical solutions are obtained at a few *chosen points* within the domain. They are therefore called *discrete* solutions. Numerical solutions are obtained by employing numerical methods. The latter are really an intermediary between the physics embodied in the transport equations and the computers that can unravel them by generating numerical solutions. The process of arriving at numerical solutions is thus quite different from the process by which analytical solutions are developed.

Before describing the essence of numerical methods, it is important to note that these methods, in principle, can overcome all three aforementioned imped-iments to obtaining analytical solutions. In fact, the history of CFD shows that numerical methods have been evolved precisely to overcome the impediments in the order of their mention. Thus, the earliest numerical methods dealt with one-dimensional equations for which analytical solutions may or may not be possible. Methods for two-dimensional transport equations, however, had to incorporate sub-stantially new features. In spite of these new features, many methods applicable to two-dimensional coupled equations could not be extended to three-dimensional equations. Similarly, the earlier methods were derived for transport equations cast in only orthogonal co-ordinates (Cartesian, cylindrical polar, or spherical). Later, however, as computations over complex domains were attempted, the equations were cast in completely arbitrary curvilinear (ξ_1, ξ_2, ξ_3) coordinates. This led to development of an important branch of CFD, namely, *numerical grid generation*. With this development, domains of arbitrary shape could be mapped such that the coordinate lines followed the shape of the domain boundary. Today, complex domains are mapped by yet another development called *unstructured mesh gener-ation*. In this, the domain can be mapped by a completely arbitrary distribution of points. When the points are connected by straight lines, one obtains polygons (in two dimensions) and polyhedra (in three dimensions). Several methods (as well as packages) for unstructured mesh generation are now available.

1.4 Main Task

It is now appropriate to list the main steps involved in arriving at numerical solutions to the transport equation. To enhance understanding, an example of an *idealised*

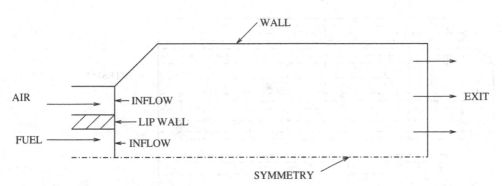

Figure 1.1. Typical two-dimensional domain.

combustion chamber of a gas-turbine engine will be considered.

1. *Given the flow situation of interest, define the physical (or space) domain of interest. In unsteady problems, the time domain is imagined.* Figure 1.1 shows the domain of interest of the idealised chember. Fuel and air streams, separated by a lip wall, enter the chamber at the *inflow* boundary. The cross section of the chamber is taken to be a perfect circle so that a *symmetry* boundary coinciding with the axis is readily identified. The enclosing *wall* is solid and the burnt products of combustion leave through the *exit* boundary. Because the situation is idealised as a two-dimensional axisymmetric domain that will involve fluid recirculation, there are four boundaries of interest: inflow, wall, symmetry, and exit.

2. *Select transport equations with appropriate diffusion and source laws. Define boundary conditions on segments of the domain boundary for each variable Φ. Also, define the fluid properties.* The boundary segments have already been identified in Figure 1.1. Now, since air and fuel mix and react chemically, equations for $\Phi = u_1, u_2, u_3$ (swirl velocity), T or h, and several mass fractions ω_k must be solved. The choice of ω_k will of course depend on the *reaction model* postulated by the analyst. Further, additional equations must be solved to capture effects of turbulence via a *turbulence model*. This matter will become clear in later chapters.

3. *Select points (called nodes) within the domain so as to map the domain with a grid. Construct control volumes around each node.* In Figure 1.2, the domain of interest is mapped by three types of grids: Cartesian, Curvilinear, and Unstructured. The hatched portions show the control volumes and the filled circles are the nodes. Note that in the Cartesian grids, the control volumes near the slanted wall are not rectangular as elsewhere. This type of difficulty is overcome in the curvilinear grids where all control volumes are quadrilaterals and the grid lines follow the contours of the domain boundary as required. The unstructured grid is completely arbitrary. Although most control volumes are triangular, one can also

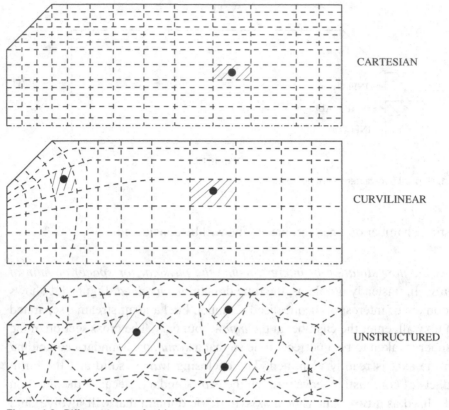

Figure 1.2. Different types of grids.

have polygons of any number of sides. This activity of specifying coordinates of nodes and of specification of control volumes is called *grid generation.*

4. *Integrate Equation 1.6 over a typical control volume so as to convert the partial differential equation into an algebraic one.* This is unlike the analytical solutions in which the original PDEs are converted to ordinary ones. Thus, if there are NV variables of interest and the number of nodes chosen is NP, one obtains a set of $NV \times NP$ algebraic equations. The process of converting PDEs into algebraic equations is called *discretisation.*

5. *Devise a numerical method to solve the set of algebraic equations.* This can be done sequentially, so that NP equations are solved for each Φ in succession. Alternatively, one may solve the entire set of $NV \times NP$ equations simultaneously. The construction of the overall calculation sequence is called an *algorithm.*

6. *Devise a computer program to implement the numerical method on a computer.* Different numerical methods require different amounts of computer storage and different amounts of computer time to arrive at a solution. Aspects such as economy in terms of number of arithmetic operations, convergence rate, and stability of the numerical method are thus important.

7. "*Interpret the solution*:" The numerical solution results in values of each Φ at each node. Such a Φ field provides the distribution of Φ over the domain. The task now is to interpret the solution to retrieve quantities of engineering interest such as the friction factor, a Nusselt number at the wall, or average concentrations of CO, fuel, and NO_x at the exit from a combustion chamber. Sometimes the field may be curve-fitted to take the appearance of an analytical solution. Similarly, the derived quantities may also be curve-fitted to take the appearance of an experimentally derived correlation for ready use in further design work.

8. "*Display of results*:" Since a numerical solution is obtained at discrete points, the solution comprises numbers that can be printed in tabular forms. The inconvenience of reading numbers can be circumvented by plotting results on a graph or by displaying the Φ fields by means of contour or vector plots. Fortunately, such graphic displays can now be made using computers. This activity is called *postprocessing* of results. The commercial success of computer codes often depends on the quality and flexibility of their *postprocessors*.

The primary focus of this book is to explain procedures for executing these steps. Computer code developers and researchers adopt a variety of practices to implement the procedures depending on their background, familiarity, and notions of convenience. Clearly biases are involved.

In this book, emphasis is laid on physical principles. In fact, the attitude is one of relearning fluid mechanics and heat and mass transfer by obtaining numerical (as opposed to restrictive analytical) solutions. The book is not intended to provide a survey of all numerical methods; rather, the objective is to introduce the reader to a few specific methods and procedures that have been found to be robust in a wide variety of situations of a specific class. The emphasis is on skill development, skills required for problem formulation, computer code writing, and interpretation of results.

1.5 A Note on Navier–Stokes Equations

The law of conservation of mass for the bulk fluid together with Newton's second law of motion constitutes the main laws governing fluid motion. As shown in Appendix A, the equations of motion are written in *differential form* and, therefore, assume existence of a fluid continuum. In this section, attention is drawn to an often overlooked requirement that assumes considerable importance in the context of CFD in which numerical solutions are obtained at *discrete points* rather than at every point in space as in a continuum.

Attention is focussed primarily on the normal stress expressions given in Appendix A (see Equations A.15). As presented in Schlichting [65], the normal stresses are given by

$$\sigma_x = -p + \sigma_x' = -p + q + \tau_{xx} = -p + q + 2\mu \frac{\partial u}{\partial x}, \tag{1.7}$$

$$\sigma_y = -p + \sigma_y' = -p + q + \tau_{yy} = -p + q + 2\mu \frac{\partial v}{\partial y}, \tag{1.8}$$

$$\sigma_z = -p + \sigma_z' = -p + q + \tau_{zz} = -p + q + 2\mu \frac{\partial w}{\partial z}. \tag{1.9}$$

In these normal stress expressions, σ' is called the deviotoric stress and the significance of quantity q in its definition requires elaboration. Schlichting [65] and Warsi [86], for example, define a space-averaged pressure $\overline{p}$ as

$$\overline{p} = -\frac{1}{3}(\sigma_x + \sigma_y + \sigma_z). \tag{1.10}$$

Now, an often overlooked *requirement* of the Stokes's relations is that, in a continuum, $\overline{p}$ must equal the point value of pressure p and the latter, in turn, must equal the thermodynamic pressure p_{th}. Thus,

$$\overline{p} = p = p_{th} = p - q - \frac{2}{3}\mu \nabla \cdot V. \tag{1.11}$$

In the context of this requirement, we now consider different flow cases to derive the significance of q.

1. *Case 1 ($V = 0$)*: In this *hydrostatic case,*

$$\overline{p} = p - q. \tag{1.12}$$

But in this case, p can only vary linearly with x, y, and z and, therefore, the point value of p exactly equals its space-averaged value $\overline{p}$ in both continuum as well as discretised space and hence $q = 0$ exactly.

2. *Case 2 ($\mu = 0$ or $\nabla \cdot V = 0$)*: Clearly when $\mu = 0$ (inviscid flow) or $\nabla \cdot V = 0$ (constant-density incompressible flow) Equation 1.12 again holds. But, in this case, since fluid motion is considered, p can vary arbitrarily with x, y, and z and, therefore, p may not equal $\overline{p}$ in a discrete space. To understand this matter, consider a case in which pressure varies arbitrarily in the x direction, whereas its variation in y and z directions is constant or linear (as in a hydrostatic case). Such a variation is shown in Figure 1.3. Now consider a point P. According to Stokes's requirement p_P must equal $\overline{p}_P$ in a continuum. However, in a discretised space, the values of pressure are available at points E and W only, and if these points are equidistant from P then $\overline{p}_P = 0.5(p_W + p_E)$. Now, this $\overline{p}_P$ *will not equal* p_P, as seen from the figure, and therefore the requirement of the Stokes's relations is not met.

However, without violating the continuum requirement, we may set

$$q = \lambda_1 (p - \overline{p}), \tag{1.13}$$

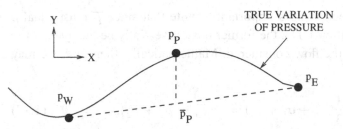

Figure 1.3. One-dimensional variation of pressure and stokes's requirement.

where λ_1 is an arbitrary constant. In most textbooks, where a continuum is assumed, λ_1 is trivially set to zero.

3. *Case 3 ($\mu \neq 0$ and $\nabla \cdot V \neq 0$)*: This case represents either compressible flow where density is a function of both temperature and pressure or incompressible flow with temperature-dependent density. Thus,

$$\overline{p} = p - \left(q + \frac{2}{3} \mu \nabla \cdot V \right). \tag{1.14}$$

In this case, Stokes's requirement will be satisfied if we set

$$q = \lambda_1 (p - \overline{p}) + \lambda \nabla \cdot V, \tag{1.15}$$

where λ is the well-known second viscosity coefficient whose value is set to $-(2/3)\mu$ even in a continuum.

It is instructive to note the reason for setting $\lambda = -(2/3)\mu$. For, if this were not done, it would amount to

$$(1 - \lambda_1)(p - \overline{p}) \nabla \cdot V = \left(\lambda + \frac{2}{3} \mu \right) (\nabla \cdot V)^2. \tag{1.16}$$

Clearly, therefore, the system will experience *dissipation* (or reversible work done at finite rate since $\nabla \cdot V$ is associated with the rate of volume change) even in an isothermal flow [65, 86]. This is, of course, highly improbable.[2]

Thus, the Stokes's relations require modifications in a continuum when compressible flow is considered, and a physical explanation for this modification can be found from thermodynamics. Now, the same interpretation can be afforded to the $\lambda_1 (p - \overline{p})$ part of q in Equation 1.13 or 1.15. This term represents a necessary modification in a discretised space. This is an important departure from the forms of normal stress expressions given in standard textbooks on fluid mechanics. It will be shown in Chapter 5 that recognition of the need to include this term is central to prediction of *smooth* pressure distributions via CFD in discrete space [17].

---

[2] Schlichting [65] shows this improbability by considering the case of an isolated sphere of a compressible isothermal gas subjected to uniform normal stress. Now if λ is not set to $-(2/3)\mu$, the gas will undergo oscillations.

Before leaving this section, it is important to note that since $\overline{p}$ must equal p in a continuum (see Equation 1.11), the former must essentially be the *hydrostatic* pressure, irrespective of the flow considered. Mathematically, therefore, we may define $\overline{p}$ as

$$\overline{p} = -\frac{1}{3}(\sigma_x + \sigma_y + \sigma_z) = \frac{1}{3}(\overline{p}_x + \overline{p}_y + \overline{p}_z), \tag{1.17}$$

where $\overline{p}_x$ is a solution to $\partial^2 p/\partial x^2 = 0$, $\overline{p}_y$ is a solution to $\partial^2 p/\partial y^2 = 0$, and $\overline{p}_z$ is a solution to $\partial^2 p/\partial z^2 = 0$.

In effect, therefore, the equations of motion (also called the Navier–Stokes equations) valid for both continuum and discrete space must read as

$$\rho\frac{Du}{Dt} = -\frac{\partial(p-q)}{\partial x} + \frac{\partial \tau_{xx}}{\partial x} + \frac{\partial \tau_{yx}}{\partial y} + \frac{\partial \tau_{zx}}{\partial z}, \tag{1.18}$$

$$\rho\frac{Dv}{Dt} = -\frac{\partial(p-q)}{\partial y} + \frac{\partial \tau_{xy}}{\partial x} + \frac{\partial \tau_{yy}}{\partial y} + \frac{\partial \tau_{zy}}{\partial z}, \tag{1.19}$$

$$\rho\frac{Dw}{Dt} = -\frac{\partial(p-q)}{\partial z} + \frac{\partial \tau_{xz}}{\partial x} + \frac{\partial \tau_{yz}}{\partial y} + \frac{\partial \tau_{zz}}{\partial z}, \tag{1.20}$$

where q is given by Equation 1.13 for incompressible (viscous or inviscid) flow and by Equation 1.15 for compressible flow. In spite of this recognition, the equations are further discussed (in conformity with standard textbooks) for a continuum only with $\lambda_1 = 0$, but the existence of finite λ_1 will be discovered in Chapter 5 where solutions in discrete space are developed.

1.6 Outline of the Book

The book is divided into nine chapters. Chapter 2 deals with one-dimensional (1D) conduction in steady and unsteady forms. In this chapter, the main ingredients of a numerical procedure are elaborately introduced so that familiarity is gained through very simple algebra. Chapter 3 deals with the 1D conduction–convection equation. This somewhat artificial equation is considered to inform the reader about the nature of difficulty introduced by convection terms. The cures for the difficulty developed in this chapter are used in all subsequent chapters dealing with solution of transport equations.

Chapter 4 deals with convective transport through *boundary layers*. This is an important class of flows encountered in fluid dynamics and heat and mass transfer. The early CFD activity relied heavily on solution of two-dimensional (2D) *parabolic* equations (a subset of the complete transport equations) appropriate to boundary layer flows. In this chapter, issues of grid adaptivity and turbulence modelling are introduced for external wall boundary layers and free-shear layers and for internal (ducted) boundary layer development.

Chapter 5 deals with solution of complete transport equations on Cartesian grids. Only 2D flow situations that may involve regions of fluid recirculation are considered. The transport equations now take the *elliptic* form. In essence, this chapter introduces all ingredients required to understand CFD practice. In this sense, the chapter provides a firm foundation for development of solution procedures employing *curvilinear* and *unstructured* grids. The latter developments are described in Chapter 6.

Chapters 7–9 deal with special topics in CFD. In Chapter 7, the reader is introduced to the topic of *phase change*. In engineering practice, heat and mass transfer are often accompanied by solid-to-liquid, liquid-to-vapour, and/or solid-to-vapour (and vice versa) transformations. This chapter, however, deals only with solidification/melting phenomena in one dimension to develop understanding of the main difficulties associated with obtaining numerical solutions. Chapter 8 deals with the topic of *numerical grid generation* and methods for curvilinear and unstructured grid generation are introduced. Finally, in Chapter 9, methods for enhancing the *rate of convergence* of iterative numerical procedures are introduced.

There are three appendices. Appendix A provides the derivation of the transport equations. In Appendix B, a computer code for solving 1D conduction problems is given. This code is based on material of Chapter 2. Appendix C provides a computer code for 2D conduction–convection problems in Cartesian coordinates. This code is based on material of Chapter 5. Familiarity with the use of these codes, it is hoped, will provide readers with sufficient exposure to enable development of their own codes for boundary layer flows (Chapter 4) , for employing curvilinear and unstructured grids (Chapter 6), for phase change (Chapter 7), and for numerical grid generation (Chapter 8).

At the end of each chapter, exercise problems are given to enhance learning. Also, in each chapter, sample problems are solved and results are presented to aid their interpretation.

EXERCISES

1. Express full forms of the S_{u_i} terms in Equation 1.3 for $i = 1, 2$, and 3. Show that if μ and ρ are constant then, for an incompressible fluid, $S_{u_i} = 0$.

2. Consider Equations 1.1–1.5. Assuming SI units, verify that units of each term in a given equation are identical.

3. Show that summing of each term in Equation 1.2 over all species of the mixture results in the mass conservation equation (1.1) for the mixture.

4. Consider the plug-flow thermo-chemical reactor (PFTCR) shown in Figure 1.4. To analyse such a reactor, the following assumptions are made: (a) All Φs vary only along the length (say, x) of the reactor. (b) Axial diffusion and conduction

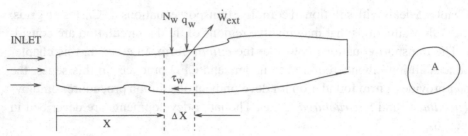

Figure 1.4. Schematic of a plug-flow reactor.

are neglected. (c) Heat (q_w W/m^2), mass (N_w kg/m^2-s), and work ($\dot{W}_{ext}$ W/m^3) through the reactor walls may be present. (d) The cross-sectional area A and perimeter P vary with x.

Following the practice adopted in Appendix A, apply the fundamental laws to a control volume $A \Delta x$. Hence, show that

$$A \frac{\partial \rho_m}{\partial t} + \frac{\partial \dot{m}}{\partial x} = N_w P \quad \text{(Bulk Mass)},$$

$$A \frac{\partial (\rho_m u)}{\partial t} + \frac{\partial (\dot{m} u)}{\partial x} = -A \frac{\partial p}{\partial x} + (N_w u - \tau_w) P \quad \text{(Momentum)},$$

$$A \frac{\partial (\rho_m \omega_k)}{\partial t} + \frac{\partial (\dot{m} \omega_k)}{\partial x} = R_k A \quad \text{(Species)},$$

$$A \frac{\partial (\rho_m h)}{\partial t} + \frac{\partial (\dot{m} h)}{\partial x} = (Q''' - \dot{W}_{ext}) A + A \frac{DP}{Dt} + N_w P \frac{u^2}{2}$$

$$+ (q_w + N_w h_w) P \quad \text{(Energy)},$$

where $\dot{m} = \rho_m A u$ and h_w is the specific enthalpy of the injected fluid.

5. Consider the well-stirred thermo-chemical reactor (WSTCR) shown in Figure 1.5. A WSTCR may be likened to a *stubby* PFTCR having *fixed volume* $V_{cv} = A \Delta x$ so that in all the PFTCR equations

$$\frac{\partial \Psi}{\partial x} = \frac{\Delta \Psi}{\Delta x} = \frac{\Psi_2 - \Psi_1}{\Delta x}.$$

Further, in a WSTCR, it is assumed that all Ψs take values of state 2 as soon as the material and energy flow into the reactor. Assuming uniform pressure ($p_1 = p_2$), show that

$$V_{cv} \frac{\partial \rho_m}{\partial t} = \dot{m}_1 - \dot{m}_2 + \dot{m}_w V_{cv} \quad \text{(Bulk Mass)},$$

$$V_{cv} \frac{\partial (\rho_m u)}{\partial t} = (\dot{m} u)_1 - (\dot{m} u)_2$$

$$+ (\dot{m}_w u - \dot{W}_{shear}) V_{cv} \quad \text{(Momentum)},$$

$$V_{cv} \frac{\partial (\rho_m \omega_k)}{\partial t} = (\dot{m} \omega_k)_1 - (\dot{m} \omega_k)_2 + (R_k + \dot{m}_{k,w}) V_{cv} \quad \text{(Species)},$$

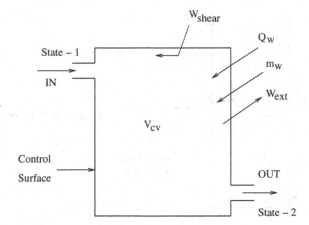

Figure 1.5. Schematic of a well-stirred reactor.

$$V_{cv} \frac{\partial(\rho_m h)}{\partial t} = (\dot{m}\, h)_1 - (\dot{m}\, h)_2 + \dot{m}_w \left(h_w + \frac{u^2}{2} \right)$$

$$+ \left(\dot{Q}_w + Q''' - \dot{W}_{ext} + \frac{\partial p}{\partial t} \right) V_{cv} \quad \text{(Energy)},$$

where $\dot{Q}_w = q_w\, P\, \Delta x / V_{cv}$ is the wall heat transfer per unit volume, $\dot{W}_{shear} = t_w\, P\, \Delta x / V_{cv}$ is the work due to wall shear, and $\dot{m}_w = \sum \dot{m}_{k,w} = N_w\, P\, \Delta x / V_{cv}$ is the mass injection through the boundary per unit volume.

6. The well-known thermodynamic *open system* having fixed volume V_{cv} is the same as the WSTCR. To derive the familiar form, consider flow of a pure-substance so that the species equation is redundant and $\rho_m = \rho$. Further, neglect viscous dissipation, radiation, and chemical heats. Also, let $m_w = 0$. Hence, show that

$$\dot{M}_{cv} = \frac{d\, M_{cv}}{dt} = \dot{m}_1 - \dot{m}_2, \tag{1.21}$$

$$\dot{E}_{cv} = \frac{d\, E_{cv}}{dt} = \dot{Q}_w - \dot{W}_{ext} + (\dot{m}\, h)_1 - (\dot{m}\, h)_2, \tag{1.22}$$

where $M_{cv} = \rho\, V_{cv}$, $E_{cv} = M_{cv}\, e$, and the symbol e stands for specific internal energy.

7. Consider a constant-volume and constant-mass (i.e., $\dot{m}_1 = \dot{m}_1 = \dot{m}_w = 0$) WSCTR with $\dot{Q}_w = \dot{W}_{ext} = 0$. Neglect heat generation due to viscous dissipation and radiation so that $Q''' = \dot{Q}_{chem} + dp/dt$. For such a reactor, show that the species and energy equations are given by

$$\rho_m \frac{d\, \omega_k}{dt} = R_k \quad \text{and} \quad \rho_m \frac{d\, e}{dt} = \dot{Q}_{chem}.$$

Typically, R_k is a function of temperature T. How will you determine T?

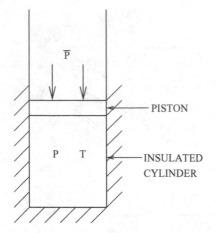

Figure 1.6. Equilibrium of an isothermal gas.

8. Consider a constant-pressure and constant-mass reactor so that volume change is permitted. Assume $Q_w = 0$. Hence, show that

$$\frac{d\, M_{cv}\omega_k}{d\, t} = R_k\, V_{cv} \quad \text{and} \quad \frac{d\, H_{cv}}{d\, t} = \dot{Q}_{chem}\, V_{cv},$$

where $V_{cv} = M_{cv}\, R_u\, T / (p\, M_{mix})$, R_u is the universal gas constant, the mixture molecular weight $M_{mix} = (\sum_k \omega_k/M_k)^{-1}$, $T = H_{cv}/(M_{cv}\, C_{p_{mix}})$, and $H_{cv} = \rho_m\, V_{cv}\, h$.

9. Consider a 2D natural convection problem in which the direction of gravity is aligned with the negative x_2 direction. Use the definition of the coefficient of cubical expansion $\beta = -\rho_{ref}^{-1}\, \partial\rho/\partial T$ and express the B_2 term in Equation 1.3 in terms of β. Now, examine whether it is possible to redefine pressure as, say, $p^* = p + \rho_{ref}\, g\, x_2$ in Equations 1.3 for $i = 1$ and 2. If so, recognise that $\rho_{ref}\, g\, x_2$ is nothing but a hydrostatic variation of pressure.

10. Consider a frictionless piston–cylinder assembly containing isothermal gas as shown in Figure 1.6. The assembly is perfectly insulated. Now, consider the unlikely circumstance in which the external pressure $\overline{p}$ is not equal to internal pressure p. Discuss the consequences if the gas temperature is to remain constant.

2 1D Heat Conduction

2.1 Introduction

A wide variety of practical and interesting phenomena are governed by the 1D heat conduction equation. Heat transfer through a composite slab, radial heat transfer through a cylinder, and heat loss from a long and thin fin are typical examples. By 1D, we mean that the temperature is a function of only one space coordinate (say x or r). This indeed is the case in steady-state problems. However, in unsteady state, the temperature is also a function of time. Thus, although there are two relevant independent variables (or dimensions), by convention, we refer to such problems as 1D *unsteady*-state problems. The extension *dimensional* thus always refers to the number of relevant space coordinates.

The 1D heat conduction equation derived in the next section is equally applicable to some of the problems arising in *convective* heat transfer, in diffusion mass transfer, and in fluid mechanics, if the dependent and independent variables of the equation are appropriately interpreted. In the last section of this chapter, therefore, problems from these neighbouring fields will be introduced. Our overall objective in this chapter is to develop a single computer program that is applicable to a wide variety of 1D problems.

2.2 1D Conduction Equation

Consider the 1D domain shown in Figure 2.1, in which the temperature varies only in the x direction although cross-sectional area A may vary with x.

The temperature over the cross section is thus assumed to be uniform. We shall now invoke the first law of thermodynamics and apply it to a typical control volume of length Δx. The law states that (Rate of energy in) − (Rate of energy out) + (Rate of generation of energy) = (Rate of change of Internal energy), or

$$Q_x - Q_{x+\Delta x} + q''' A \,\Delta x = \frac{\partial}{\partial t} \left[\rho \, A \, \Delta x \, C \, T \right] \qquad \text{W}, \qquad (2.1)$$

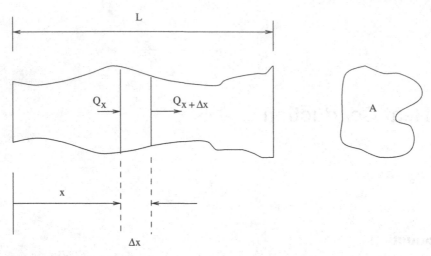

Figure 2.1. Typical 1D domain.

where q''' (W/m^3) is the *volumetric* heat generation rate, C denotes specific heat (J/kg-K), and Q (W) represents the rate at which energy is conducted. Further, it is assumed that the control volume $\Delta V = A(x) \times \Delta x$ does not change with time. Similarly, the density ρ is also assumed constant with respect to time but may vary with x. Therefore, dividing Equation 2.1 by ΔV, we get

$$\frac{Q_x - Q_{x+\Delta x}}{A\,\Delta x} + q''' = \rho\,\frac{\partial(C\,T)}{\partial t}. \tag{2.2}$$

Now, letting $\Delta x \to 0$, we obtain

$$-\frac{1}{A}\frac{\partial Q}{\partial x} + q''' = \rho\,\frac{\partial(C\,T)}{\partial t}. \tag{2.3}$$

This partial differential equation contains two dependent variables, Q and T. The equation is rendered solvable by invoking Fourier's law of heat conduction. Thus,

$$Q = -kA\,\frac{\partial T}{\partial x}, \tag{2.4}$$

where k is the thermal conductivity of the domain medium. Substituting Equation 2.4 in Equation 2.3 therefore yields

$$\frac{\partial}{\partial x}\left[kA\,\frac{\partial T}{\partial x}\right] + q'''\,A = \rho\,A\,\frac{\partial(C\,T)}{\partial t}. \tag{2.5}$$

It will be instructive to make the following comments about Equation 2.5.

1. The equation is most general. It permits variation of medium properties ρ, k, and C with respect to x and/or t.
2. The equation permits variation of cross-sectional area A with x. Thus, the equation is applicable to the case of a conical fin, for example. Similarly, the equation

PRACTISE A

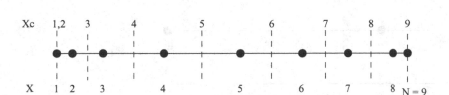

PRACTISE B

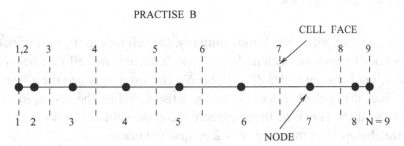

Figure 2.2. Grid layout practises.

is also applicable to the case of cylindrical radial conduction if it is recognised that $A = 2 \times \pi \times r$, and if x is replaced by r.

3. The equation also permits variation of q''' with T or x. Thus, if an electric current is passed through the medium, q''' will be a function of electrical resistance and the latter will be a function of T. Similarly, in case of a fin losing heat to the surroundings due to convection, q''' will be negative and it will be a function of the heat transfer coefficient h and perimeter P.

4. Equation 2.5 is to be solved for boundary conditions at $x = 0$ and $x = L$ (say). Thus, $0 \leq x \leq L$ specifies the domain of interest.[1]

2.3 Grid Layout

As mentioned in Chapter 1, numerical solutions are generated at a few *discrete* points in the domain. Selection of coordinates of such points (also called nodes) is called grid layout. Two practises are possible (see Figure 2.2).

Practise A

In this practise, the locations of nodes (shown by filled circles) are first chosen and then numbered from 1 to N. Note that the chosen locations need not be equispaced. Now the control volume faces (also called the cell faces) are placed *midway* between the nodes. When this is done, a difficulty arises at the near-boundary nodes 2 and $N - 1$. For these nodes, the cell face to the west of node 2

---

[1] Numerical solutions are always obtained for a domain of finite size. In many problems, the boundary condition is specified at $x = \infty$. In this case, L is assumed to be sufficiently large but finite.

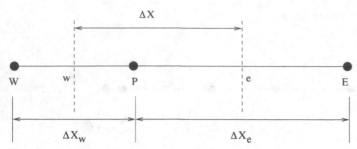

Figure 2.3. Typical node P – Practise A.

is assumed to *coincide* with node 1 and, similarly, the cell face to the east of node $N - 1$ is assumed to coincide with node N. As such, there is no cell face between nodes 1 and 2, nor between nodes $N - 1$ and N. The space between the adjacent cell faces defines the *control volume*. In this practise therefore the nodes, in general, will not be at the *centre* of their respective control volumes. Also note that if N nodes are chosen, then there are $N - 2$ control volumes.

Practise B
In this practise, the location of cell faces is first chosen and then the grid nodes are placed *at the centre* of the control volumes thus formed. Note again that the chosen locations of the cell faces need not be equispaced. Both practises have their advantages and disadvantages that become apparent only as one encounters multi-dimensional situations. Yet, a choice must be made. In this chapter, much of the discussion is carried out using practise A, but it will be shown that a generalised code can be written to accommodate either practise.

2.4   Discretisation

Having chosen the grid layout, our next step is to convert the PDE (2.5) to an algebraic one. This process of conversion is called discretisation. Here again, there are two possible approaches:

1. a Taylor series expansion (TSE) method or
2. an integration over a control volume (IOCV) method.

In both methods, a typical node P is chosen along with nodes E and W to east and west of P, respectively (see Figure 2.3). The cell face at e is midway between P and E, likewise, the cell face at w is midway between P and W.

Before describing these methods, it is important to note an important aspect of discretisation. Equation 2.5 is a partial differential equation. The time derivative on the right-hand side (RHS), therefore, must be evaluated at a fixed x. We choose this fixed location to be node P. The left-hand side (LHS) of Equation 2.5, however, contains a partial second derivative with respect to x and, therefore, this derivative must be evaluated at a fixed time. The choice of this fixed time, however, is not so straightforward because over a time step Δt, one may evaluate the LHS at time t, or $t + \Delta t$, or at an intermediate time between t and $t + \Delta t$. In general, therefore, we may write Equation 2.5 as

$$\psi\,(LHS)_P^n + (1 - \psi)(LHS)_P^o = RHS|_P \tag{2.6}$$

where ψ is a weighting factor, superscript n refers to the new time $t + \Delta t$, and superscript o refers to the old time t. If we choose $\psi = 1$ then the discretisation is called *implicit*, if $\psi = 0$ then it is called *explicit*, and if $0 < \psi < 1$, it is called *semi-implicit* or *semi-explicit*. Each choice has a bearing on economy and convenience with which a numerical solution is obtained. The choice of ψ is therefore made by the numerical analyst depending on the problem at hand. The main issues involved will become apparent following further developments.

2.4.1 TSE Method

To employ this method, Equation 2.5 is first written in a *nonconservative* form. Thus,

$$LHS|_P = kA\,\frac{\partial^2 T}{\partial x^2} + \frac{\partial(kA)}{\partial x}\,\frac{\partial T}{\partial x} + q'''\,A, \tag{2.7}$$

$$RHS|_P = \rho A\,\frac{\partial(CT)}{\partial t}. \tag{2.8}$$

Equation 2.7 contains first and second derivatives of T with respect to x. To represent these derivatives we employ a Taylor series expansion:

$$T_E = T_P + \Delta x_e\,\frac{\partial T}{\partial x}\bigg|_P + \frac{\Delta x_e^2}{2}\,\frac{\partial^2 T}{\partial x^2}\bigg|_P + \cdots, \tag{2.9}$$

$$T_W = T_P - \Delta x_w\,\frac{\partial T}{\partial x}\bigg|_P + \frac{\Delta x_w^2}{2}\,\frac{\partial^2 T}{\partial x^2}\bigg|_P + \cdots. \tag{2.10}$$

From these two expressions, it is easy to show that

$$\frac{\partial T}{\partial x}\bigg|_P = \frac{\Delta x_w^2\,T_E - \Delta x_e^2\,T_W + (\Delta x_e^2 - \Delta x_w^2)\,T_P}{\Delta x_e\,\Delta x_w\,(\Delta x_e + \Delta x_w)}, \tag{2.11}$$

$$\frac{\partial^2 T}{\partial x^2}\bigg|_P = \frac{\Delta x_w\,T_E + \Delta x_e\,T_W - (\Delta x_e + \Delta x_w)\,T_P}{\Delta x_e\,\Delta x_w\,(\Delta x_e + \Delta x_w)/2}. \tag{2.12}$$

Note that, in Equations 2.9 and 2.10, terms involving derivative orders greater than 2 are ignored. Therefore, Equations 2.11 and 2.12 are called second-order-accurate representations of first- and second-order derivatives with respect to x.

Now to evaluate the time derivative, we write

$$(C\,T)_P^n = (C\,T)_P^o + \Delta t \, \left.\frac{\partial (C\,T)}{\partial t}\right|_P + \cdots, \tag{2.13}$$

or

$$\left.\frac{\partial (C\,T)}{\partial t}\right|_P = \frac{(C\,T)_P^n - (C\,T)_P^o}{\Delta t}. \tag{2.14}$$

In Equation 2.13, derivatives of order higher than 1 are ignored; therefore, Equation 2.14 is only a first-order-accurate representation of the time derivative.[2]

Inserting Equations 2.11 and 2.12 in Equation 2.7 and Equation 2.14 in Equation 2.8 and employing Equation 2.6, we can show that

$$\left[\left.\frac{\rho\,\Delta V\,C^n}{\Delta t}\right|_P + \psi\,(AE + AW)\right] T_P^n = \psi\,\left[AE\,T_E^n + AW\,T_W^n\right] + S, \tag{2.15}$$

with

$$AE = \frac{2}{\Delta x_e}\left[(kA)_P + \frac{\Delta x_w}{2}\left.\frac{d\,(kA)}{d\,x}\right|_P\right]\frac{\Delta x}{(\Delta x_e + \Delta x_w)}, \tag{2.16}$$

$$AW = \frac{2}{\Delta x_w}\left[(kA)_P - \frac{\Delta x_e}{2}\left.\frac{d\,(kA)}{d\,x}\right|_P\right]\frac{\Delta x}{(\Delta x_e + \Delta x_w)}, \tag{2.17}$$

$$S = \left[\psi\,q_P^{\prime\prime\prime,n} + (1 - \psi)q_P^{\prime\prime\prime,o}\right]\Delta V + (1 - \psi)\left[AE\,T_E^o + AW\,T_W^o\right]$$
$$+ \left[\left.\frac{\rho\,\Delta V\,C^o}{\Delta t}\right|_P - (1 - \psi)(AE + AW)\right] T_P^o, \tag{2.18}$$

where $\Delta V = A\,\Delta x$. Note that if the cell faces were midway between adjacent nodes, $2\Delta x = \Delta x_e + \Delta x_w$. Before leaving the discussion of the TSE method, we make the following observations:

1. Calcuation of coefficients AE and AW requires evaluation of the derivative $d\,(kA)/d\,x\,|_P$. This derivative can be evaluated using expressions such as (2.11) in which T is replaced by kA.
2. For certain variations of (kA) and choices of Δx_e and Δx_w, AE and/or AW can become negative.
3. For certain choices of Δt, the multiplier of T_P^o in Equation 2.18 can become negative.
4. In steady-state problems, $\Delta t = \infty$ and T^o has no meaning. Therefore, in such problems, ψ always equals 1.

---

[2] Clearly, it is possible to represent the time derivative to a higher-order accuracy. However, the resulting expression will involve reference to T^n, T^0, T^{00}, and so on.

From the point of view of obtaining stable and convergent numerical solutions, observations 2 and 3 are significant. The associated matter will become clear in a later section.

2.4.2 IOCV Method

In this method, the RHS and LHS of Equation 2.5 are *integrated* over a control volume Δx and over a time step Δt. Thus,

$$\text{Int}(\text{LHS}) = \int_t^{t'} \int_w^e \frac{\partial}{\partial x} \left[kA \frac{\partial T}{\partial x} \right] dx \, dt + \int_t^{t'} \int_w^e q''' A \, dx \, dt, \quad (2.19)$$

where $t' = t + \Delta t$. It is now *assumed* that the integrands are constant over the time interval Δt. Further, q''' is assumed constant over the control volume and since the second-order derivative is evaluated at a fixed time, we may write

$$\text{Int}(\text{LHS}) = \left[kA \frac{\partial T}{\partial x} \bigg|_e - kA \frac{\partial T}{\partial x} \bigg|_w \right] \Delta t + q_P''' A \, \Delta x \, \Delta t. \quad (2.20)$$

It is further assumed that T varies *linearly* with x between adjacent nodes. Then

$$\frac{\partial T}{\partial x} \bigg|_e = \frac{T_E - T_P}{\Delta x_e}, \qquad \frac{\partial T}{\partial x} \bigg|_w = \frac{T_P - T_W}{\Delta x_w}. \quad (2.21)$$

Note that when the cell faces are midway between the nodes, these representations of the derivatives are second-order accurate (see Equation 2.11). Using Equation 2.21 therefore gives

$$\text{Int}(\text{LHS}) = \left[\frac{kA}{\Delta x} \bigg|_e (T_E - T_P) + \frac{kA}{\Delta x} \bigg|_w (T_W - T_P) \right] \Delta t$$

$$+ q_P''' A \, \Delta x \, \Delta t. \quad (2.22)$$

Similarly,

$$\text{Int}(\text{RHS}) = \rho A \int_t^{t'} \int_w^e \frac{\partial (CT)}{\partial t} dx \, dt$$

$$= (\rho A \Delta x)_P \left[(CT)^n - (CT)^0 \right]_P. \quad (2.23)$$

Substituting Equations 2.22 and 2.23 into the integrated version of Equation 2.6, therefore, we can show that

$$\left[\frac{\rho \, \Delta V \, C^n}{\Delta t} \bigg|_P + \psi \, (AE + AW) \right] T_P^n = \psi \left[AE \, T_E^n + AW \, T_W^n \right] + S, \quad (2.24)$$

where

$$AE = \left. \frac{kA}{\Delta x} \right|_{\mathrm{e}} \tag{2.25}$$

$$AW = \left. \frac{kA}{\Delta x} \right|_{\mathrm{w}} \tag{2.26}$$

$$\begin{aligned}
S = &\left[\psi q_{\mathrm{P}}^{\prime\prime\prime,\mathrm{n}} + (1 - \psi) q_{\mathrm{P}}^{\prime\prime\prime,\mathrm{o}} \right] \Delta V \\
&+ (1 - \psi) \left[AE\, T_{\mathrm{E}}^{\mathrm{o}} + AW\, T_{\mathrm{W}}^{\mathrm{o}} \right] \\
&+ \left[\left. \frac{\rho \Delta V\, C^{\mathrm{o}}}{\Delta t} \right|_{\mathrm{P}} - (1 - \psi)(AE + AW) \right] T_{\mathrm{P}}^{\mathrm{o}}.
\end{aligned} \tag{2.27}$$

Note that Equation 2.24 has the same form as Equation 2.15, but there are important differences:

1. Coefficients AE and AW can never be negative since $kA/\Delta x$ can only assume positive values.
2. AE and AW are also amenable to physical interpretation; they represent conductances.
3. Again, in steady-state problems, $\psi = 1$ because $\Delta t = \infty$. In unsteady problems, for certain choices of Δt, however, the multiplier of $T_{\mathrm{P}}^{\mathrm{o}}$ can still be negative. This observation is in common with the TSE method.

2.5 Stability and Convergence

Before discussing the issues of stability and convergence, we recognize that there will be one equation of the type (2.24) [or (2.15)] for each node P. To minimize writing, we designate each node by a running index $i = 1, 2, 3, \ldots, N$, where $i = 1$ and $i = N$ are boundary nodes. Thus, Equations 2.24 are written as

$$AP_i\, T_i = \psi \left[AE_i\, T_{i+1} + AW_i\, T_{i-1} \right] + S_i, \quad i = 2, 3, \ldots, N - 1, \tag{2.28}$$

where superscript n is now dropped for convenience. In these equations, AP_i represents multiplier of T_{P} in Equation 2.24.

It will be shown later that this equation set can be written in a matrix form $[A][T] = [S]$, where $[A]$ is the coefficient matrix and $[T]$ and $[S]$ are column vectors. This set can be solved by a variety of *direct* and *iterative* methods. The methods yield *converged* solutions only when the condition for convergence (also known as Scarborough's criterion [64]) is satisfied. To put it simply, the criterion states that

Condition for Convergence

$$\frac{\psi \left[|AE_i| + |AW_i| \right]}{|AP_i|} \leq 1 \qquad \text{for all nodes,} \tag{2.29}$$

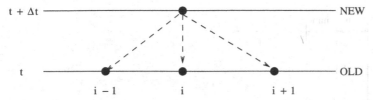

Figure 2.4. Explicit procedure.

$$\frac{\psi\left[|AE_i| + |AW_i|\right]}{|AP_i|} < 1 \qquad \text{for at least one node.} \qquad (2.30)$$

Condition for Stability

In unsteady problems, the stability of the calculation procedure, however, requires that the coefficient of T_i^o contained in the S_i term always be positive.[3] It will be shown in the next section that this implies a restriction on the permissible size of the time step.

2.5.1 Explicit Procedure $\psi = 0$

In this case, Equation 2.28 will read as

$$\left[\frac{\rho \Delta V_i C_i^n}{\Delta t}\right] T_i = AE_i T_{i+1}^o + AW_i T_{i-1}^o + q_i^{'''\,,o} A_i \Delta x$$

$$+ \left[\frac{\rho \Delta V_i C_i^o}{\Delta t} - (AE_i + AW_i)\right] T_i^o. \qquad (2.31)$$

Equation 2.31 shows that the values of T_i at a new time step are now calculable explicitly in terms of values T_{i-1}^o, T_i^o, and T_{i+1}^o. Terms containing T_{i+1} and T_{i-1} do not appear on the RHS. Therefore, the equation is explicit and no iterations are required. This situation is also depicted in Figure 2.4. Thus, starting with known initial temperature distribution at $t = 0$, one can evaluate temperatures at each new time step. Such a solution procedure is called a *marching* solution procedure. It is very easy to devise computer code for a marching procedure.

In an explicit procedure, the issue of convergence is irrelevant but the stability of the calculation procedure requires that the coefficient of T_i^o always be positive. From Equation 2.31 it is clear that this requirement is satisfied when

$$\Delta t < \left[\frac{\rho \Delta V_i C_i^o}{AE_i + AW_i}\right]_{\min}. \qquad (2.32)$$

---

[3] This condition of positiveness is strictly meant for the case of $\psi = 0$. For $\psi = 1$, the condition is automatically satisfied. For $0 < \psi < 1$, however, the condition again holds but can be violated without impairing stability of the solution procedure. This is discussed in Section 2.5.2.

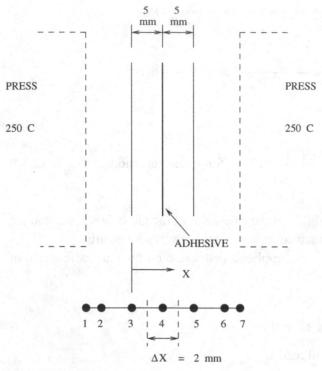

Figure 2.5. Bonding of plastic sheets – Problem 1.

Mathematically more rigorous arguments concerning this stability condition will be discussed in the next chapter. Here we consider a problem[4] to develop essential understanding.

Problem 1 [80]

Two plastic sheets, each 5 mm thick, are to be bonded together with a thin layer of adhesive that fuses at 140°C. For this purpose, they are pressed between two surfaces at 250°C (see Figure 2.5). Determine the time for which the two sheets should be pressed together, if the initial temperature of the sheets (and the adhesive) is 30°C. For plastic sheets, $k = 0.25$ W/m-K, $C = 2,000$ J/kg-K, and $\rho = 1,300$ kg/m^3.

Solution

In this problem, we measure x from the edge of one of the sheets as shown. We divide the domain of 10 mm such that $\Delta x = 2$ mm. This will yield seven grid nodes, as shown in Figure 2.5. Note that the distance between nodes 1 and 2 and that between 6 and 7 will be 1 mm. In this problem, area A is constant and may be assigned value of 1 m^2 (say). Also, since there is no internal heat generation,

---

[4] The USER file for this problem is given in Appendix B.

Table 2.1: Explicit procedure with $\Delta t = 10$ s (stable).

Time	0 mm	1 mm	3 mm	5 mm	7 mm	9 mm	10 mm
0	250	30	30	30	30	30	250
10	250	135.7	30	30	30	135.7	250
20	250	165.3	55.43	30	55.43	165.3	250
30	250	179.6	75.72	42.22	75.22	179.6	250
40	250	188.5	92.5	58.33	92.5	188.5	250
50	250	195.0	107.4	74.82	107.4	195.0	250
60	250	200.4	120.6	90.5	120.6	200.4	250
70	250	205.1	132.6	105.0	132.6	205.1	250
80	250	209.3	143.4	118.3	143.4	209.3	250
90	250	213.0	153.2	130.3	153.2	213.0	250
100	250	216.4	162.1	141.3	162.1	216.4	250

$q''' = 0$. We solve this problem by an explicit method ($\psi = 0$) and employ the IOCV method.[5]

We now note that $\rho \, A_i \, \Delta x_i \, C = 1{,}300 \times 1 \times 0.002 \times 2{,}000 = 5{,}200$, $A W_2 = 0.25 \times 1/0.001 = 250$, $A W_i = 0.25 \times 1/0.002 = 125$ for $i = 3$ to $N - 1$, $A E_{N-1} = 0.25 \times 1/0.001 = 250$, and $A E_i = 0.25 \times 1/0.002 = 125$ for $i = 2$ to $N - 2$. Therefore, the applicable discretised equations are

$$\frac{5{,}200}{\Delta t} T_2 = 250 \, T_1^\circ + 125 \, T_3^\circ + \left(\frac{5{,}200}{\Delta t} - 375 \right) T_2^\circ, \tag{2.33}$$

$$\frac{5{,}200}{\Delta t} T_i = 125 \left(T_{i-1}^\circ + T_{i+1}^\circ \right) + \left(\frac{5{,}200}{\Delta t} - 250 \right) T_i^\circ, \tag{2.34}$$

for $i = 3$, 4, and 5 and

$$\frac{5{,}200}{\Delta t} T_{N-1} = 125 \, T_{N-2}^\circ + 250 \, T_N^\circ + \left(\frac{5{,}200}{\Delta t} - 375 \right) T_{N-1}^\circ. \tag{2.35}$$

Finally, the boundary conditions are $T_1 = 250$ and $T_N = 250$. These conditions apply because it is assumed that when the sheets are pressed, the thermal contact between the sheets and the pressing surface is perfect.

This set of discretised equations dictates that Δt must be less than $5{,}200/375 = 13.87$ s (see Equation 2.32). We therefore carry out two sets of computations, one in which $\Delta t = 10$ s (see Table 2.1) and another in which $\Delta t = 20$ s, so that the stability condition is violated (see Table 2.2). In both cases, computations are stopped when T_4 ($x = 5$ mm) exceeds 140°C.

Table 2.1 clearly shows monotonic evolution of temperature within the sheets and thus accords with our expectation. The time for which the two sheets should

[5] Note that because in this problem kA is constant, the coefficients AE_i, AW_i, and AP_i will be identical in both the IOCV and TSE methods.

Table 2.2: Explicit procedure with $\Delta t = 20$ s (unstable).

Time	0 mm	1 mm	3 mm	5 mm	7 mm	9 mm	10 mm
0	250	30	30	30	30	30	250
20	250	241.5	30	30	30	241.5	250
40	250	148.0	131.7	30	131.7	148.0	250
60	250	238.3	90.63	127.8	90.63	238.3	250
80	250	178.6	179.5	92.05	179.5	178.6	250
100	250	247.7	137.0	176.1	137.0	247.7	250

be pressed together can be determined by interpolation as $(t - 90)/(100 - 90) = (140 - 130.33)/(141.31 - 130.33)$ or at $t = 98.8$ s. This calculated time, of course, need not be considered accurate. Its accuracy can be ensured by repeating calculations with increasingly smaller Δx (increasingly greater number of nodes) and by taking ever smaller values of Δt. Further, note that the temperature distributions at any time t are symmetric about $x = 5$ mm. This is because of the symmetry of the boundary and the initial condition. Now, unlike in Table 2.1, the results presented in Table 2.2 show zigzag or nonmonotonic evolution of temperature. For example, at any x, the temperature first rises (as expected) and then falls (against expectation). In fact, the reader is advised to carry the computations well beyond 100 s or with larger values of Δt. Then, it will be found that the evolved temperatures will show even more unexpected trends. That is, the interior temperatures will *exceed* the bounds of 30°C and 250°C. Clearly, this is in violation of the second law of thermodynamics. Results of Table 2.2 are, therefore, unacceptable.

2.5.2 Partially Implicit Procedure $0 < \psi < 1$

In this case, if the condition of positiveness of the coefficient of T_P^o is invoked then Δt must obey the following constraint:

$$\Delta t < \left[\frac{\rho \, \Delta V_i \, C_i^o}{(1 - \psi)(AE_i + AW_i)} \right]_{\min}. \tag{2.36}$$

However, computations of the previous problem will show that stable (monotonically evolving) solutions can be obtained even with

$$\Delta t < \left[\frac{\rho \, \Delta V_i \, C_i^o}{(1 - 2\psi)(AE_i + AW_i)} \right]_{\min} \quad \text{for} \quad \psi < 0.5, \tag{2.37}$$

and, for $\psi \geq 0.5$, Δt can be chosen *without any restriction*. Clearly, therefore, condition (2.36), though valid, is too restrictive on the time step. The reader will appreciate this matter by solving Exercise 29. The more rigorous proof

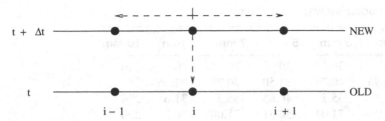

Figure 2.6. Implicit procedure.

can be developed by carrying out the *stability* analysis described in the next chapter.

2.5.3 Implicit Procedure $\psi = 1$

In this case, Equation 2.28 will read as

$$\left[\frac{\rho \, \Delta V_i \, C_i^n}{\Delta t} + AE_i + AW_i \right] T_i = AE_i \, T_{i+1} + AW_i \, T_{i-1} + q_i''' \, \Delta V_i$$

$$+ \frac{\rho \, \Delta V_i \, C_i^o}{\Delta t} \, T_i^o. \qquad (2.38)$$

This equation is implicit because the RHS also contains reference to temperatures at the new time step. Further, note that the multiplier of T_i^o is always positive and, therefore, Equation 2.38 is unconditionally stable irrespective of the time step. The situation of Equation 2.38 is shown in Figure 2.6. Because of its implicitness, Equation 2.38 must be solved iteratively, so that we may write

$$AP_i \, T_i^{l+1} = AE_i \, T_{i+1}^{l+1} + AW_i \, T_{i-1}^{l+1} + S_i, \quad i = 2, 3, \ldots, N-1, \quad (2.39)$$

where l is the iteration number.

Now, in the IOCV method, the condition of convergence (2.29) is always satisfied because the AP coefficient is the largest (see Equation 2.38) and condition (2.30) is satisfied at the boundary node. Also, AE and AW are always positive (see Equations 2.25 and 2.26).

The overall procedure can thus be described through the following steps:

1. Specify T_i^o for $i = 1$ to N and set $T_i = T_i^o$.
2. Begin a new time step. Choose Δt.
3. Solve Equation 2.39 to obtain T_i^{l+1}.
4. Check convergence by calculating the fractional change $FC_i = (T_i^{l+1} - T_i^l)/T_i^l$.
5. If $FC_{i,\max} >$ convergence criterion (CC) go to step 3 by setting $T_i^l = T_i^{l+1}$; else, go to step 6.
6. Set $T_i^o = T_i$ and go to step 2.

Table 2.3: Implicit procedure with $\Delta t = 10$ s.

Time	0 mm	1 mm	3 mm	5 mm	7 mm	9 mm	10 mm
0	250	30	30	30	30	30	250
10	250	92.96	40.79	33.50	40.79	92.96	250
20	250	131.6	55.5	40.65	55.5	131.6	250
30	250	156.2	71.04	50.51	71.04	156.2	250
40	250	172.6	86.07	62.06	86.07	172.6	250
50	250	184.1	100.1	74.41	100.1	184.1	250
60	250	192.5	112.9	86.92	112.9	192.5	250
70	250	199.1	124.7	99.19	124.7	199.1	250
80	250	204.4	135.4	110.9	135.4	204.4	250
90	250	208.9	145.2	122.1	145.2	208.9	250
100	250	213.3	154.2	132.0	154.2	213.3	250
110	250	216.1	162.2	142.1	162.2	216.1	250

The specification of procedural steps is called an *algorithm*. To illustrate the algorithm, we again consider Problem 1. Then, using the IOCV method, the equations to be solved are

$$\left(\frac{5{,}200}{\Delta t} + 375\right) T_2 = 250\, T_1 + 125\, T_3 + \frac{5{,}200}{\Delta t}\, T_2^o, \qquad (2.40)$$

$$\left(\frac{5{,}200}{\Delta t} + 250\right) T_i = 125\,(T_{i-1} + T_{i+1}) + \frac{5{,}200}{\Delta t}\, T_i^o \quad i = 3, \ldots, N-2,$$
$$(2.41)$$

$$\left(\frac{5{,}200}{\Delta t} + 375\right) T_{N-1} = 125\, T_{N-2} + 250\, T_N + \frac{5{,}200}{\Delta t}\, T_{N-1}^o. \qquad (2.42)$$

It is now possible to cast our algorithm in the form of a computer program. This matter is taken up in a later section. Here, results of computations with $\Delta t = 10$ and 20 s are presented in Tables 2.3 and 2.4, respectively.

Table 2.4: Implicit procedure with $\Delta t = 20$ s.

Time	0 mm	1 mm	3 mm	5 mm	7 mm	9 mm	10 mm
0	250	30	30	30	30	30	250
20	250	121.6	55.52	42.51	55.52	121.6	250
40	250	164.7	84.10	62.90	84.10	164.7	250
60	250	184.5	109.9	84.94	109.9	184.5	250
80	250	201.2	131.9	108.5	131.9	201.2	250
100	250	210.4	150.4	129.1	150.4	210.4	250
120	250	217.2	166.0	147.1	166.0	217.2	250

From the computed results, we make the following observations:

1. The temperature evolutions are monotonic irrespective of the time step since there is no restriction on the time step in the implicit procedure.
2. With $\Delta t = 10$ s, the time for pressing is evaluated at 107.81 s and with $\Delta t = 20$ s at 112.09 s. Again these times are not necessarily accurate. Accuracy can only be established by repeating computations with ever smaller values of Δt and Δx till the evaluated total time is independent of the choices made.
3. Comparison of results in Table 2.3 with those in Table 2.1 shows that temperature evolutions calculated by the implicit procedure are more realistic. Note, for example, that T_4 in the explicit procedure does not even recognise that heating has started for the first 20 s. Of course, this lacuna can be nearly eliminated by taking smaller time steps.
4. For the same time step, the explicit procedure reaches $T_4 = 140$ in 10 time steps. The implicit procedure has, however, required 11 time steps. In addition, at each time step, a few iterative calculations have been carried out. Thus, in this example, the implicit procedure involves more arithmetic operations than the explicit procedure. This, however, is not a general observation. When Δx and Δt are reduced to obtain accurate solutions, or when coefficients AE and AW are not constant but functions of temperature (through temperature-dependent conductivity, for example), or when $q''' = q'''(T)$ is present, one may find that an implicit procedure may yield more economic solutions than the explicit procedure because the former enjoys freedom over the size of the time step.

2.6 Making Choices

In the previous two sections, we have introduced TSE and IOCV methods as well as explicit and implicit procedures. Here, we offer advice on the best choice of combination, keeping in mind the requirements of multidimensional problems (including convection) to be discussed in later chapters. Further, we also keep in mind that coefficients AE and AW are in general not constant. This makes the discretised equations nonlinear.

1. Note that the TSE method casts the governing equations in non-conservative form whereas the IOCV method uses the as-derived conservative form. As we shall observe later, this matter is of considerable physical significance when convective problems are considered.
2. In the TSE method, coefficients AE and AW carry little physical meaning. In the IOCV method, they represent conductances.
3. In the TSE method, Scarborough's criterion may be violated. In the IOCV method, this can never happen.

4. The question of invoking explicit procedure arises only when unsteady-state problems are considered. The implicit procedure, in contrast, can be invoked for both unsteady-state as well as steady-state problems. In fact, in steady-state problems ($\Delta t = \infty$) the implicit procedure is the only one possible.[6]

5. The explicit procedure imposes restriction on the largest time step to obtain stable solutions. The implicit procedure does not suffer from such a restriction.

In view of these comments, the best choice is to employ the IOCV method with an implicit procedure. Throughout this book, therefore, this combination will be preferred.

2.7 Dealing with Nonlinearities

Now that we have accepted a combination of IOCV with the implicit procedure, we restate the main governing discretised equation (equations 2.38 and 2.39) but in a slightly altered form:

$$(AP_i + Sp_i) T_i^{l+1} = AE_i \, T_{i+1}^{l+1} + AW_i \, T_{i-1}^{l+1} + Su_i, \quad i = 2, 3, \ldots, N - 1,$$

$$\tag{2.43}$$

$$AP_i = AE_i + AW_i, \tag{2.44}$$

$$AE_i = \left. \frac{kA}{\Delta x} \right|_{i+1/2}, \tag{2.45}$$

$$AW_i = \left. \frac{kA}{\Delta x} \right|_{i-1/2}, \tag{2.46}$$

$$Su_i = \frac{\rho \, \Delta V_i \, C_i^{\circ}}{\Delta t} \, T_i^{\circ}, \quad Sp_i = \frac{\rho \, \Delta V_i \, C_i^{n}}{\Delta t}. \tag{2.47}$$

In these equations, the q''' term is deliberately ignored because it is a *problem-dependent* term. The altered form shown in Equation 2.43 will be useful in dealing with nonlinearities. Also, a generalised computer code can be constructed around Equation 2.43 in such a way that preserves the underlying physics. The nonlinearities can emanate from three sources:

1. if q''' is a function of T
2. if conductivity k is a function of T or changes abruptly, as in a composite material and/or
3. boundary conditions at $x = 0$ and $x = L$.

---

[6] Some analysts employ an explicit procedure even for a steady-state problem. In this case, calculations proceed by introducing a *false* or imaginary time step. Hence, such procedures are called false transient procedures.

In the following, we discuss methods for dealing with nonlinearities through modification of Su_i and Sp_i.

2.7.1 Nonlinear Sources

Consider a pin fin losing heat to its surroundings under *steady state* by convection with heat transfer coefficient h. Then, q''' will be given by

$$q_i''' = -\frac{h_i\, P_i\, \Delta x_i\, (T_i - T_\infty)}{A_i\, \Delta x_i},\qquad (2.48)$$

where P_i is the local fin perimeter. Therefore,

$$q_i'''\, \Delta V_i = -h_i\, P_i\, \Delta x_i\, (T_i - T_\infty).\qquad (2.49)$$

When this equation is included in Equation 2.43, it is obvious that T_i will now appear on both sides of the equation. One can therefore write the total source term as

$$\text{Source term} = Su_i + h_i\, P_i\, \Delta x_i\, (T_\infty - T_i).\qquad (2.50)$$

This prescription can be accommodated by *updating* Su_i and Sp_i as

$$Su_i = Su_i + h_i\, P_i \Delta x_i\, T_\infty,$$
$$Sp_i = Sp_i + h_i\, P_i\, \Delta x_i,\qquad (2.51)$$

where Su_i and Sp_i on the RHSs are the original quantities given in Equation 2.47.

Note that, in this case, the updated Sp_i is positive and, therefore, there is no danger of rendering $AP_i + Sp_i$ negative. Thus, Scarborough's criterion cannot be violated. However, if we considered dissipation of heat due to an electric current or chemical reaction (as in setting of cement) then, because heat is generated within the medium, $q_i''' = a + b\, T_i^m$, where b is positive. In this case, $Su_i = Su_i + a\, \Delta V_i$ and $Sp_i = Sp_i - b\, T_i^{m-1}\, \Delta V_i$. But now, there is a danger of violating Scarborough's criterion and, therefore, one simply sets $Su_i = Su_i + q_i'''\, \Delta V_i$ and Sp_i is *not* updated.

Accounting for the source term in the manner of Equation 2.51 is called source term linearization [49]. We shall discover further advantages of this form when dealing with the application of boundary conditions.

2.7.2 Nonlinear Coefficients

Coefficients AE_i and AW_i can become functions of temperature owing to thermal conductivity as in $k = a + b\,T + c\,T^2$. Thus, $k_{i+1/2}$ in AE_i (see Equation 2.45),

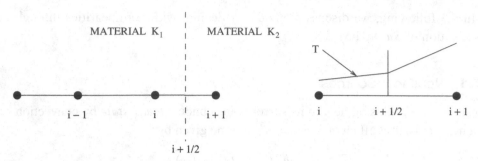

Figure 2.7. Interpolation of conductivity.

for example, may be evaluated in two ways:

$$k_{i+1/2} = a + b\,T_{i+1/2} + c\,T_{i+1/2}^2, \qquad T_{i+1/2} = 0.5\,(T_i + T_{i+1}) \quad (2.52)$$

or

$$k_{i+1/2} = 0.5\,[k(T_i) + k(T_{i+1})]. \quad (2.53)$$

Both of these representations are pragmatically acceptable but neither can be justified on the basis of the physics of conductance. To illustrate this point, let us consider a composite medium consisting of two materials with constant conductivities k_1 and k_2 (see Figure 2.7). In this case, we lay the grid nodes i and $i + 1$ in such a way that the cell face $i + 1/2$ *coincides* with the location where the two materials are joined. Thus, there is a discontinuity in conductivity at the $i + 1/2$ location.

Now, in spite of the discontinuity, the heat transfer $Q_{i+1/2}$ on either side of $i + 1/2$ must be the same. Therefore,

$$Q_{i+1/2} = k_1\,A_{i+1/2}\,\frac{T_i - T_{i+1/2}}{x_{i+1/2} - x_i}, \qquad k_1 = k_i, \quad (2.54)$$

$$Q_{i+1/2} = k_2\,A_{i+1/2}\,\frac{T_{i+1/2} - T_{i+1}}{x_{i+1} - x_{i+1/2}}, \qquad k_2 = k_{i+1}. \quad (2.55)$$

Eliminating $T_{i+1/2}$ from these equations gives

$$Q_{i+1/2} = A_{i+1/2}\left[\frac{x_{i+1/2} - x_i}{k_i} + \frac{x_{i+1} - x_{i+1/2}}{k_{i+1}}\right]^{-1}(T_i - T_{i+1}). \quad (2.56)$$

We recall, however, that our discretised equation was derived on the basis of *linear* temperature variation between nodes i and $i + 1$ (see Equation 2.21). This implies that

$$Q_{i+1/2} = \left.\frac{A}{\Delta x}\right|_{i+1/2} k_{i+1/2}\,(T_i - T_{i+1}). \quad (2.57)$$

Comparing Equations 2.56 and 2.57, leads to

$$k_{i+1/2} = \Delta x_{i+1/2} \left[\frac{x_{i+1/2} - x_i}{k_i} + \frac{x_{i+1} - x_{i+1/2}}{k_{i+1}} \right]^{-1}. \tag{2.58}$$

If the cell face were midway between the nodes then this equation would read as

$$k_{i+1/2} = 2 \left[\frac{1}{k_i} + \frac{1}{k_{i+1}} \right]^{-1}. \tag{2.59}$$

These equations suggest that the conductivity at a cell face should be evaluated by a harmonic mean to accord with the physics of conductance. We shall regard this as a general practise and extend it to the case when thermal conductivity varies with temperature. Thus, instead of using either Equation 2.52 or 2.53, Equation 2.58 will be used with k_i and k_{i+1} evaluated in terms of temperatures T_i and T_{i+1}, respectively. Further, note that if conductivity is constant, $k_{i+1/2} = k_i = k_{i+1}$.

2.7.3 Boundary Conditions

In practical problems, three types of boundary conditions are encountered:

1. Boundary temperatures T_1 and/or T_N are specified.
2. Boundary heat fluxes q_1 and/or q_N are specified.
3. Boundary heat transfer coefficients h_1 and/or h_N are specified.

Our interest in this section lies in prescribing these boundary conditions by employing Su and Sp for the near-boundary nodes.

Boundary Temperature Specified

For the purpose of illustration, consider the $i = 2$ node, where T_1 is specified. Then, Equation 2.43 will read as

$$(AP_2 + Sp_2) T_2^{l+1} = AE_2 T_3^{l+1} + AW_2 T_1^{l+1} + Su_2, \tag{2.60}$$

where Su_2 and Sp_2 are already updated to account for any source term. Equation 2.60 can be left as it is but we alter it via a three-step procedure in which we set

$$Su_2 = Su_2 + AW_2 T_1,$$

$$Sp_2 = Sp_2 + AW_2,$$

$$AW_2 = 0.0. \tag{2.61}$$

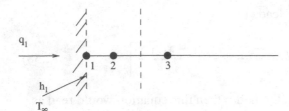

Figure 2.8. Flux boundary condition.

With this specification, AP_2 will now equal AE_2 because AW_2 is set to zero, but the coefficient of T_2^{l+1} remains intact because Sp_2 has been updated. Thus, the boundary condition specification is accomplished by *snapping* the boundary connection in the main discretised equation.

Heat Flux Specified

Let heat flux q_1 be specified at $x = 0$ (see Figure 2.8) Then, temperature T_1 is unknown and heat transfer will be given by

$$Q_1 = A_1 q_1 = AW_2 (T_1 - T_2), \tag{2.62}$$

$$T_1 = \frac{A_1 q_1}{AW_2} + T_2. \tag{2.63}$$

From Equation 2.60, it is clear that one can apply the boundary condition by employing the following sequence:

1. Calculate T_1 from Equation 2.63.
2. Update $Su_2 = Su_2 + A_1 q_1$ and $Sp_2 = Sp_2 + 0$.
3. Set $AW_2 = 0$.

The q_N-specified boundary condition can be similarly dealt with by altering AE_{N-1} and Su_{N-1}.

Heat Transfer Coefficient Specified

In this case, let h_1 be the specified heat transfer coefficient (see Figure 2.8 again) and let T_∞ be the fluid temperature adjacent to the surface at $x = 0$. Then,

$$Q_1 = A_1 q_1 = A_1 h_1 (T_\infty - T_1) = AW_2 (T_1 - T_2). \tag{2.64}$$

Therefore,

$$T_1 = \frac{T_2 + (A_1 h_1 / AW_2) T_\infty}{1 + (A_1 h_1 / AW_2)}. \tag{2.65}$$

In this case, the boundary condition can be implemented via the following steps:

1. Calculate T_1 from Equation 2.65.
2. Update

$$Sp_2 = Sp_2 + \left[\frac{1}{A_1 h_1} + \frac{1}{AW_2} \right]^{-1} \text{ and } Su_2 = Su_2 + \left[\frac{1}{A_1 h_1} + \frac{1}{AW_2} \right]^{-1} T_\infty.$$

3. Set $AW_2 = 0$.

Thus, for all types of boundary conditions, we are able to find appropriate Su and Sp augmentations and then set the boundary coefficient of the near-boundary node (AW_2 in our examples) to zero. The usefulness of this practise will become apparent when we consider the issue of convergence enhancement of the iterative solution procedures of 2D equations in Chapter 9.

2.7.4 Underrelaxation

In a nonlinear problem, if k and/or q''' are strong functions of temperature then, in an iterative procedure, as the temperature field changes, the coefficients AP, AE, and AW and the source S may change very rapidly from iteration to iteration. In such highly nonlinear problems, the iterative solution may yield oscillatory or erratic convergence or may even diverge. Therefore, it is desirable to restrict the changes in temperature implied by Equation 2.43. Such a restriction is called *underrelaxation*. It can be effected by rewriting Equation 2.43 as

$$T_i^{l+1} = \frac{\alpha \left[AE_i \, T_{i+1}^{l+1} + AW_i \, T_{i-1}^{l+1} + Su_i \right]}{AP_i + Sp_i} + (1 - \alpha) \, T_i^l, \qquad (2.66)$$

where $0 < \alpha \leq 1$. If $\alpha = 1$, no underrelaxation will be effected. If $\alpha = 0$, no change will be effected, therefore, this case is not of interest. The underrelaxation can be effected without altering the structure of Equation 2.43 by simply augmenting Su and Sp *before* every iteration. Thus,

$$Su_i = Su_i + \frac{(1 - \alpha)}{\alpha} (AP_i + Sp_i) \, T_i^l, \qquad (2.67)$$

$$Sp_i = Sp_i + \frac{(1 - \alpha)}{\alpha} (AP_i + Sp_i). \qquad (2.68)$$

If the coefficients AE_i and AW_i were constants and not functions of T then it is also possible to take $1 \leq \alpha < 2$. This is called *overrelaxation*. Typically, compared to the case of $\alpha = 1$, the convergence rate with overrelaxation is faster up to a certain optimum α_{opt}, but for $\alpha > \alpha_{opt}$, the convergence rate again slows down, so much so that it may be even slower than that with $\alpha = 1$. The magnitude of α_{opt} is problem dependent.

2.8 Methods of Solution

When coefficients AE_i, AW_i, and AP_i are calculated and Su_i and Sp_i are suitably updated to account for the effects of source linearization, boundary conditions, and underrelaxation, we are ready to solve the set of equations (2.43) at an iteration level $l + 1$. There are two extensively used methods for solving such equations.

2.8.1 Gauss–Seidel Method

The Gauss–Seidel (GS) method is extremely simple to implement on a computer. The main steps are as follows:

1. At a given iteration level l, calculate coefficients AE, AW, AP, Su, and Sp using temperature T^l for $i = 2$ to $N - 1$
2. Hence, execute a DO loop:

```
100     FCMX = 0
        DO 1 I = 2, N-1
        TL = T(I)
        ANUM = AE(I)*T(I+1) + AW(I)*T(I-1) + SU(I)
        ADEN = AE(I) + AW(I) + SP(I)
        T(I) = ANUM / ADEN
        FC = (T(I) - TL) / TL
        IF (ABS(FC).GT.FCMX) FCMX = ABS(FC)
1       CONTINUE
```

3. If FCMX > CC, go to step 1.

The method is also called a *point-by-point* method because each node i is visited in succession. The method is very reliable but requires a large number of iterations and hence considerable computer time, particularly when N is large.

2.8.2 Tridiagonal Matrix Algorithm

In the tridiagonal matrix algorithm (TDMA), Equation 2.43 is rewritten as

$$T_i = a_i\, T_{i+1} + b_i\, T_{i-1} + c_i, \tag{2.69}$$

where

$$a_i = \frac{AE_i}{AP_i + Sp_i}, \qquad b_i = \frac{AW_i}{AP_i + Sp_i}, \qquad c_i = \frac{Su_i}{AP_i + Sp_i}. \tag{2.70}$$

Note that since $Sp_i \geq 0$, a_i and b_i can only be fractions. Equation 2.69 represents $(N - 2)$ simultaneous algebraic equations. In matrix form, these equations can be written as $[A]\,[T] = [C]$, where the coefficient matrix $[A]$ will appear as shown

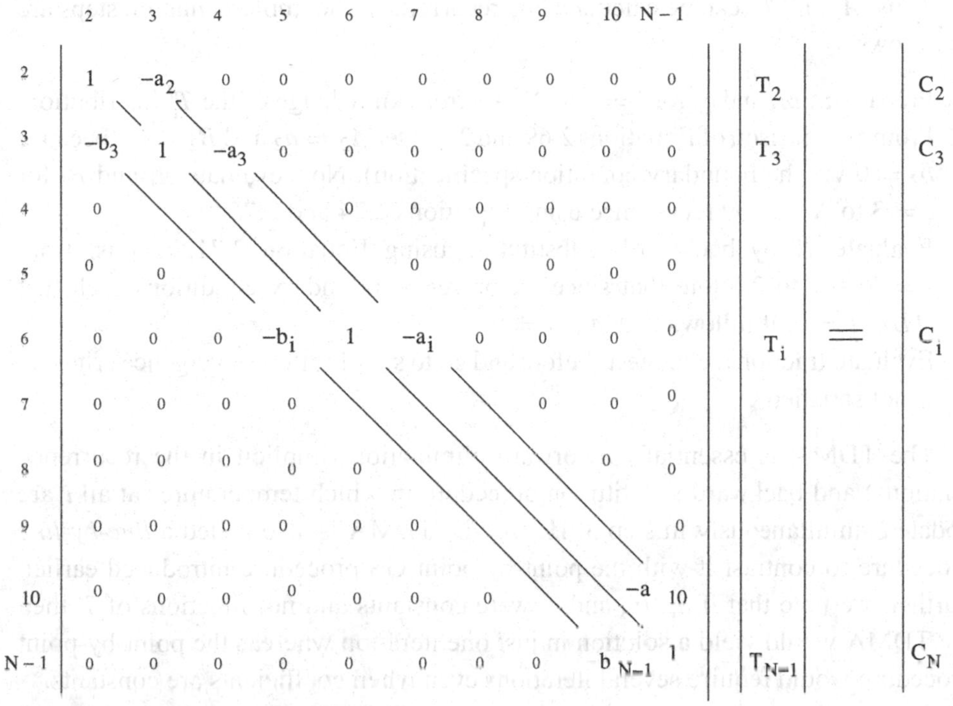

	2	3	4	5	6	7	8	9	10	N-1				
2	1	$-a_2$	0	0	0	0	0	0	0	0		T_2		C_2
3	$-b_3$	1	$-a_3$	0	0	0	0	0	0	0		T_3		C_3
4	0				0	0	0	0	0	0				
5	0	0			0	0	0	0	0					
6	0	0	0	$-b_i$	1	$-a_i$	0	0	0	0		T_i	$=$	C_i
7	0	0	0	0			0	0	0					
8	0	0	0	0	0			0	0					
9	0	0	0	0	0	0	0		0					
10	0	0	0	0	0	0	0		$-a$	10				
N-1	0	0	0	0	0	0	0	0	b_{N-1}	1		T_{N-1}		C_{N-1}

Figure 2.9. Diagonally dominant matrix [A].

in Figure 2.9. Notice that the coefficient of T_i occupies the diagonal position of the matrix with $-a_i$ and $-b_i$ occupying the neighbouring diagonal positions. All other elements of the matrix are zero. The matrix [A] thus has diagonally dominant tridiagonal structure. This structure can be exploited as follows. Let

$$T_i = A_i T_{i+1} + B_i, \qquad i = 2, \ldots, N-1. \tag{2.71}$$

Then

$$T_{i-1} = A_{i-1} T_i + B_{i-1}. \tag{2.72}$$

Now, substituting this equation in Equation 2.69, we can show that

$$T_i = \left[\frac{a_i}{1 - b_i A_{i-1}}\right] T_{i+1} + \left[\frac{b_i B_{i-1} + c_i}{1 - b_i A_{i-1}}\right]. \tag{2.73}$$

Comparison of Equation 2.73 with Equation 2.71 shows that

$$A_i = \frac{a_i}{1 - b_i A_{i-1}}, \tag{2.74}$$

$$B_i = \frac{b_i B_{i-1} + c_i}{1 - b_i A_{i-1}}. \tag{2.75}$$

Thus, A_i and B_i can be calculated by recurrence. The implementation steps are as follows:

1. Prepare a_i, b_i, and c_i for $i = 2$ to $N - 1$ from knowledge of the T_i^l distribution.
2. From comparison of Equations 2.69 and 2.71, set $A_2 = a_2$ and $B_2 = c_2$ (because $b_2 = 0$ via the boundary condition specification). Now evaluate A_i and B_i for $i = 3$ to $N - 1$ by recurrence using Equations 2.74 and 2.75.
3. Evaluate T_i by backwards substitution using Equation 2.71, that is, from $i = N - 1$ to 2. Note that since we prescribe boundary conditions such that $AE_{N-1} = 0$, it follows that $A_{N-1} = 0$.
4. Evaluate fractional change as before and go to step 1 if the convergence criterion is not satisfied.

The TDMA is essentially a forward elimination (implicit in the recurrence relations) and backward substitution procedure in which temperatures at all i are updated simultaneously in step 3. Hence, the TDMA is also called a *line-by-line* procedure to contrast it with the point-by-point GS procedure introduced earlier. Further, we note that if a_i, b_i, and c_i were constants and not functions of T then the TDMA would yield a solution in just one iteration whereas the point-by-point procedure would require several iterations even when coefficients are constants.

2.8.3 Applications

To illustrate performance of the methods just described, we consider two steady-state problems.[7]

Problem 2 – Rectangular Fin [80]
A rectangular fin of length 2 cm, thickness 2 mm, and breadth 20 cm is attached to a plane wall as shown in Figure 2.10. The wall temperature $T_w = 225°C$ and ambient temperature $T_\infty = 25°C$. For the fin material, $k = 45$ W/m-K and the operating $h = 15$ W/m²-K. Determine the heat loss from the fin and its effectiveness. Assume the tip heat loss to be negligible.

Solution
The exact solution to this problem is

$$\frac{T - T_\infty}{T_w - T_\infty} = \frac{\cosh m (L - x)}{\cosh m L}, \qquad Q_{loss} = \sqrt{h \, P \, k A} \, (T_w - T_\infty) \tanh (m \, L),$$

$$(2.76)$$

where $m = \sqrt{h \, P / k A}$. In our problem, perimeter $P = 2 \times 20 = 40$ cm, area $A = 20 \times 0.2 = 4$ cm², and $L = 2$ cm. Therefore, $m = 18.257 \text{m}^{-1}$ and $Q_{loss} = 23$ W.

---

[7] The USER files for these problems are given in Appendix B.

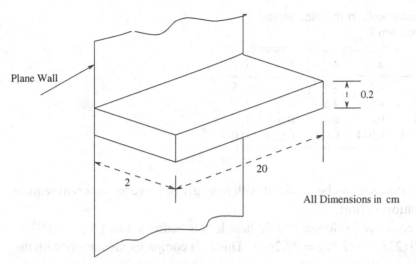

Figure 2.10. Rectangular fin – Problem 2.

To obtain a numerical solution, let us take $N = 7$ so that we have five control volumes of length $\Delta x = 0.4$ cm. Thus, we have a uniform grid. Using definitions (2.25) and (2.26), it follows that $AW_2 = 15 \times 4 \times 10^{-4}/0.002 = 9$ and $AW_i = 4.5$ for $i = 3$ to 6. Similarly, $AE_i = 4.5$ for $i = 2$ to 5 and $AE_6 = 9$. The boundary conditions are $T_1 = 225$ and $q_7 = 0$ (negligible tip loss).

Further, $Su_i = h_i P \Delta x_i T_\infty = 15 \times 0.4 \times 0.004 \times 25 = 0.6$ and $Sp_i = 15 \times 0.4 \times 0.004 = 0.024$. Now, from an equation such as (2.63), $T_7 = 0 + T_6 = T_6$. Thus, our discretised equations are

$$T_1 = 225,$$

$$[9 + 4.5 + 0.024]\, T_2 = 4.5\, T_3 + 9\, T_1 + 0.6,$$

$$[4.5 + 4.5 + 0.024]\, T_i = 4.5\, T_{i+1} + 4.5\, T_{i-1} + 0.6, \quad i = 3, 4, 5,$$

$$[4.5 + 0.024]\, T_6 = 4.5\, T_5 + 0.6,$$

$$T_7 = T_6.$$

In this problem, the conductivity, area, perimeter, and heat transfer coefficient are constants. Therefore, coefficients AE_i and AW_i do not change with iterations. Thus, after carrying out the developments of Section 2.7.3, it is possible to construct a coefficient table. The relevant quantities are shown in Table 2.5.

The solutions obtained using the GS method are shown in Table 2.6. No underrelaxation is used. Entries for $l = 0$ indicate the initial guess for temperatures (assuming a linear variation). At subsequent iterations, maximum fractional change (FCMX) reduces monotonically from 0.01 at $l = 1$ to 0.000092 at $l = 24$. The convergence criterion was set at 10^{-4}. The converged solution compares favourably with the exact solution although only five control volumes have been

Table 2.5: Coefficients in the discretised
equation – Problem 2.

i	2	3	4	5	6
AW_i	0	4.5	4.5	4.5	4.5
AE_i	4.5	4.5	4.5	4.5	0
Su_i	2025.6	0.6	0.6	0.6	0.6
Sp_i	9.024	0.024	0.024	0.024	0.024

used. Greater accuracy can be obtained with finer grids; however, this will require
more computational effort.

From the converged solution, the fin heat loss is estimated as $Q_{loss} = AW_2 \times (T_1 - T_2) = 9(225 - 222.42) = 23.26$ W. This also compares favourably with the
exact solution already mentioned.

Table 2.7 shows the execution of the same problem using TDMA. The table
shows values of A_i and B_i derived from Table 2.5 and Equations 2.74 and 2.75. Since
these are constants, solution is now obtained in only one iteration. Also, the initial
guess becomes irrelevant. The estimated heat loss is $Q_{loss} = 9(225 - 222.45) = 22.967$ W.

Thus, compared to GS, the TDMA procedure is considerably faster. Experience
shows that this conclusion is valid even in nonlinear problems. For this reason, the
TDMA is the most preferred solution procedure in generalised codes.

Problem 3 – Annular Composite Fin

Consider an annular fin put on a tube (of outer radius $r_1 = 1.25$ cm), as shown
in Figure 2.11. The fin is made from two materials: The inner material has radius
$r_2 = 2.5$ cm and conductivity $k_2 = 200$ W/m-K and the outer material extends
to radius $r_3 = 3.75$ cm and has conductivity $k_3 = 40$ W/m-K. The fin thickness
$t = 1$ mm. The tube wall (and hence the fin base) temperature is $T_0 = 200°$C. The

Table 2.6: Solution by Gauss–Seidel method – Problem 2.

l	FCMX	0 cm	0.2 cm	0.6 cm	1.0 cm	1.4 cm	1.8 cm	2.0 cm
0		225	223	219	215	211	207	205
1	0.01	225	222.65	218.31	214.15	210.08	209.1	209.1
2	0.0034	225	222.42	217.77	213.44	210.77	209.78	209.78
3	0.0021	225	222.24	217.32	213.54	211.16	210.18	210.18
⋮	⋮	⋮	⋮	⋮	⋮	⋮	⋮	⋮
22	0.00012	225	222.41	218.28	215.22	213.19	212.19	212.19
23	0.00011	225	222.41	218.30	215.24	213.21	212.21	212.21
24	0.000092	225	222.42	218.31	215.25	213.23	212.23	212.23
Exact	–	225	222.58	218.52	215.51	213.49	212.49	212.37

Table 2.7: Solution by TDMA – Problem 2.

x (cm)	0	0.2	0.6	1.0	1.4	1.8	2
A_i	—	0.333	0.598	0.711	0.772	0.0	—
B_i	—	149.78	89.628	63.776	49.357	212.375	—
$l = 1$	225	222.45	218.40	215.38	213.37	212.37	212.37
Exact	225	222.58	218.52	215.51	213.49	212.49	212.37

fin surface experiences heat transfer coefficient $h = 20$ W/m^2-K and the ambient temperature is $T_\infty = 25°C$. Assuming conduction to be radial, estimate the heat loss from the fin and the fin effectiveness. Neglect heat loss from the fin tip.

Solution
In this problem, if the origin $x = 0$ is assumed to coincide with the base of the fin, then at any radius r, area $A = 2\pi r t = 2\pi (r_1 + x)t$ and perimeter $P = 2 \times (2\pi r) = 2 \times [2\pi (r_1 + x)]$. The multiplication factor 2 in P arises because the fin loses heat from both its faces. Further, since the fin material is a composite, grids must be laid such that the cell face coincides with the location of the discontinuity in conductivity. Therefore, we adopt practise B and specify cell-face coordinate (x_e) values. Choosing $N = 8$ and equal cell-face spacings, we have six control volumes of size $\Delta x = (r_3 - r_1)/(N - 2) = 0.4167$ cm. This grid specification provides three control volumes in each material. The boundary conditions at the fin base and fin tip are $T(1) = 200$ and $q_N = 0$, respectively. Finally, the heat loss from the fin is accounted for in the manner of Equations 2.51.

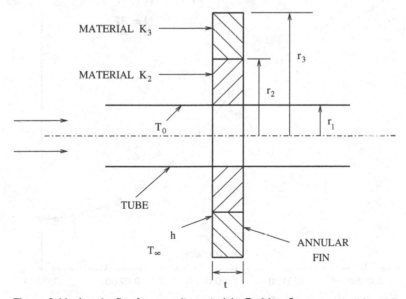

Figure 2.11. Annular fin of composite material – Problem 3.

Table 2.8: Solution by TDMA ($N = 8$) – Problem 3.

$x \times 10^3$	0	2.083	6.25	10.417	14.58	18.75	22.917	25.0
$A \times 10^5$	7.845	7.845	10.5	13.1	15.7	18.3	20.9	23.6
T	200	196.7	192.43	189.4	183.38	177.39	174.63	174.63

The predicted temperature distribution in the fin is shown in Table 2.8 and plotted (open circles) in Figure 2.12. From the table, the heat loss $Q = -k_2 A \, \partial T / \partial x \,|_{x=0} = -200 \times 7.845 \times 10^{-5}(196.7 - 200)/2.083 \times 10^{-3} = 24.86$ W. To evaluate fin effectiveness, the maximum possible heat loss from the fin is evaluated from $2 \times h \times \pi \,(r_3^2 - r_1^2) \times (T_0 - T_\infty) = 27.49$ W. Therefore, the predicted effectiveness $\Phi = 24.86/27.49 = 0.9046$.

To carry out the grid-independence study, computations are repeated for $N = 16$ and $N = 32$. These results are also plotted in Figure 2.12. The figure shows that results for $N = 16$ (open squares) and $N = 32$ (solid line) almost coincide. Thus, in this problem, results obtained with $N = 16$ may be considered quite accurate for engineering purposes. This is also corroborated by the computed Q and Φ for the two grids. For $N = 16$, the computed results are $Q = 24.933$ and $\Phi = 0.907$; for $N = 32$, they are $Q = 24.941$ and $\Phi = 0.9073$. Note also the change in the

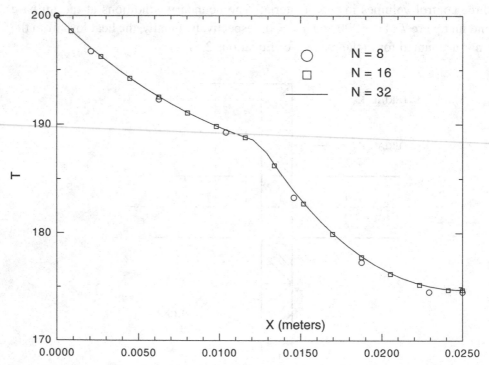

Figure 2.12. Variation of temperature with X – Problem 3.

slope of the temperature profile at the point of discontinuity ($x = 0.0125$ m) in conductivity. Finally, by assigning different values to k_2, k_3, r_2, r_3, and t, it would be possible to carry out a parametric study to aid optimisation of fin volume and economic cost in a separate design study.

2.9 Problems from Related Fields

Quite a few problems from the fields of fluid mechanics, convective heat transfer, and diffusion mass transfer are governed by equations that bear similarity with Equation 2.5. Only the dependent variable, the coefficients, and the source term need to be interpreted appropriately. We discuss such problems next.

Fully Developed Laminar Flow

Steady, fully developed laminar flow in a tube is governed by

$$\frac{\partial}{\partial r}\left(\mu \, 2\pi \, r \, \frac{\partial u}{\partial r}\right) - 2\pi \, r \, \frac{d\,p}{d\,z} = 0, \tag{2.77}$$

where u is velocity parallel to the tube axis and the pressure gradient is a negative constant. Since velocity u is directed in the z direction, it can be treated as a scalar with respect to the r direction. Comparison with Equation 2.5 shows that $T \equiv u$, $\partial x \equiv \partial r, A \equiv 2\pi r, k \equiv \mu,$ and $q''' \equiv -d\,p/d\,z$. For a circular tube, $u = 0$ at $r = R$ (tube radius) and $\partial u/\partial r = 0$ at the tube axis $r = 0$. Equation 2.77 is also applicable to an annulus with boundary conditions $u = 0$ at $r = R_i$ and $r = R_o$. Similarly, the equation is applicable to flow between parallel plates if we set $A = 2\pi r = 1$ and $\partial x \equiv \partial r \equiv \partial y$, where y is measured from the symmetry axis.

Fully Developed Turbulent Flow

In this case, if Boussinesq approximation is considered valid then the axial velocity is governed by

$$\frac{\partial}{\partial r}\left\{(\mu + \mu_t)\, 2\pi \, r \, \frac{\partial u}{\partial r}\right\} - 2\pi \, r \, \frac{d\,p}{d\,z} = 0, \tag{2.78}$$

where the turbulent viscosity $\mu_t = \rho \, l_m^2 \left|\frac{\partial u}{\partial r}\right|$ with

$$l_m = \begin{cases} \kappa \, y \left[1 - \exp\left(-\dfrac{y^+}{26}\right)\right] & \text{for} \quad \dfrac{y}{R} < y_l, \\[4mm] 0.085 \, R & \text{for} \quad \dfrac{y}{R} > y_l, \end{cases} \tag{2.79}$$

where $\kappa = 0.41$, $y = R - r$, $y_l \simeq 0.2$, and $y^+ = y \sqrt{\tau_w/\rho}\,/\nu$ with τ_w the shear stress at the wall (i.e., $\tau_w = \mu \, \partial u/\partial y\,|_{y=0}$). Clearly, Equation 2.78 can be solved iteratively by estimating the turbulent viscosity distribution from the velocity gradient.

Fully Developed Heat Transfer

The equation governing laminar fully developed heat transfer in a tube is given by

$$\frac{\partial}{\partial r}\left(k\, 2\pi r\, \frac{\partial T}{\partial r}\right) - 2\pi r\, \rho\, C_p\, u_{\text{fd}}\, \frac{\partial T}{\partial z} = 0, \tag{2.80}$$

where $u_{\text{fd}} = 2\bar{u}\,(1 - r^2/R^2)$ or can be taken from the numerical solution of Equation 2.77. Evaluation of $\partial T/\partial z$ can be carried out from the boundary conditions at the tube wall as follows.

Constant Wall Heat Flux: From the overall heat balance and from the condition of fully developed heat transfer [33], it can be shown that

$$\frac{\partial T}{\partial z} = \frac{dT_b}{dz} = \frac{2\,q_w}{\rho\, C_p\, \bar{u}\, R}. \tag{2.81}$$

Therefore, Equation 2.80 can be written as

$$\frac{\partial}{\partial r}\left(k 2\pi r\, \frac{\partial T}{\partial r}\right) - 8\pi\, \frac{r}{R}\left(1 - \frac{r^2}{R^2}\right) q_w = 0. \tag{2.82}$$

Thus, if ∂r is replaced by ∂x, A by $2\pi r$, and q''' by $-4(1 - r^2/R^2)q_w/R$, Equation 2.82 is same as the steady-state form of Equation 2.5.

Constant Wall Temperature: In this case, the condition of fully developed heat transfer implies that

$$\frac{\partial T}{\partial z} = (T_w - T_b)^{-1}\frac{dT_b}{dz} = (T_w - T_b)^{-1}\frac{2k\,\partial T/\partial r\,|_{r=R}}{\rho\, C_p \bar{u}\, R}, \tag{2.83}$$

where T_b is the mixed-mean or bulk temperature. Thus, by setting $q''' = -4k/R\,(1 - r^2/R^2)(T_w - T_b)^{-1}\,\partial T/\partial r\,|_{r=R}$, Equation 2.80 is same as Equation 2.5. However, T_b and $\partial T/\partial r\,|_{r=R}$ must be evaluated at each iteration. The bulk temperature T_b is evaluated as

$$T_b = \frac{\int_0^R \rho\, C_p\, u\, T\, 2\pi\, r\, dr}{\int_0^R \rho\, C_p\, u\, 2\pi\, r\, dr}. \tag{2.84}$$

Thermal Entry Length Solutions

Consider laminar flow between two parallel plates separated by distance $2b$. When $Pr \gg 1$, it is possible to obtain the variation of the heat transfer coefficient h with axial distance z by solving the following differential equation:

$$\frac{\partial}{\partial y}\left(k\, \frac{\partial T}{\partial y}\right) = \rho\, C_p\, u_{\text{fd}}\, \frac{\partial T}{\partial z}, \tag{2.85}$$

where

$$u_{fd} = \frac{3}{2}\,\bar{u}\left(1 - \frac{y^2}{b^2}\right) \tag{2.86}$$

and y is measured from the symmetry axis. The initial condition is $T = T_i$ at $z = 0$ and the symmetry boundary condition is $\partial T/\partial y = 0$ at $y = 0$. At $y = b$, however, $T = T_w$ if both walls are at constant wall temperature, or, if constant wall heat flux is specified, then $k\,\partial T/\partial y\,|_b = q_w$. For this problem, if we set $y \equiv x$, $z \equiv t$, $q''' = 0$, $A = 1$, and $C_p\,u_{fd} \equiv 1.5\,\bar{u}\,(1 - y^2/b^2)\,C_p$ then Equation 2.85 is the same as Equation 2.5 in which the unsteady term is retained.

Diffusion Mass Transfer

In a binary mixture of species i and j, the equation (in spherical coordinates) governing radial diffusion of j in a *stationary* medium i is given by

$$\frac{\partial}{\partial r}\left\{\frac{\rho_m\,D\,4\pi\,r^2}{(1 - \omega_j)}\,\frac{\partial \omega_j}{\partial r}\right\} = \rho_m\,4\pi r^2\,\frac{\partial \omega_j}{\partial t}, \tag{2.87}$$

where ω_j is the mass fraction of j in the mixture and D is the mass diffusivity. Thus, if we set $\partial r = \partial x$, $A - 4\pi\,r^2$, $k = \rho_m\,D/(1 - \omega_j)$, $C_p = 1$, $T = \omega_j$, and, $q''' = 0$ then this equation is the same as Equation 2.5. To solve the equation, one will need boundary conditions at $r = r_i$ and $r = r_o$ and the initial condition at $t = 0$. Estimation of penetration depth during surface hardening of materials, estimation of leakage flow of gases from storage vessels, or estimation of burning rate of volatile fuel in still surroundings are some of the mass transfer problems of interest. The reader is referred to the unified formulation of the mass transfer problem by Spalding [72] and to the book by Gupta and Srinivasan [26].

EXERCISES[8]

1. Show that the derivative expressions in Equation 2.21 are second-order accurate if the cell face is midway between adjacent nodes.

2. A slab of thickness $2b$ is initially at temperature T_0. At $t = 0$, the boundary temperatures at $x = -b$ and $+b$ are raised to T_b and maintained there. The exact solution for evolution of temperature in this case is given by

$$\frac{T - T_b}{T_0 - T_b} = 2\sum_{n=1}^{\infty} \frac{\sin(\lambda_n\,b)}{\lambda_n\,b}\cos(\lambda_n\,x)\exp\left(-\alpha\,\lambda_n^2\,t\right),$$

where $\lambda_n\,b = (2n - 1)\pi/2$. Hence, considering the data of Problem 1 in the text, write a computer program to determine the value of t for the centerline

---

[8] All numerical problems given in these exercises can be solved by the generalised computer code given in Appendix B.

temperature to reach $140°C$. What is the minimum value of n required to obtain an accurate estimate of t?

3. Repeat Problem 1 from the text using both explicit and implicit methods by choosing $N = 7, 12$, and 22. Determine the largest allowable time step in the explicit case. Compare your solution for the time required for adhesion with the exact solution determined in the previous problem.

4. Evaluate Su_{N-1} and Sp_{N-1} for an unsteady problem when T_N is specified as a function of time. Assume an arbitrary value of ψ.

5. Consider a time-varying heat-flux-specified condition at $i = 1$. Hence, derive Su_2 and Sp_2 for arbitrary ψ. Confirm the validity of the three-step procedure following Equation 2.63 for $\psi = 1$.

6. Repeat Exercise 5 for a time-varying heat transfer coefficient boundary condition. Hence, confirm the validity of the procedure following Equation 2.65 for $\psi = 1$.

7. Confirm the correctness of Equations 2.67 and 2.68.

8. Verify the entries in Tables 2.5 and 2.7 by carrying out the necessary calculations.

9. Develop a TDMA routine in which the postulated equation is

$$T_i = A_i T_{i-1} + B_i.$$

10. Consider a slab of width $b = 20$ cm. At $x = 0$, $T = 100°C$ and at $x = b$, $q = 1$ kW/m^2. The heat generation rate is $q''' = 1,000 - 5T$ W/m^3. Calculate the steady-state temperature distribution with and without source-term linearisation. Compare the number of iterations required in the two cases for $N = 22$ and 42. Also calculate the heat flux at $x = 0$ and T_b and check the overall heat balance. Take $k = 1$ W/m-K. Use TDMA.

11. Consider a nuclear fuel rod of length L and diameter D. The two ends of the rod are maintained at T_0. The internal heat generation rate is $q''' = a \sin(\pi x/L)$, where x is measured from one end of the rod and a is an arbitrary constant. The rod loses heat by convection (coefficient h) to a coolant fluid at T_∞.

   (a) Nondimensionalise the steady-state heat conduction equation and identify the dimensionless parameters. [Hint: Define $\theta = (T - T_\infty)/(T_0 - T_\infty)$, $x^* = x/L$, $P_1 = a L^2/k(T_0 - T_\infty)$, and $P_2 = 4h L^2/(k D)$.]

   (b) Compute the temperature distribution in the rod and compare with the exact solution for $0 < P_1, P_2 < 10$. Use source-term linearisation and TDMA. Carry out an overall heat balance from the computed results

   (c) Solve the problem for $P_1 = P_2 = 10$ using different underrelaxation parameters $0 < \alpha < 2$ for $N = 22$ and $N = 42$. Determine α_{opt} in each case. Use uniform grid spacing and the GS procedure.

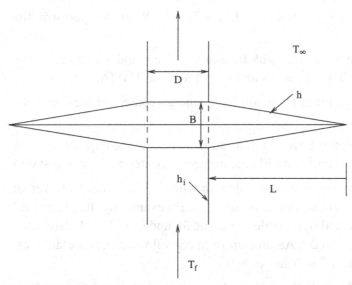

Figure 2.13. Circumferential fin.

12. Exploit the symmetry in Exercise 11 at $L/2$ and compute the temperature distribution over $0 \le x \le L/2$. Compare the value of $T_{L/2}$ with the exact solution.

13. Consider the fin shown in Figure 2.13. The following are given: $T_\infty = 25°C$, $T_t = 200°C$, $B = 2$ mm, $L = 6$ mm, tube diameter $D = 4$ mm, $k_{fin} = 40$ W/m-K, $h = 20$ W/m²-K, and $h_i = 200$ W/m²-K.

 (a) Write the appropriate differential equation for steady-state heat transfer and the boundary conditions to determine the temperature distribution in the fin.

 (b) Discretise the equation assuming six nodes (four control volumes) and list AE, AW, Su, Sp, and AP for each node.

 (c) Evaluate the effectiveness of the fin.

14. Consider a rod of circular cross section ($L = 10$ cm, $d = 1$ cm, $k = 1$ W/m-K, $\rho = 2,000$ kg/m³, and $C = 850$ J/kg-K). The rod is perfectly insulated around its periphery. At $t = 0$, the rod is at 25°C. For $t > 0$, $T_{x=0} = 25°C$ and $T_{x=L} = 25 + t(s)°C$. Compute temperature distribution in the rod as a function of x and t over a period of 15 min using $\psi = 0, 0.5$, and 1. Also determine $q_{x=0}$ as a function of time and plot the variation. Take $N = 22$ and $\Delta t = 5$ s in each case.

15. Consider a rod of circular cross section ($L = 10$ cm, $d = 1$ cm, $k = 1$ W/m-K, $\rho = 2,000$ kg/m³, and $C = 850$ J/kg-K). The rod is initially at 600°C. The temperatures at the two ends of the rod are suddenly reduced to 100°C and maintained at that temperature. The rod is also cooled by natural convection

to surroundings at 25°C. If $h = 3 (T_{rod} - T_\infty)^{0.25}$ W/m²-K, perform the following:

(a) Compute the variation of h with time at $x = 5$ cm and $x = 9$ cm over a period of 1 min. Take $\Delta t = 1$ s and $\psi = 1$ and use TDMA.

(b) Compute the percentage reduction in the energy content of the rod at the end of 1 min.

(c) Extend the calculation beyond 1 min and estimate the time required to reach near steady state. (Hint: You will need to specify a criterion for steady state.)

16. Consider an unsteady conduction problem in which T_1 is given. However, at $x = L$, the heat transfer coefficient is specified. By examining the discretised equation for a general node i, for node $i = 2$, and for node $i = N - 1$, determine the stability constraint on Δt. Assume uniform control volumes, constant area, and conductivity with $q''' = 0$ and $\psi = 0$.

17. A semi-infinite solid is initially at 25°C. At $t = 0$, the solid surface $(x = 0)$ is suddenly exposed to $q_w = 10$ kW/m². A thermocouple is placed at $x = 1$ mm to apparently measure the surface temperature. Compute the temperature distribution in the solid as a function of x and t and estimate the error in the thermocouple reading as a function of time. Carry out computations up to 1 s. Given are the following: $k = 80$ W/m-K, $\rho = 7,870$ kg/m³, and $C = 450$ J/kg-K. [Hint: The boundary condition at $x = \infty$ is $T_L = 25$°C at all times. Choose sufficiently large L (say 1 cm) and execute with $\Delta t = 0.01$ s.]

18. A laboratory built in the Antarctic has a composite wall made up of plaster board (10 mm), fibreglass insulation (100 mm), and plywood (20 mm). The inside room temperature is maintained at $T_i = 293$ K throughout. The plywood is exposed to an outside temperature T_o that varies with time t (in hours) as

$$T_o = \begin{cases} 273 + 5\sin\left(\dfrac{\pi}{12}t\right) & \text{for} \quad 0 \le t \le 12\,\text{h}, \\ 273 + 30\sin\left(\dfrac{\pi}{12}t\right) & \text{for} \quad 12 \le t \le 24\,\text{h}. \end{cases}$$

(a) Compute the heat loss to the outside over a typical 24-h period (i.e., under periodic steady state) in J/m².

(b) Plot the variation of interface temperatures between the plasterboard and the fibreglass and between the fibreglass and the plywood as a function of time. Assume: $h_i = 15$ W/m²-K and $h_o = 60$ W/m²-K. Material properties are given in Table 2.9.

19. Solve for fully developed laminar flow in a concentric annular $(r^* = R_i/R_o = 0.6)$ duct. Compare the predicted velocity profile with the exact solution [33]

$$\frac{u}{\bar{u}} = \frac{2}{A}\left[1 - \left(\frac{r}{R_o}\right)^2 + B\ln\left(\frac{r}{R_o}\right)\right],$$

Table 2.9: Properties of the wall materials.

Material	ρ (kg/m^3)	C (J/kg-K)	k (W/m-K)
Plasterboard	1000	1380	0.15
Fibreglass	30	850	0.038
Plywood	545	1200	0.1

where $B = (r^{*^2} - 1)/\ln r^*$ and $A = 1 + r^{*^2} - B$. Hence, compare the predicted friction factor based on a hydraulic diameter $D_h = 2(R_o - R_1)$ with

$$(f\,Re)_{D_h} = \frac{16}{A}\left(1 - r^{*^2}\right).$$

20. Solve Equation 2.78 for turbulent flow in a circular tube and compare your
results with the expressions [33]

$$\frac{u}{u_\tau} = \begin{cases} y^+, & y^+ \leq 11.6 \\ 2.5\ln\left[y^+ \dfrac{1.5(1 + r/R)}{1 + 2(r/R)^2}\right] + 5.5, & y^+ > 11.6. \end{cases}$$

Also compare the predicted friction factor f with $f = 0.079\,Re^{-0.25}$ for $Re <$
2×10^4 and with $f = 0.046\,Re^{-0.2}$ for $Re > 2 \times 10^4$. Plot the variation of total
(laminar plus turbulent) shear stress with radius r. Is it linear? (Hint: Make sure
that the first node away from the wall is at $y^+ \sim 1$.)

21. Engine oil enters a tube ($D = 1.25$ cm) at uniform temperature $T_{in} = 160°$C.
The oil mass flow rate is 100 kg/h and the tube wall temperature is maintained at
$T_w = 100°$C. If the tube is 3.5 m long, calculate the bulk temperature of oil at exit
from the tube. The properties of the oil are $\rho = 823$ kg/m^3, $C_p = 2,351$ J/kg-K,
$v = 10^{-5}$m$^2/s$, and $k = 0.134$ W/m-K. Plot the axial variation of Nusselt number Nu_x and bulk temperature $T_{b,x}$ and compare with the exact solution given
in Table 2.10.

Table 2.10: Thermal entry length solution – $T_w =$ constant [33].

$(x/R)/(Re\,Pr)$	Nu_x	$(T_w - T_b)/(T_w - T_{in})$
0	∞	1.0
0.001	12.80	0.962
0.004	8.03	0.908
0.01	6.0	0.837
0.04	4.17	0.628
0.08	3.77	0.459
0.10	3.71	0.396
0.20	3.66	0.190
∞	3.66	0.0

22. It is proposed to remove NO from exhaust gases of an internal combustion engine by passing them over a catalyst surface. It is assumed that chemical reactions involving NO are very slow so that NO is neither generated nor destroyed in the gas phase. At the catalyst surface, however, NO is absorbed at the rate of $\dot{m}'' = K\rho_m\omega_0$, where the rate constant $K = 0.075$ m/s and ω_0 is the mass fraction of NO at the catalyst surface. In the exhaust gases ($T = 500°C$, $p = 1$ bar, $M = 30$) the *mole* fraction of NO is $X_{NO} = 0.002$. Now, it is assumed that NO diffuses to the catalyst surface over a *stagnant* layer of 1 mm with *effective* diffusivity $= 3 \times D$, where $D = 10^{-4}$ m^2/s. Determine the steady-state absorption rate (kg/m^2-s) of NO and its mass fraction at the surface.

23. The mass fraction of carbon in a low-carbon steel rod (2 cm diameter) is 0.002. To case-harden the rod it is preheated to 900°C and packed in a carburising mixture at 900°C. The mass fraction of carbon at the rod surface is now 0.014 and is maintained at this value. Calculate the time required for the carbon mass fraction to reach 0.008 at a depth of 1 mm from the rod surface. Assume radial diffusion only. In this case, cross-sectional area $A = 2\pi r$. However, since the penetration depth is only 10% of the rod radius, one may take $A = 2\pi R =$ constant (i.e., assume plane diffusion). Compare the time required in the two cases. Take the diffusivity of carbon in steel to be $D = 5.8 \times 10^{-10}$ m^2/s.

24. Gaseous H$_2$ at 10 bar and 27°C is stored in a 10-cm inside diameter spherical tank having a 2-mm-thick wall. If diffusivity of H$_2$ in steel is $D = 0.3 \times 10^{-12}$ m^2/s and solubility $S = 9 \times 10^{-3}$ kmol/m^3-bar, estimate the time required for the tank pressure to reduce to 9.9 bar. Also, plot the time variation of tank pressure p_{H_2} and the instantaneous hydrogen loss rate. Take $\rho_{steel} = 8,000$ kg/m^3. The density of hydrogen at the *inner surface* of the tank is given by $\rho_{H_2,i} = Sp_{H_2} M_{H_2}$. Is an exact solution possible for this problem?

25. Consider steady-state heat transfer through the composite slab shown in Figure 2.14. Assume $k_1 = 0.05(1 + 0.008\,T)$, $k_2 = 0.05(1 + 0.0075\,T)$, and $k_3 = 2$ W/m-K, where T is in degrees centigrade. Calculate the rate of heat transfer and the temperatures of the two interfaces. Ignore radiation.

26. Repeat Exercise 25 including the effect of radiation. The emissivities at $x = 0$ and $x = 17$ cm are 0.1 and 0.8, respectively. In this problem, one must use the concept of effective heat transfer coefficient $h_{eff} = h + h_{rad}$. Thus, at $x = 0$, for example,

$$h_{eff} = 50 + 0.1\sigma\left(T_\infty + T_{x=0}\right)\left(T_\infty^2 + T_{x=0}^2\right),$$

where the Stefan–Boltzmann constant $\sigma = 5.67 \times 10^{-8}$ W/m^2–K^4, and T_∞ $T_{x=0}$ are in Kelvin.

27. Consider fully developed turbulent heat transfer in a circular tube under constant wall heat flux conditions. Equations 2.80 and 2.81 are again applicable

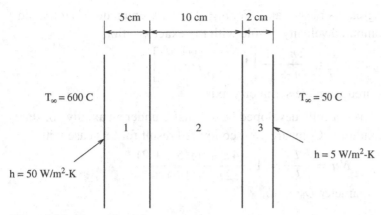

Figure 2.14. Composite slab.

but the fully developed velocity profile is determined from Exercise 19. Also, k in Equation 2.80 is replaced by $(k + k_t)$, where $k_t = C_p \mu_t / Pr_t$. Calculate the Nusselt number Nu for different Reynolds numbers at Prandtl numbers $Pr = 1$, 10, and 100. Take $Pr_t = 0.85 + 0.039(Pr + 1)/Pr$. Compare your result with following correlations: (a) $Nu_1 = 0.023\, Re^{0.8}\, Pr^{0.4}$ and (b) $Nu_2 = 5 + 0.015\, Re^m\, Pr^n$, where $m = 0.88 - 0.24(4 + Pr)^{-1}$ and $n = 0.333 + 0.5 \exp(-0.6\, Pr)$.

28. Consider laminar fully developed flow and heat transfer in a circular tube under constant wall heat flux conditions. The fluid is highly viscous. Therefore, Equation 2.80 must be augmented to account for viscous dissipation $\mu\,(\partial u/\partial r)^2$. Calculate Nu and compare your result with $Nu = 192/(44 + 192\,Br)$, where the Brinkman number $Br = \mu\,\bar{u}^2 / (q_w\, D)$. In this problem, Equation 2.81 must be modified as follows:

$$\frac{\partial T}{\partial z} = \frac{d T_b}{dz} = \frac{2\,(q_w + 4\mu\bar{u}^2/R)}{\rho\, Cp\,\bar{u}\, R}.$$

Explain why.

29. Repeat Problem 1 from the text using $\psi = 0.3$ and $\psi = 0.7$. Choose $N = 7$. Determine the largest allowable time step using constraints (2.36) and (2.37). Compare your solution for the time required for adhesion with the exact solution determined in Exercise 2.

30. Consider fully developed laminar flow of a *non-Newtonian* fluid between two parallel plates $2b$ apart. For such a fluid, the shear stress is given by

$$\tau_{yx} = \mu \left| \frac{\partial u}{\partial y} \right|^{n-1} \frac{\partial u}{\partial y},$$

where n may be greater or less than 1. For $n = 1$, a Newtonian fluid is retrieved. Compare the computed velocity profile with the exact solution

$$\frac{u}{\bar{u}} = \frac{2n + 1}{n + 1} \left[1 - \left(\frac{y}{b}\right)^{(n+1)/n} \right],$$

where y is measured from the symmetry axis.

31. In Exercise 30 consider fully developed heat transfer under an axially constant wall heat flux condition. Compare your computed result for this case with

$$Nu = \frac{h\, D_h}{k} = 12 \frac{(4n + 1)(5n + 2)}{32n^2 + 17n + 2},$$

where hydraulic diameter $D_h = 4b$.

3 1D Conduction–Convection

3.1 Introduction

Consider a 1D domain ($0 \leq x \leq L$) through which a fluid with a velocity u is flowing. Then, the steady-state form of the first law of thermodynamics can be stated as

$$\frac{\partial q_x}{\partial x} = S,$$

(3.1)

where

$$q_x = q_x^{\text{conv}} + q_x^{\text{cond}} = \rho\, C_p\, u\, T - k\frac{\partial T}{\partial x}.$$

(3.2)

These equations are to be solved for two boundary conditions, $T = T_0$ at $x = 0$ and $T = T_L$ at $x = L$. It is further assumed that ρu is a constant as are properties C_p and k.

Our interest in this chapter is to examine certain discretisational aspects associated with Equation 3.1. This is because in computational fluid dynamics (momentum transfer) and in convective heat and mass transfer, we shall recurringly encounter representation of the total flux in the manner of Equation 3.2. Note that if $u = 0$, only conduction is present and the discretisations carried out in Chapter 2 readily apply. However, difficulty is encountered when *convective* flux is present. The objective here is to understand the difficulty and to learn about commonly adopted measures to overcome it. In the last section of this chapter, stability and convergence aspects of explicit and implicit procedures for an *unsteady* equation in the presence of conduction and convection are considered.

3.2 Exact Solution

Because our interest lies in examining the discretisational aspects associated with convective–conductive flux, we take the special case of $S = 0$. For this case, an

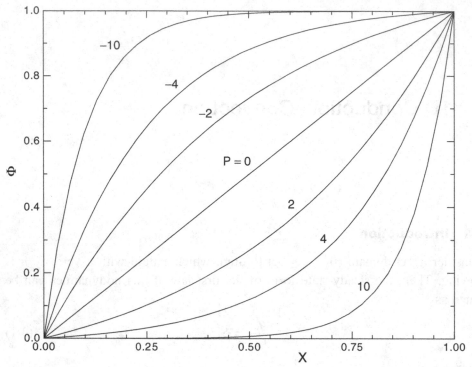

Figure 3.1. Effect of P – exact solution.

elegant closed-form solution is possible. Thus, we define

$$\Phi = \frac{T - T_0}{T_L - T_0}, \tag{3.3}$$

$$X = \frac{x}{L}, \tag{3.4}$$

$$P = \frac{\rho\, C_p\, u}{k/L} = \frac{\text{Convective flux}}{\text{Conduction flux}}, \tag{3.5}$$

where P is called the *Peclet* number. Therefore, Equations 3.1 and 3.2 can be written as

$$\frac{\partial}{\partial X}\left[P\,\Phi - \frac{\partial \Phi}{\partial X} \right] = 0 \tag{3.6}$$

with $\Phi = 0$ at $X = 0$ and $\Phi = 1$ at $X = 1$. The exact solution is

$$\frac{\Phi - \Phi_{X=0}}{\Phi_{X=1} - \Phi_{X=0}} = \Phi = \frac{\exp(P\,X) - 1}{\exp(P) - 1}. \tag{3.7}$$

The solution is plotted in Figure 3.1 for both positive and negative values of P. Negative P implies that the fluid flow is from $x = L$ to $x = 0$ (or u is negative).

It will be instructive to note the tendencies exhibited by the solution.

1. Figure 3.1 shows that irrespective of the value of P, Φ always lies between 0 and 1. This means that Φ at any x is *bounded* between its extreme values.
2. When $P = 0$, the conduction solution is obtained and, as expected, the solution is linear.
3. At $X = 0.5$ (i.e., at the midpoint)

$$\Phi(0.5) = \frac{\exp(0.5\,P) - 1}{\exp(P) - 1}. \tag{3.8}$$

It is seen from the figure that as $P \to +\infty$, $\Phi(0.5) \to 0$ and as $P \to -\infty$, $\Phi(0.5) \to 1$. Thus, at large values of $|P|$, the midpoint solution tends to a value at the *upstream* extreme.

This last comment is particularly important because a large $|P|$ implies dominance of convection over conduction. As we will shortly discover, the main difficulty in obtaining numerical solution to Equation 3.6 is also associated with large $|P|$.

3.3   Discretisation

Equation 3.6 will now be discretised using the IOCV method. Then with reference to Figure 2.3 of Chapter 2, we have

$$\int_{w}^{e} \frac{\partial}{\partial X} \left[P\,\Phi - \frac{\partial \Phi}{\partial X} \right] dX = 0, \tag{3.9}$$

or

$$P\,\Phi_e - \frac{\partial \Phi}{\partial X}\bigg|_e - P\,\Phi_w + \frac{\partial \Phi}{\partial X}\bigg|_w = 0. \tag{3.10}$$

Now, as in the case of conduction, it will be assumed that Φ varies linearly between adjacent nodes. Also, though not essential, we shall assume a uniform grid so that $\Delta X_e = \Delta X_w = \Delta X$. Thus, since the cell face is midway between adjacent nodes,

$$\Phi_e = \frac{1}{2}(\Phi_E + \Phi_P), \qquad \Phi_w = \frac{1}{2}(\Phi_W + \Phi_P) \tag{3.11}$$

and

$$\frac{\partial \Phi}{\partial X}\bigg|_e = \frac{\Phi_E - \Phi_P}{\Delta X} \qquad \frac{\partial \Phi}{\partial X}\bigg|_w = \frac{\Phi_P - \Phi_W}{\Delta X}. \tag{3.12}$$

This practise of representing cell-face value and cell-face gradient is called the central difference scheme (CDS). Substituting Equations 3.11 and 3.12 in Equation 3.10, we have

$$\frac{P}{2}(\Phi_E - \Phi_W) - \frac{1}{\Delta X}[\Phi_E - 2\,\Phi_P + \Phi_W] = 0. \tag{3.13}$$

Clearly, the first term represents the net convection whereas the second term represents the net conduction. However, note that, unlike in the conduction term, Φ_P *does not* appear in the convection term.

Equation 3.13 will now be rewritten in the familiar discretised form to read as

$$AP\,\Phi_P = AE\,\Phi_E + AW\,\Phi_W, \tag{3.14}$$

where

$$AE = \left(1 - \frac{P_c}{2}\right), \tag{3.15}$$

$$AW = \left(1 + \frac{P_c}{2}\right), \tag{3.16}$$

$$AP = AE + AW = 2, \tag{3.17}$$

and

$$P_c = P\,\Delta X = \frac{u\,L}{\alpha}\frac{\Delta x}{L} = \frac{u\,\Delta x}{\alpha}, \tag{3.18}$$

where $\alpha = k/(\rho\,C_p)$ is the thermal diffusivity and P_c is called the *cell Peclet* number. If we now invoke Scarborough's criterion, it is clear that Equation 3.14 will be convergent only when AE and AW are positive. This implies that the condition for convergence is

$$|P_c| \le 2. \tag{3.19}$$

Thus, when convection is very large compared to conduction, to satisfy condition (3.19), one will need to employ very small values of ΔX or a very fine mesh. However, this can prove to be very expensive.

The more relevant question, however, is, Why do AE and/or AW turn negative when convection is dominant? The answer to this question can be found in Equation 3.11, where, contrary to the advice provided by the exact solution, the cell-face values are *linearly* interpolated between the values of Φ at the adjacent nodes. Note that when $P_c > 2$ and large, the exact solution gives $\Phi_e \to \Phi_P$ and $\Phi_w \to \Phi_W$. Similarly, when $P_c < -2$, $\Phi_e \to \Phi_E$ and $\Phi_w \to \Phi_P$. In Equation 3.11, we took no cognizance of either the direction of flow (sign of P_c) or its magnitude.

To obtain economic convergent solutions, therefore, one must write

$$\Phi_e = \psi\,\Phi_P + (1 - \psi)\,\Phi_E, \qquad \Phi_w = \psi\,\Phi_W + (1 - \psi)\,\Phi_P, \tag{3.20}$$

where ψ is sensitized to the sign and the magnitude of P_c. Note that, in Equation 3.11, we took $\psi = 0.5$, an absolute constant.

3.4 Upwind Difference Scheme

The upwind difference scheme (UDS) was originally proposed in [8] but later independently developed by Runchal and Wolfshtein [60] among others. The scheme simply senses the sign of P_c but *not its magnitude*. Thus, instead of Equation 3.11, we write

$$P \Phi_e = \frac{1}{2} [P + |P|] \Phi_P + \frac{1}{2} [P - |P|] \Phi_E, \qquad (3.21)$$

$$P \Phi_w = \frac{1}{2} [P + |P|] \Phi_W + \frac{1}{2} [P - |P|] \Phi_P. \qquad (3.22)$$

These expressions show that when $P > 0$, $\Phi_e = \Phi_P$ and $\Phi_w = \Phi_W$. Similarly, when $P < 0$, $\Phi_e = \Phi_E$ and $\Phi_w = \Phi_P$. That is, the cell-face values always pick up the *upstream* values of Φ irrespective of the magnitude of P, hence, giving rise to the name of this interpolation scheme as the upwind difference scheme.[1] Substituting these equations in Equation 3.10, we can show that Equation 3.14 again holds with

$$AE = 1 + \frac{1}{2} (|P_c| - P_c), \qquad (3.23)$$

$$AW = 1 + \frac{1}{2} (|P_c| + P_c), \qquad (3.24)$$

and $AP = AE + AW$. Equations 3.23 and 3.24 show that, irrespective of the magnitude or sign of P (or P_c), AE and AW can never become negative. Also, AP remains dominant. Therefore, obstacles to convergence are removed for all values of P_c. This was not the case with CDS.[2]

---

[1] Physically, the UDS can be understood as follows: Imagine standing at the middle of a long corridor at one end of which there is an icebox (at T_{ice}) and at the other end a firebox (at T_{fire}). Then, neglecting radiation, the temperature experienced by you will be $T_m = 0.5 (T_{ice} + T_{fire})$ when the air in the corridor is *stagnant* and heat transfer is only by conduction. Now, imagine that there is air-flow over the firebox flowing through the corridor in the direction of the icebox. You will now experience T_m that weighs more in favour of T_{fire} than T_{ice}. The reverse would be the case if the airflow was from the icebox end and towards the firebox end. The UDS takes an extreme view of both situations and sets $T_m = T_{fire}$ in the first case and $T_m = T_{ice}$ in the second case.

[2] Incidentally, with respect to Equation 3.20, we may generalise AE and AW coefficients for both CDS and UDS in terms of ψ as

$$AE = 1 - (1 - \psi) P_c, \qquad AW = 1 + \psi P_c, \qquad (3.25)$$

$$\psi = 0.5 \ \ (CDS), \qquad \psi = \frac{1}{2} \left(1 + \frac{|P_c|}{P_c} \right) \ \ (UDS). \qquad (3.26)$$

Table 3.1: Φ_P values for $\Phi_E = 1$ and $\Phi_W = 0$.

P_c	Exact	CDS	UDS	HDS	Power
10	0.454e−4	−2	0.0833	0.0	0.0
8	0.335e−3	−1.5	0.100	0.0	0.40e−4
6	0.247e−2	−1.0	0.125	0.0	0.17e−2
4	0.018	−0.5	0.167	0.0	0.0187
2	0.119	0.0	0.25	0.0	0.123
1	0.269	0.25	0.333	0.25	0.271
0	0.5	0.5	0.5	0.5	0.5
−1	0.731	0.75	0.667	0.75	0.729
−2	0.881	1.0	0.75	1.0	0.981
−4	0.982	1.5	0.833	1.0	1.0
−6	0.998	2.0	0.875	1.0	1.0
−8	1.0	2.5	0.900	1.0	1.0
−10	1.0	3.0	0.917	1.0	1.0

3.5 Comparison of CDS, UDS, and Exact Solution

To compare the exact solution with CDS and UDS formulas, let $L = 2\,\Delta x$. Then, it can be shown that (see Equation 3.7)

$$\Phi = \left[1 - \frac{\exp\left(2\,P_c\,x^*\right) - 1}{\exp\left(2\,P_c\right) - 1} \right] \Phi_W + \left[\frac{\exp\left(2\,P_c\,x^*\right) - 1}{\exp\left(2\,P_c\right) - 1} \right] \Phi_E, \quad (3.27)$$

where x is measured from node W and $x^* = x/(2\,\Delta x)$. Therefore, $\Phi_P\,(x^* = 0.5)$ is given by

$$\Phi_P = \left[1 - \frac{\exp\left(P_c\right) - 1}{\exp\left(2\,P_c\right) - 1} \right] \Phi_W + \left[\frac{\exp\left(P_c\right) - 1}{\exp\left(2\,P_c\right) - 1} \right] \Phi_E, \quad \text{(Exact)}.$$

$$(3.28)$$

The corresponding CDS and UDS formulas are

$$\Phi_P = \frac{1}{2}\left(1 - \frac{P_c}{2} \right) \Phi_E + \frac{1}{2}\left(1 + \frac{P_c}{2} \right) \Phi_W \quad \text{(CDS)}, \quad (3.29)$$

$$\Phi_P = \left[\frac{1 - 0.5\,(P_c - |P_c|)}{2 + |P_c|} \right] \Phi_E + \left[\frac{1 + 0.5\,(P_c + |P_c|)}{2 + |P_c|} \right] \Phi_W \quad \text{(UDS)}.$$

$$(3.30)$$

In general, Φ_E and Φ_W may have any value. However, to simplify matters, we take the case of $\Phi_E = 1$ and $\Phi_W = 0$ and study the behaviour of Φ_P with P_c. Values computed from Equations 3.28–3.30 are tabulated in Table 3.1 and plotted in Figure 3.2. Two points are worth noting:

1. The CDS goes out of bounds for $|P_c| > 2$. For this range, the CDS is also not convergent as was noted earlier. It is a reasonable approximation to the exact

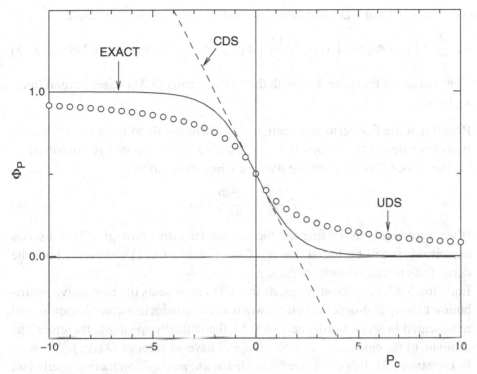

Figure 3.2. Comparison of CDS and UDS with exact solution.

solution when $|P_c| \to 0$. In spite of this, mathematically speaking, CDS is taken as the best *reference case* to compare all other differencing approximations because the CDS representation evaluates both convective and conductive contributions with the *same* approximation. That is, the spatial variation of Φ is assumed to be linear between adjacent grid nodes.

2. Although UDS is convergent at all values of P_c and nearly approximates the exact solution for $|P_c| \to \infty$, it is not a very good approximation to the exact solution at moderate values of $|P_c|$. Also, UDS deviates from CDS for $|P_c| < 2$.

3.6 Numerical False Diffusion

It was already noted that CDS is mathematically consistent. We consider the CDS formula (3.13) again and write it as

$$\frac{P_c}{2}(\Phi_E - \Phi_W) - [\Phi_E - 2\Phi_P + \Phi_W] = 0 \quad \text{(CDS)}. \quad (3.31)$$

Now, consider UDS formula (3.30) for $P_c > 0$ (say):

$$P_c(\Phi_P - \Phi_W) - [\Phi_E - 2\Phi_P + \Phi_W] = 0 \quad \text{(UDS)}. \quad (3.32)$$

To compare CDS and UDS formulas, we modify Equation 3.32 to read as[3]

$$\frac{P_c}{2}(\Phi_E - \Phi_W) - \left(1 + \frac{|P_c|}{2}\right)[\Phi_E - 2\Phi_P + \Phi_W] = 0 \quad \text{(UDS)}. \quad (3.33)$$

Comparison of Equation 3.33 with the CDS formula (3.31) raises several interesting issues:

1. Recall that the first term in Equation 3.31 corresponds to the convective contribution whereas the second term corresponds to the conductive contribution. Further, since P is constant, we may view Equation 3.6 as

$$P\frac{\partial \Phi}{\partial X} - \frac{\partial^2 \Phi}{\partial X^2} = 0. \quad (3.34)$$

   If we discretise both the first and the second derivative through a Taylor series expansion, it will be found that the CDS formula (3.31) represents both the derivatives to second-order accuracy.

2. Equation 3.32, in contrast, suggests that UDS represents the convective contribution to only first-order accuracy, whereas the conductive contribution is still represented to second-order accuracy. Mathematically speaking, therefore, the estimate of the convective contribution will have an error of $O(\Delta x)$.

3. In Equation 3.33, this error is reflected in the *augmented* conduction coefficient because the convective term is now written to second-order accuracy as in the CDS formula. Mathematically speaking, therefore, it may be argued that the second-order-accurate UDS formula represents discretisation with augmented or *false* conductivity $k_{false} = \rho C_p |u| \Delta x / 2$. In fact, it can be shown that Equation 3.33 is nothing but a CDS representation of

$$\frac{\partial}{\partial x}\left[\rho C_p u T - \left(k + \frac{\rho C_p |u| \Delta x}{2}\right)\frac{\partial T}{\partial x}\right] = 0. \quad (3.35)$$

Thus, if the last comment is given credence, then clearly the UDS represents distortion of reality and is therefore a poor choice. Yet, the closeness of the UDS result to the exact solution shown in Figure 3.2 suggests that the so-called false conductivity is indeed needed. In fact, it is this false conductivity that *reduces* the value of the *effective* Peclet number and thereby ensures convergence of the UDS formula for all Peclet numbers.

Patankar [49] has therefore argued that to form a proper view of false diffusion, it is necessary to compare the UDS with the exact solution rather than with the second-order-accurate CDS formula. This is yet another example where the TSE method is found wanting.

Of course, this is not to suggest that the UDS formula is the best representation of reality. The properties embodied in the UDS formula suggest that one can derive other variants that will sense not only the sign of P_c but also its magnitude. Further

---

[3] Equation 3.33 can also be derived for $P_c < 0$.

considerations associated with false diffusion in multidimensional flows will be discussed in Chapter 5.

3.7 Hybrid and Power-Law Schemes

Spalding [75] derived a hybrid difference scheme (HDS) such that, in Equation 3.20, ψ is given by

$$\psi = \frac{1}{P_c} \left[P_c - 1 + \max \left(-P_c, 1 - \frac{P_c}{2}, 0 \right) \right] \quad \text{(HDS)}. \quad (3.36)$$

Similarly, Patankar [49] argued that the best representation is the exact solution itself (see Equation 3.28). However, this will require evaluation of exponential terms and this is not economically attractive in practical computing. Therefore, he chose to mimic Equation 3.28 through a power-law scheme, which implies that

$$\psi = [P_c - 1 + \max(0, -P_c)] / P_c$$
$$+ \max \left\{ 0, (1 - 0.1|P_c|)^5 \right\} / P_c \quad \text{(Power law)}. \quad (3.37)$$

With these two expressions for ψ, it is now possible to construct AE and AW coefficients (see Equation 3.25) for the HDS and power-law schemes. The resulting implications for Φ_P are tabulated in Table 3.1. Notice that for $|P_c| \leq 2$, the HDS results match exactly with those of the CDS. For $|P_c| > 2$, the HDS assumes that $|P_c| = \infty$ or, in other words, conduction flux is set to zero. This may be considered too drastic but it nonetheless ensures positivity of coefficients for all values of P_c. The results from the power-law scheme, of course, do mimic the exact solution quite well.

3.8 Total Variation Diminishing Scheme

The difference schemes discussed so far are found to be adequate when the spatial variation of Φ is expected to be smooth and continuous. Often, however, the Φ variation is almost discontinuous (as across a shock). To capture such variation, extremely small values of Δx become necessary, resulting in uneconomic computations. However, if *coarse* grids are employed then UDS, HDS, or power-law schemes produce *smeared* shock predictions.

Total variation diminishing (TVD) schemes enable sharper shock predictions on coarse grids. In these schemes, in addition to magnitude and sign of P_c, the nature of the variation of Φ in the neighbourhood of node P is also sensed. Thus, instead of Equations 3.21 and 3.22, we write

$$P \Phi_e = \frac{1}{2} (P + |P|) \left[f_e^+ \Phi_E + (1 - f_e^+) \Phi_W \right]$$
$$+ \frac{1}{2} (P - |P|) \left[f_e^- \Phi_P + (1 - f_e^-) \Phi_{EE} \right], \quad (3.38)$$

$$P \, \Phi_w = \frac{1}{2}(P + |P|) \left[f_w^+ \, \Phi_P + (1 - f_w^+) \, \Phi_{WW} \right]$$

$$+ \frac{1}{2}(P - |P|) \left[f_w^- \, \Phi_W + (1 - f_w^-) \, \Phi_E \right], \qquad (3.39)$$

where the fs are the appropriate weighting functions to be determined from

$$f = f(\xi) = f \left(\frac{\Phi_U - \Phi_{UU}}{\Phi_D - \Phi_{UU}} \right) \qquad (3.40)$$

with suffix D referring to downstream, U to upstream, and UU to upstream of U. The f_e^+, for example, is thus a function of $(\Phi_P - \Phi_W)/(\Phi_E - \Phi_W)$ and f_e^- is a function of $(\Phi_E - \Phi_{EE})/(\Phi_P - \Phi_{EE})$. Here, EE refers to the node east of node E and WW to the node west of node W.

It is interesting to note that if f equals its associated ξ then Equations 3.38 and 3.39 readily retrieve the UDS formula. Therefore, writing

$$f(\xi) = \xi + f_c(\xi) \qquad (3.41)$$

we can show that

$$P \, \Phi_e = P \, \Phi_e |_{UDS} + \frac{1}{2}(P + |P|) f_{ce}^+ (\Phi_E - \Phi_W)$$

$$- \frac{1}{2}(P - |P|) f_{ce}^- (\Phi_{EE} - \Phi_P), \qquad (3.42)$$

$$P \, \Phi_w = P \, \Phi_w |_{UDS} + \frac{1}{2}(P + |P|) f_{cw}^+ (\Phi_P - \Phi_{WW})$$

$$- \frac{1}{2}(P - |P|) f_{cw}^- (\Phi_E - \Phi_W). \qquad (3.43)$$

Substituting the last two equations in Equation 3.10, we can show that

$$AP \, \Phi_P = AE \, \Phi_E + AW \, \Phi_W + S_{TVD}, \qquad (3.44)$$

where AE, AW, and AP are the same as those for the UDS and the additional source term S_{TVD} contains the f_c terms in Equations 3.42 and 3.43, which the reader can easily derive. The $f_c(\xi)$ functions for some variants of TVD schemes are tabulated in Table 3.2.

To appreciate the implications of the TVD scheme, consider the case in which $P_c > 0$. Then, from Equation 3.42, $P \, \Phi_e = P \, \Phi_P + P f_{ce}^+ (\Phi_E - \Phi_W)$ and $\xi = (\Phi_P - \Phi_W)/(\Phi_E - \Phi_W)$. Therefore, using the Lin–Lin scheme, for example, we get

$$\Phi_e = \begin{cases} \Phi_P, \xi \ni (0, 1), \\[2mm] 2 \, \Phi_P - \Phi_W, \xi \in (0, 0.3), \\[2mm] \dfrac{3}{4} \Phi_P + \dfrac{3}{8} \Phi_E - \dfrac{1}{8} \Phi_W, \xi \in (0.3, 5/6), \\[2mm] \Phi_E, \xi \in (5/6, 1.0). \end{cases} \qquad (3.45)$$

Table 3.2: Function $f_c(\xi)$.

Scheme	Range of ξ	f_c
Second-order UPWIND	$-\infty < \xi < \infty$	$\xi/2$
QUICK [42]	$-\infty < \xi < \infty$	$3/8 - \xi/4$
HLPA [90]	$\xi \ni [0,1]$	0
	$\xi \in [0,1]$	$\xi(1-\xi)$
Lin–Lin [43]	$\xi \ni [0,1]$	0
	$\xi \in [0,0.3]$	ξ
	$\xi \in [0.3, 5/6]$	$3/8 - \xi/4$
	$\xi \in [5/6, 1]$	$1 - \xi$

Thus, for positive P_c, whereas UDS will always return $\Phi_e = \Phi_P$, the TVD scheme returns different values of Φ_e depending on the value of ξ (or shape of the local Φ profile). In fact, as the last expression shows, even a *downwind* value may be returned. The TVD schemes thus typically switch among upwind, central-like, and downwind (DDS) schemes.

3.9 Stability of the Unsteady Equation

We now consider the unsteady conduction–convection equation

$$\rho C_p \frac{\partial T}{\partial t} + \rho C_p u \frac{\partial T}{\partial x} = k \frac{\partial^2 T}{\partial x^2}, \tag{3.46}$$

where all properties and u (positive) are constant. Now, let $X = x/\lambda$, $\tau = \alpha t/\lambda^2$, and $P = u\lambda/\alpha$, where λ is an arbitrary length scale to be further defined shortly. Then, Equation 3.46 will read as

$$\frac{\partial T}{\partial \tau} + P \frac{\partial T}{\partial X} = \frac{\partial^2 T}{\partial X^2}. \tag{3.47}$$

3.9.1 Exact Solution

If at $t = 0$, with $T = T_0 \sin(X)$, the exact solution to Equation 3.47 is

$$T = T_0 \exp(-\tau) \sin(X - P\tau). \tag{3.48}$$

The solution represents a *wave* that moves $P \Delta\tau$ to the right in each time interval $\Delta\tau$. The *amplitude* of the wave is $T_0 \exp(-\tau)$. Thus, over a time interval $\Delta\tau$, the amplitude ratio (or the amplitude decay factor) AR is given by

$$AR = \frac{T_0 \exp[-(\tau + \Delta\tau)]}{T_0 \exp(-\tau)} = \exp(-\Delta\tau). \tag{3.49}$$

To understand the relevance of AR, let T_P be the temperature at X_P after the *first* time step. Then, from Equation 3.48, it follows that

$$\frac{T_P}{T_0 \sin(X_P + \epsilon)} = \exp(-\Delta\tau) = AR, \qquad (3.50)$$

where the wave *propagation speed* ϵ is given by

$$\epsilon_{\text{exact}} = -P\,\Delta\tau = -\frac{u\,\Delta t}{\lambda}. \qquad (3.51)$$

Finally, we note that the arbitrary length scale λ is nothing but the *wavelength* and the propagation speed depends on λ. This dependence on λ is called *dispersion*.

3.9.2 Explicit Finite-Difference Form

Since $P > 0$, using UDS, the explicit discretised form of Equation 3.47 will read as

$$T_P = AE\,T_E^o + AW\,T_W^o + \{1 - (AE + AW)\}\,T_P^o, \qquad (3.52)$$

where

$$AE = \frac{\Delta\tau}{\Delta X^2}, \qquad AW = \frac{\Delta\tau}{\Delta X^2} + P\,\frac{\Delta\tau}{\Delta X}. \qquad (3.53)$$

Now, consider the first time step. Then, $T_P^o = T_0 \sin(X_P)$, $T_E^o = T_0 \sin(X_P + \Delta X)$, and $T_W^o = T_0 \sin(X_P - \Delta X)$. Therefore, after some manipulation, it can be shown that

$$\frac{T_P}{T_0 \sin(X_P)} = [1 - (AE + AW)(1 - \cos\Delta X)] \times \left[1 + \frac{\tan\epsilon_{ED}}{\tan(X_P)}\right], \qquad (3.54)$$

where

$$\tan\epsilon_{ED} = \frac{(AE - AW)\sin(\Delta X)}{1 - (AE + AW)(1 - \cos\Delta X)}. \qquad (3.55)$$

In these equations, the suffix ED denotes explicit differencing. Now, consider the identity

$$\sin(X_P + \epsilon_{ED}) = \sin(X_P)\cos(\epsilon_{ED})\left[1 + \frac{\tan\epsilon_{ED}}{\tan(X_P)}\right]. \qquad (3.56)$$

Substituting Equation 3.56 in Equation 3.54, it follows that

$$AR_{ED} = \frac{T_P}{T_0 \sin(X_P + \epsilon_{ED})} = \frac{1 - (AE + AW)(1 - \cos\Delta X)}{\cos\epsilon_{ED}}. \qquad (3.57)$$

Now, let us consider tendencies of AR_{ED} and $\tan\epsilon_{ED}$ for fine ($\Delta X \to 0$) and coarse ($\Delta X \to \pi$) grids.[4] These are shown in Table 3.3.

---

[4] Note that $1 - \cos\Delta X = 2\sin^2(\Delta X/2)$.

Table 3.3: Comparison of exact and explicit-differencing solutions.

	Exact	Fine grid	Coarse grid
Wave speed	$-P\,\Delta\tau$	$\epsilon_{ED} \to -P\,\Delta\tau$	$\epsilon_{ED} \to 0$
AR	$\exp(-\Delta\tau)$	$\dfrac{1-0.5\,(AE+AW)\,\Delta X^2}{\cos\epsilon_{ED}}$	$\dfrac{1-2\,(AE+AW)}{\cos\epsilon_{ED}}$

The table shows that, for fine grids, ϵ_{ED} behaves in a correct manner but, for coarse grids, ϵ_{ED} does not demonstrate the expected dependence on λ. Therefore, for reasonable *accuracy*, $\Delta X \ll 1$, which implies that one must live with dispersion. Now, *instability* occurs when absolute amplitude ratio exceeds 1. Thus, for *stability*,

$$|AR| = \left| \frac{T_P}{T_0 \sin(X_P + \epsilon)} \right| < 1. \tag{3.58}$$

From Table 3.3, therefore, we must have

$$\left| 1 - 4\frac{\Delta\tau}{\Delta X^2} - 2P\frac{\Delta\tau}{\Delta X} \right| < 1 \quad \text{(coarse grid)},$$

$$\left| 1 - \Delta\iota\left(1 + P\frac{\Delta X}{2}\right) \right| < \cos\epsilon_{ED} \quad \text{(fine grid).} \tag{3.59}$$

These equations show that, to meet the stability requirement, $\Delta\tau$ must be limited to a small value. In pure conduction ($P = 0$), we had already stated these requirements and showed consequences of their violation through a worked example. For the entire range of Ps, however, it is best to observe the following conditions for stability [76]:

$$\frac{\Delta\tau}{\Delta X^2} < \frac{1}{2} \quad \text{and} \quad P\frac{\Delta\tau}{\Delta X} < 1. \tag{3.60}$$

The first condition is operative when $P \to 0$; the second when P is large.

3.9.3 Implicit Finite-Difference Form

The implicitly discretised form of Equation 3.47 will read as

$$(1 + AE + AW)\,T_P = AE\,T_E + AW\,T_W + T_P^o. \tag{3.61}$$

Therefore, substituting for T_P, T_E, T_W, and T_P^o for the first time step, we can show that

$$(1 + AE + AW)\sin(X_P + \epsilon) = AE\sin(X_P + \Delta X + \epsilon)$$
$$+ AW\sin(X_P - \Delta X + \epsilon)$$
$$+ \sin(X_P)\exp(\Delta\tau), \tag{3.62}$$

where ϵ is given by Equation 3.51. To derive an expression for $\tan \epsilon_{\text{ID}}$ (where the subscript ID stands for implicit differencing), therefore, let $X_P = 0$. Then, from Equation 3.62, it can be shown that

$$\tan \epsilon_{\text{ID}} = \frac{(AE - AW)\sin(\Delta X)}{1 + (AE + AW)(1 - \cos \Delta X)}. \tag{3.63}$$

Equation 3.63 again shows that, as $\Delta X \to 0$, $\epsilon_{\text{ID}} \to \epsilon_{\text{exact}}$. Also, notice that the denominator of this equation with a plus sign before $(AE + AW)$ is not the same as the denominator in Equation 3.55. The plus sign indicates that the propagation wave will be more severely damped than in the explicit procedure and this damping will be greater for large ΔX (small wavelength) than for small ΔX. Now, to derive an expression for AR_{ID}, let $X_P = \pi/2$. Then, using Equations 3.62 and 3.63, we can show that

$$AR_{\text{ID}} = \frac{T_P}{T_0 \sin(x_P + \epsilon_{\text{ID}})} = \frac{\cos \epsilon_{\text{ID}}}{1 + (AE + AW)(1 - \cos \Delta X)}. \tag{3.64}$$

Again, this expression is different from Equation 3.57. Equation 3.64 shows that when ΔX and ϵ_{ID} are small, $AR_{\text{ID}} = \left[1 + (AE + AW)\Delta X^2/2\right]^{-1} = (1 + \Delta \tau)^{-1} \sim 1 + \Delta \tau \to \exp(-\Delta \tau)$ as required. When $\Delta X = \pi$ (i.e., for a coarse grid), however, $AR_{\text{ID}} = \cos \epsilon_{\text{ID}}/[1 + 2(AE + AW)]$.

These remarkable results show that AR_{ID} can *never* be greater than 1 because neither AE nor AW can be negative. Thus, the implicit discretisation is *unconditionally* stable and there is no restriction on the time step. Again, in pure conduction ($P = 0$), we had demonstrated this result in Chapter 2 through a worked example. The implicit discretisation is thus *safe*. The only disadvantage is that the discretised equation must be solved iteratively rather than by a *marching* procedure, which is possible in an explicit scheme.

The conclusions arrived at in this section apply equally to variables other than T, to nonuniform grids, to Φ-dependent coefficients, and to multiple dimensions.

EXERCISES

1. Derive Equation 3.7.

2. Show that the CDS formula (3.31) is second-order accurate for both the first and the second derivatives.

3. Show that the UDS formula (3.32) represents convection to only first-order accuracy.

4. Show that the UDS formula is a CDS representation of Equation 3.35.

5. Show correctness of the HDS (3.36) and power-law (3.37) expressions by recalculating the Φ_P values shown in Table 3.1.

6. Consider the steady 1D conduction–convection problem discussed in this chapter. Assume a *nonuniform* grid (i.e., $\Delta x_e \neq \Delta x_w$). Hence, derive expressions for AE, AW, and AP using the power-law scheme. If $\Phi_E = 1$ and $\Phi_W = 0$, calculate Φ_P for $P_{c_e} = u \, \Delta x_e / \alpha = -10, -5, -1, 0, 1, 5$, and 10 when $\Delta x_e / \Delta x_w = 1.2$. [Hint: Start with Equation 3.20 with $\psi_e = F(P_{c_e})$ and $\psi_w = F(P_{c_w})$.]

7. Show that if f in Equation 3.40 equals its associated ξ, Equations 3.38 and 3.39 will yield the UDS formula. Hence, derive Equations 3.42 and 3.43 and the expression for the S_{TVD} term in Equation 3.44.

8. Use $\Phi_E = 1$ and $\Phi_W = 0$ and determine the variation of Φ_P with P_c for the TVD scheme when $\Phi_{EE} = 5$ and $\Phi_{WW} = -0.1$. Assume $-200 < P_c < 200$ and use the Lin–Lin and HLPA schemes. Assume a uniform grid. Compare your results with those given in Table 3.1 and comment on the result. (Hint: Iterations are required.)

9. Show that for a general differencing scheme, the false conductivity is given by $k_{false} = \rho \, C_p \, u \, \Delta x \, (\psi - 0.5)$, where ψ is defined by Equation 3.20. Hence, compare k_{false} for UDS and HDS and comment on the result. Assume a uniform grid.

10. Runchal [61] developed a controlled numerical diffusion with internal feedback (CONDIF) scheme capable of sensing the shape of the local Φ profile. According to this scheme, AE and AW in Equation 3.14 are given by

$$AE = 1 + \left(1 + \frac{1}{R}\right)\left[\frac{|P_c| - P_c}{4}\right], \qquad AW = 1 + (1 + R)\left[\frac{|P_c| + P_c}{4}\right],$$

where

$$R = \frac{\partial \Phi / \partial X |_e}{\partial \Phi / \partial X |_w} = \frac{(\Phi_E - \Phi_P) \, \Delta X_w}{(\Phi_P - \Phi_W) \, \Delta X_e}.$$

Further, the values of R are constrained as follows: If $R < 1/R_{max}$ then $R = 1/R_{max}$; if $R > R_{max}$ then $R = R_{max}$. Typical values assigned to R_{max} vary between 4 and 10. Assuming a uniform grid, show that

(a) If $R = 1$, the CONDIF scheme is the same as the UDS.

(b) CONDIF represents both convection and diffusion terms to second-order accuracy irrespective of the sign and the magnitude of the Peclet number.

(c) Taking $\Phi_W = 0$ and $\Phi_E = 1$, compare values of Φ_P for $|P_c| < 20$ with the exact solution given in Table 3.1. Carry out this comparison for $R_{max} = 4$ and 10.

11. Derive Equations 3.55, 3.57, 3.63, and 3.64.

12. Starting with Equation 3.59, show the correctness of Equations 3.60.

13. Verify that $T = T_0 \exp(-\tau) \sin(X)$ is an exact solution to the *unsteady* heat conduction equation $\partial T / \partial \tau = \partial^2 T / \partial X^2$.

14. It is desired to investigate stability of the equation in Exercise 13 for different values of weighting factor ψ (see Equation 2.6) so that the equation will read as

$$\frac{\partial T}{\partial \tau} = \psi \frac{\partial^2 T}{\partial X^2} + (1 - \psi) \frac{\partial^2 T^o}{\partial X^2}.$$

(a) Obtain a discretised analogue of this equation and substitute the exact solution for temperatures at P, E, and W. Set $X_P = \pi/2$ and show that

$$\exp(-\Delta \tau) = \frac{1 - 4A(1 - \psi)\sin^2(\Delta X/2)}{1 + 4A\psi \sin^2(\Delta X/2)},$$

where $A = AE = AW = \Delta \tau/(\Delta X)^2$.

(b) Hence, show that AR for any X_P is given by

$$AR = \frac{T_P}{T_P^o} = \exp(-\Delta \tau).$$

(c) For stability, $|AR| < 1$. Hence, show that for $\psi < 0.5$, the solution is stable when $A < 0.5/(1 - 2\psi)$ whereas, for $0.5 \le \psi \le 1$, the solution is unconditionally stable.

4   2D Boundary Layers

4.1   Governing Equations

It will be fair to say that the early developments in CFD and heat and mass transfer began with calculation of boundary layers. The term *boundary layer* is applied to long and thin flows: long in the streamwise direction and thin in the transverse direction. The term applies equally to flows attached to a solid boundary (*wall boundary layers*) as well as to jets or wakes (*free-shear layers*).

Calculation of boundary layer phenomena received a considerable boost following the development of a robust numerical procedure by Patankar and Spalding [50]. This made phenomena that were either impossible or too cumbersome to calculate by means of earlier methods (similarity, nonsimilarity, and integral) amenable to fast and economic computation. The procedure, for example, permitted use of variable properties, allowed for completely arbitrary variations of boundary conditions in the streamwise direction, and led to several new explorations of diffusion and source laws. Thus, calculation of free or forced flames or wall fires could be carried out by considering the detailed chemistry of chemical reactions. Similarly, calculation of turbulent flows (and development of turbulence models, in particular) could be brought to a substantial level of maturity through newer explorations of diffusion and source laws governing transport of variables that characterise turbulence. Computer programs based on the Patankar–Spalding procedure are available in [50, 77, 10]. There are also other methods, for example, the Keller–Box method described in [35].

The emphasis in this chapter is on describing the Patankar–Spalding procedure using simple notation. The procedure generalises all two-dimensional boundary layer phenomena by introducing the coordinate system shown in Figure 4.1. This system permits consideration of

1. axisymmetric as well as plane flows,
2. wall boundary layers as well as free-shear layers, and
3. internal (or ducted) as well as external boundary layers.

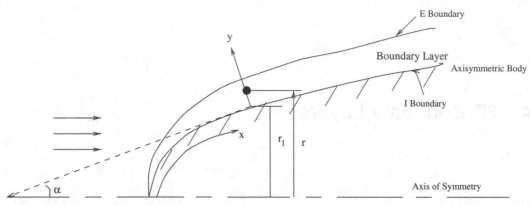

Figure 4.1. The generalised coordinate system.

Following the generalised manner of presentation introduced in Chapter 1, the equations governing *steady* two-dimensional boundary layer phenomena can be written as

$$\frac{\partial(\rho\,u\,r\,\Phi)}{\partial x} + \frac{\partial(\rho\,v\,r\,\Phi)}{\partial y} = \frac{\partial}{\partial y}\left[r\,\Gamma_\Phi\,\frac{\partial\Phi}{\partial y}\right] + r\,S_\Phi, \qquad (4.1)$$

where Φ stands for u (streamwise velocity), w (azimuthal velocity), T (temperature), h (specific enthalpy), and ω_k (mass fraction). The meanings of Γ_Φ and S_Φ are given in Table 4.1. The source terms of the u and w equations assume axisymmetry and $\partial p/\partial r \to 0$ so that $\partial p/\partial x = dp/dx$. In writing the energy equation in terms of T, we assume the specific heat to be constant. Note that in the presence of mass transfer, ρ and Γ represent mixture properties and, in turbulent flows, the suffix eff (for *effective*) must be attached to Γ. Later, we shall find that Φ may also represent further scalar variables such as turbulent kinetic energy k and its dissipation rate ϵ. Independent variables x and y are shown in Figure 4.1 and are applicable to both axisymmetric and plane flows. In the latter, $r = 1$. It will be shown later that r, y, and angle $\alpha(x)$ are connected by an algebraic relation.

Table 4.1: Generalized representation of boundary layer equation.

Φ	Γ_Φ	S_Φ
1	0	0
u	μ	$-dp/dx + B_x$
w	μ	0
ω_k	$\rho\,D_k$	R_k
T	k/C_p	Q'''/C_p
h	k/C_p	Q'''

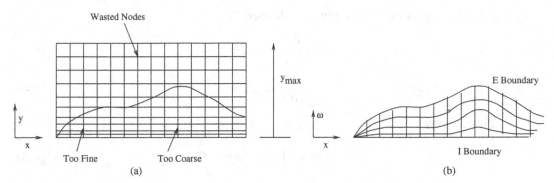

Figure 4.2. Notion of adaptive grid.

Equation 4.1 is to be solved with appropriate boundary conditions at I (inner) and E (external) boundaries and an initial condition at $x = x_0$ (say) for each Φ. Although the I boundary with radius $r_I(x)$ is shown as a wall boundary, it may well be an axis of symmetry with $r_I(x) = 0$. Similarly, although the E boundary is shown as a *free* boundary, it may be a wall boundary. Thus, the specification of the three types of flows mentioned here can be sensed through appropriate designation of I and E boundaries as free, wall, or symmetry boundaries.

Finally, we note that Equation 4.1 is *parabolic*. This implies that the values of Φ at a given x are influenced only by Φ – values *upstream* of x; values *downstream* of x have no influence. Our task now is to discretise Equation 4.1.

4.2 Adaptive Grid

It is well known from boundary layer theory that, in general, boundary layer thicknesses of velocity and other scalar variables can grow or shrink in an arbitrary manner in the streamwise direction. Also, for a given domain length L (say) in the x direction, the maximum values of thicknesses for different Φs are a priori not known. This makes the choice of y_{max} [see Figure 4.2(a)] difficult if the (x, y) coordinate system is used. Further, in this system, for a given number of nodes in the y direction, the boundary layer region of interest may be occupied by too few grid nodes, resulting in *wasted* nodes. Similarly, in some other regions, there may be more nodes than necessary for accuracy. What one would ideally like is a grid that expands and contracts with the changes in boundary layer thickness preserving the *same number* of grid nodes in the transverse direction at each axial location. Such a grid (called an *adaptive* grid) is shown in Figure 4.2(b) with coordinates x and ω, where ω is defined as

$$\omega = \frac{\psi - \psi_I}{\psi_E - \psi_I}, \quad 0 \le \omega \le 1, \tag{4.2}$$

and where ψ is the stream function defined by

$$\frac{\partial \psi}{\partial x} = -\rho \, v \, r, \tag{4.3}$$

$$\frac{\partial \psi}{\partial y} = \rho \, u \, r. \tag{4.4}$$

Thus, at any x

$$\psi = \int \rho \, u \, r \, dy + C, \tag{4.5}$$

where C is a constant. The y coordinate is thus related to ψ and the latter, in turn, is related to ω via Equation 4.2. Suffixes I and E, of course, refer to inner and external boundaries.

4.3 Transformation to (x, ω) Coordinates

Our task now is to transform Equation 4.1 from the (x, y) coordinate system to the (x, ω) coordinate syatem. To do this, we shall follow the sequence $(x, y) \rightarrow (x, \psi) \rightarrow (x, \omega)$. Making use of the mass conservation equation ($\Phi = 1$), we can write Equation 4.1 in *nonconservative* form as

$$\rho \left[u \frac{\partial \Phi}{\partial x} + v \frac{\partial \Phi}{\partial y} \right] = \frac{1}{r} \frac{\partial}{\partial y} \left[r \, \Gamma \frac{\partial \Phi}{\partial y} \right] + S. \tag{4.6}$$

Now, the transformation $(x, y) \rightarrow (x, \psi)$ implies that

$$\left. \frac{\partial}{\partial x} \right|_y = \frac{\partial \psi}{\partial x} \left. \frac{\partial}{\partial \psi} \right|_x + \left. \frac{\partial}{\partial x} \right|_\psi, \tag{4.7}$$

$$\left. \frac{\partial}{\partial y} \right|_x = \frac{\partial \psi}{\partial y} \left. \frac{\partial}{\partial \psi} \right|_y = \rho \, r \, u \left. \frac{\partial}{\partial \psi} \right|_y. \tag{4.8}$$

Substituting these equations in Equation 4.6, we can show that

$$\left. \frac{\partial \Phi}{\partial x} \right|_\psi = \frac{\partial}{\partial \psi} \left[\rho \, r^2 \, u \, \Gamma \frac{\partial \Phi}{\partial \psi} \right] + \frac{S}{\rho \, u}. \tag{4.9}$$

Further, the $(x, \psi) \rightarrow (x, \omega)$ transformation implies that

$$\left. \frac{\partial \Phi}{\partial x} \right|_\psi = \left. \frac{\partial \Phi}{\partial x} \right|_\omega + \left. \frac{\partial \omega}{\partial x} \right|_\psi \left. \frac{\partial \Phi}{\partial \omega} \right|_x, \tag{4.10}$$

but, from Equation 4.2,

$$\left. \frac{\partial \omega}{\partial x} \right|_\psi = \psi_{EI}^{-1} \left[\frac{\partial \psi}{\partial x} - \frac{\partial \psi_I}{\partial x} - \omega \frac{\partial \psi_{EI}}{\partial x} \right]_\psi$$

$$= -\psi_{EI}^{-1} \left[\frac{\partial \psi_I}{\partial x} + \omega \frac{\partial \psi_{EI}}{\partial x} \right], \tag{4.11}$$

where, for convenience,

$$\psi_{EI} \equiv \psi_E - \psi_I. \tag{4.12}$$

Thus, substituting Equation 4.11 in Equation 4.10, we can write Equation 4.9 as

$$\left.\frac{\partial \Phi}{\partial x}\right|_{\omega} + (a + b\omega) \left.\frac{\partial \Phi}{\partial \omega}\right|_{x} = \frac{\partial}{\partial \psi}\left[\rho r^2 u \Gamma \frac{\partial \Phi}{\partial \psi}\right] + \frac{S}{\rho u}, \tag{4.13}$$

where

$$a \equiv -\psi_{EI}^{-1}\frac{\partial \psi_I}{\partial x}, \tag{4.14}$$

$$b \equiv -\psi_{EI}^{-1}\frac{\partial \psi_{EI}}{\partial x}. \tag{4.15}$$

Now, invoking Equation 4.2 again, we obtain

$$\frac{\partial}{\partial \psi} = \psi_{EI}^{-1}\frac{\partial}{\partial \omega}. \tag{4.16}$$

Therefore, Equation 4.13 can be written as

$$\left.\frac{\partial \Phi}{\partial x}\right|_{\omega} + (a + b\omega) \left.\frac{\partial \Phi}{\partial \omega}\right|_{x} = \frac{\partial}{\partial \omega}\left[c\frac{\partial \Phi}{\partial \omega}\right]_{x} + \frac{S}{\rho u}, \tag{4.17}$$

where

$$c \equiv \psi_{EI}^{-2}\rho r^2 u \Gamma. \tag{4.18}$$

Equation 4.17 represents Equation 4.1 in the (x, ω) coordinate system in nonconservative form. To develop the *conservative* counterpart, the equation is written as

$$\left.\frac{\partial \Phi}{\partial x}\right|_{\omega} + \frac{\partial}{\partial \omega}\left[(a + b\omega)\Phi - c\frac{\partial \Phi}{\partial \omega}\right] - \Phi\frac{\partial}{\partial \omega}(a + b\omega) = \frac{S}{\rho u}, \tag{4.19}$$

where, since a and b are not functions of ω,

$$\Phi\frac{\partial}{\partial \omega}(a + b\omega) = b\Phi. \tag{4.20}$$

Now, consider the identity

$$\psi_{EI}^{-1}\frac{\partial}{\partial x}(\psi_{EI}\Phi) = \frac{\partial \Phi}{\partial x} + \Phi\psi_{EI}^{-1}\frac{\partial \psi_{EI}}{\partial x} = \frac{\partial \Phi}{\partial x} - b\Phi. \tag{4.21}$$

Using the last two equations, we can write Equation 4.19 as

$$\frac{\partial}{\partial x}[\psi_{EI}\Phi] + \frac{\partial}{\partial \omega}\left[\psi_{EI}\left\{(a + b\omega)\Phi - c\frac{\partial \Phi}{\partial \omega}\right\}\right] = \frac{\psi_{EI}S}{\rho u}. \tag{4.22}$$

This is the required boundary layer equation in the (x, ω) coordinate system written in conservative form. It will be useful at this stage to interpret the terms in Equation 4.22. Thus, from Equations 4.12 and 4.5, it is easy to show that ψ_{EI}

represents the total streamwise mass flow rate through the boundary layer at any x. Similarly, making use of the definitions of a, b, and c and using Equation 4.16, we can show that

$$\psi_{EI} \left\{ (a + b\omega)\,\Phi - c\,\frac{\partial \Phi}{\partial \omega} \right\} = r\,\dot{m}\,\Phi - r\,\Gamma\,\frac{\partial \Phi}{\partial y}, \tag{4.23}$$

where

$$\frac{\partial}{\partial \omega} = \frac{\psi_{EI}}{\rho\,r\,u}\,\frac{\partial}{\partial y} \tag{4.24}$$

and

$$r\,\dot{m} = r\,\rho\,v = (1 - \omega)\,r_I\,\dot{m}_I + \omega\,r_E\,\dot{m}_E$$

$$\text{with} \quad \dot{m}_E = (\rho\,v)_E, \qquad \dot{m}_I = (\rho\,v)_I. \tag{4.25}$$

Thus the total transverse mass flux $\dot{m}$ at any y is a weighted sum of mass fluxes at the inner ($\dot{m}_I$) and external ($\dot{m}_E$) boundaries in the positive y direction. Equation 4.23 therefore represents the total convective–diffusive flux in the y direction. Then by substituting Equation 4.23, Equation 4.22 can be written as

$$\frac{\partial}{\partial x}\,[\psi_{EI}\,\Phi] + \frac{\partial}{\partial \omega}\left[r\,\dot{m}\,\Phi - r\,\Gamma\,\frac{\partial \Phi}{\partial y} \right] = \frac{\psi_{EI}\,S}{\rho\,u}. \tag{4.26}$$

4.4 Discretisation

Figure 4.3 shows the (x, ω) grid at streamwise location x. Suffix u refers to *upstream* and d refers to *downstream*. Note that nodes N, P, and S are not equidistant because $\Delta \omega$, in general, will not be uniform. This will become apparent in a later section. To derive the discretised version of Equation 4.26, each term in the equation will be integrated over the control volume. Thus, assuming source term S to be constant over the control volume, we have

$$\int_{x_u}^{x_d} \int_{s}^{n} \frac{\psi_{EI}\,S}{\rho\,u}\,dx\,d\omega = \int_{x_u}^{x_d} \int_{s}^{n} \frac{S}{\rho\,u}\,dx\,d\psi = \int_{x_u}^{x_d} \int_{s}^{n} S\,r\,dx\,dy$$

$$= S\,r_P\,\Delta x\,\Delta y = S\,\Delta V, \tag{4.27}$$

where

$$\Delta V = r_P\,\Delta x\,\Delta y. \tag{4.28}$$

Similarly, the streamwise convection term integrates to

$$\int_{x_u}^{x_d} \int_{s}^{n} \frac{\partial}{\partial x}\,[\psi_{EI}\,\Phi]\,dx\,d\omega = \left[(\psi_{EI}\,\Phi)^d - (\psi_{EI}\,\Phi)^u \right]_P \Delta\omega. \tag{4.29}$$

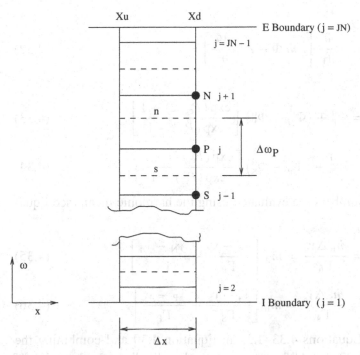

Figure 4.3 The (x, ω) grid.

Finally, the convection–diffusion term in the transverse direction integrates to

$$\int_{x_u}^{x_d} \int_s^n \frac{\partial}{\partial \omega} \left[r \, \dot{m} \, \Phi - r \, \Gamma \, \frac{\partial \Phi}{\partial y} \right] dx \, d\omega = \left\{ r \, \dot{m} \, \Phi - r \, \Gamma \, \frac{\partial \Phi}{\partial y} \right\}_n \Delta x$$

$$- \left\{ r \, \dot{m} \, \Phi - r \, \Gamma \, \frac{\partial \Phi}{\partial y} \right\}_s \Delta x. \quad (4.30)$$

Equation 4.30 implies that the net flux at the cell faces is uniform between x_u and x_d. Now, assuming *linear* variation of Φ between adjacent nodes gives

$$\left. \frac{\partial \Phi}{\partial y} \right|_n = \frac{(\Phi_N^d - \Phi_P^d)}{\Delta y_n}, \qquad \left. \frac{\partial \Phi}{\partial y} \right|_s = \frac{(\Phi_P^d - \Phi_S^d)}{\Delta y_s}, \qquad (4.31)$$

where $\Delta y_n = y_N - y_P$ and $\Delta y_s = y_P - y_S$. Note that the Φs are evaluated at x_d rather than midway between x_u and x_d. However, assuming that Δx is small, this liberty is permissible.

The next task is to evaluate convective fluxes at the cell faces. To do this, we may use any of the schemes introduced in the previous chapter but, following Patankar [52], we use the *exponential* scheme that follows from the exact solution to the equation

$$\frac{\partial}{\partial y}\left[r\,\dot{m}\,\Phi - r\,\Gamma\,\frac{\partial \Phi}{\partial y}\right] = 0. \tag{4.32}$$

Then, it follows that

$$\Phi_n = \Phi_P^d + \left(\Phi_N^d - \Phi_P^d\right)\left[\frac{\exp\left(P_{c_n}/2\right) - 1}{\exp\left(P_{c_n}\right) - 1}\right], \tag{4.33}$$

$$\Phi_s = \Phi_S^d + \left(\Phi_P^d - \Phi_S^d\right)\left[\frac{\exp\left(P_{c_s}/2\right) - 1}{\exp\left(P_{c_s}\right) - 1}\right], \tag{4.34}$$

where, the cell Peclet numbers are evaluated using the harmonic mean (see Equation 2.58):

$$P_{c_n} = \frac{\dot{m}_n\,\Delta y_n}{\Gamma_n} = \dot{m}_n\left[\frac{y_n - y_P}{\Gamma_P} + \frac{y_N - y_n}{\Gamma_N}\right], \tag{4.35}$$

$$P_{c_s} = \frac{\dot{m}_s\,\Delta y_s}{\Gamma_s} = \dot{m}_n\left[\frac{y_s - y_S}{\Gamma_S} + \frac{y_P - y_s}{\Gamma_P}\right]. \tag{4.36}$$

Thus, substituting Equations 4.33–4.36 in Equation 4.30 and combining the latter with Equations 4.27 and 4.29, we can show that the discretised version of Equation 4.26 takes the following form:

$$AP\,\Phi_P^d = AN\,\Phi_N^d + AS\,\Phi_S^d + AU\,\Phi_P^u + S\,\Delta V, \tag{4.37}$$

where

$$AN = \frac{r_n\,\dot{m}_n\,\Delta x}{\exp P_{c_n} - 1}, \tag{4.38}$$

$$AS = \frac{r_s\,\dot{m}_s\,\Delta x\,\exp P_{c_s}}{\exp P_{c_s} - 1}, \tag{4.39}$$

$$AU = \psi_{EI}^u\,\Delta\omega, \qquad AP = AU + AN + AS. \tag{4.40}$$

In deriving the AP coefficient, use is made of the mass conservation equation. Thus,

$$\int_s^n \frac{\partial}{\partial x}(\rho\,r\,u)\,dy = -\int_s^n \frac{\partial}{\partial y}(\rho\,r\,v)\,dy$$

$$= -(r_n\,\dot{m}_n - r_s\,\dot{m}_s) \tag{4.41}$$

$$= \frac{\partial}{\partial x}\int_s^n \frac{\partial\psi}{\partial y}\,dy$$

$$= \frac{\Delta\omega}{\Delta x}\left(\psi_{EI}^d - \psi_{EI}^u\right). \tag{4.42}$$

Finally, the node-indexed version of Equation 4.37 can be written as

$$AP_j\,\Phi_j = AN_j\,\Phi_{j+1} + AS_j\,\Phi_{j-1} + AU_j\,\Phi_j^u + S_j\,\Delta V_j \tag{4.43}$$

for $j = 2, 3, \ldots, JN - 1$. Note that superscript d is now dropped for convenience.

4.5 Determination of ω, y, and r

Equation 4.43 represents a set of algebraic equations at a streamwise location x_d. These equations can be solved by TDMA when values of Φ_j^u at x_u are known along with the two boundary conditions at x_d (i.e., at $j = 1$ and $j = JN$). Thus, starting with $x = x_0$ (say), one can execute a *marching* procedure taking step Δx. This situation is very much like the unsteady conduction problem in which the marching procedure is executed with time step Δt.

Thus, at $x = x_0$, the $u_j \sim y_j$ relationship is assumed to have been prescribed either from experimental data or from an analytical solution. One can use this prescription to set ω_j *once and for all*. Let

$$\omega_j = \omega_P, \quad \omega_{c,j} = \omega_s, \quad \psi_j = \psi_P, \quad \psi_{c,j} = \psi_s,$$

$$y_j = y_P, \quad y_{c,j} = y_s, \quad r_j = r_P, \quad r_{c,j} = r_s, \tag{4.44}$$

where, at $x = x_0$, $y_j (j = 1, 2, \ldots, JN)$ are known. Thus, one can set $y_{c,1} = y_{c,2} = y_1$ where y_1 refers to the I boundary and y_{JN} to the E boundary. Now, from the geometry of Figure 4.1, it follows that r_j and $r_{c,j}$ can be evaluated from the formula

$$r = r_I + y \cos(\alpha), \tag{4.45}$$

where α is function of x. This completes the grid specification at $x = x_0$.

For evaluation of ω_j, we first calulate ψ_j. Thus, setting $\psi_1 = \psi_{c,1} = \psi_I$ (say), where ψ_I is arbitrarily chosen, one can use Equation 4.5 to set all other ψ_j. The relevant discretised equations are

$$\psi_{c,j} = \psi_{c,j-1} + (\rho r u)_{j-1} (y_{c,j} - y_{c,j-1}), \qquad j = 2, 3, \ldots, JN, \tag{4.46}$$

$$\psi_j = \psi_{j-1} + 0.5 \left\{ (\rho r u)_j + (\rho r u)_{j-1} \right\} (y_j - y_{j-1}), \qquad j = 2, 3, \ldots, JN. \tag{4.47}$$

It is now a simple matter to evaluate ω_j and $\omega_{c,j}$ using definition (4.2). Thus, ω_j at y_j represents the ratio of streamwise mass flow rate from $y_1 = y_I$ to y_j to the total mass flow rate from y_I to y_E at $x = x_0$. It is now assumed that this ratio remains *intact at all values of x* and thus the ω_j distribution does not change throughout the domain in the x direction.

Note, however, that the physical distance y (and therefore r) must go on changing at different values of x as the boundary layer grows or shrinks. We thus seek the $y_j \sim \omega_j$ relationship applicable to every x.

Plane Flow

From Equations 4.4 and 4.2, it can be shown that

$$y = \psi_{\text{EI}} \int_0^\omega \frac{d\omega}{\rho u} = I(\text{say}).\tag{4.48}$$

Thus, knowing the initially set values of ω_j and $\omega_{c,j}$, y_j and $y_{c,j}$ can be estimated. Note that ψ_{EI} and ρu will change with x. Therefore, y will also change with x.

Axisymmetric Flow

In this case, from Equation 4.45, it follows that

$$\frac{\psi_{\text{EI}}}{\rho u} d\omega = (r_I + y \cos\alpha)\, dy\tag{4.49}$$

and, therefore, from Equation 4.48

$$I = r_I y + \cos\alpha \frac{y^2}{2}.\tag{4.50}$$

The solution to this quadratic equation suitable for computer implementation is

$$y = \frac{2I}{r_I + \left(r_I^2 + 2I \cos\alpha\right)^{0.5}},\tag{4.51}$$

where I is given by Equation 4.48. Now, knowing y_j and $y_{c,j}$ in this manner, r_j and $r_{c,j}$ can be evaluated using Equation 4.45.

4.6 Boundary Conditions

At the E and I boundaries, three types of boundary conditions are possible: symmetry, wall, or free stream. We discuss them in turn.

4.6.1 Symmetry

There can be no mass flux across the symmetry plane. Also, $\partial\Phi/\partial n|_b = 0$, where suffix b denotes the E or I boundary node. This implies that

$$\Phi_b = \Phi_{nb} \quad \text{and} \quad \dot{m}_b = 0,\tag{4.52}$$

where suffix nb stands for *near-boundary* node. A further consequence of the $\dot{m}_b = 0$ condition is that $\partial\psi_b/\partial x = 0$ or $\psi_b = $ constant. The boundary condition can be effected by setting $AS_2 = 0$ at the I boundary or $AN_{JN-1} = 0$ at the E boundary.

4.6.2 Wall

The term *wall* signifies a solid boundary. However, it must be remembered that when a gas flows over a liquid surface, the gas–liquid *interface* too will act like a wall. For different Φs, the wall boundary conditions are also different. We consider them in turn.

Velocity Variables $\Phi = u$ **or** w
For these variables,

$$u_b = u_{wall}, \qquad w_b = w_{wall}. \tag{4.53}$$

Thus, if the surface is rotating about the axis of symmetry (see Figure 4.1) with angular velocity Ω, then the surface fluid velocity will be $w_{wall} = r_1 \Omega$. Similarly, the streamwise velocity will always be zero unless the surface itself is moving with velocity u_{wall}. Equation 4.53, therefore, signifies the *no-slip* condition.

In some circumstances, a fluid may be injected (by blowing) into the boundary layer or the boundary layer fluid may be withdrawn (by suction) through the wall. Alternatively, in case of evaporation or surface burning, mass will be transferred into the boundary layer. In all such cases $\dot{m}_b$ is known or knowable and the consequence is

$$\psi_b(x) = \psi_b(x - \Delta x) - r_b \dot{m}_b \Delta x. \tag{4.54}$$

Thermal Variables $\Phi = T$ **or** h
For these variables, typically two types of conditions are specified. In the first, the value of the variable itself is specified. Thus,

$$T_b = T_{wall}(x), \qquad h_b = h_{wall}(x). \tag{4.55}$$

In the second, the heat flux q_b is specified. Then, at the I boundary, for example,

$$q_b = -k \left. \frac{\partial T}{\partial y} \right|_{y=0} = -\frac{k}{C_p} \left. \frac{\partial h}{\partial y} \right|_{y=0} = -\Gamma \left. \frac{\partial h}{\partial y} \right|_{y=0}. \tag{4.56}$$

The flux boundary condition is effected by adding $q_b \Delta x$ to the source term of Equation 4.43 for $j = 2$ and, further, by setting $AS_2 = 0$, the values of T_b or h_b can be extracted in the usual manner. A similar procedure is adopted if q_b is specified at the E boundary.

In a chemically reacting boundary layer, the mass transfer flux at the wall $\dot{m}_b''$ is given by

$$\dot{m}_b'' = (h_b - h_T)^{-1} \sum_k \left[\rho_m D_k \frac{\partial \omega_k}{\partial y} h_k + k_m \frac{\partial T}{\partial y} \right]_{y=0}, \tag{4.57}$$

where h_T is the enthalpy of the mixture deep inside the I boundary. If the Lewis number is taken to be unity (i.e., $Pr = Sc$) or a simple chemical reaction (SCR) is assumed with equal specific heats then this relationship can be simplified to [33]

$$\dot{m}_b'' = (h_b - h_T)^{-1} \Gamma \left. \frac{\partial h}{\partial y} \right|_{y=0}. \tag{4.58}$$

Knowing $\dot{m}_b''$, boundary condition h_b can be extracted.

Mass Transfer Variables $\Phi = \omega_k$

The most common boundary condition [33] for these variables at the I boundary, for example, is

$$\dot{m}_b'' = r_b \dot{m}_b = (\omega_{k,b} - \omega_{k,T})^{-1} \Gamma_k \left. \frac{\partial \omega_k}{\partial y} \right|_{y=0}, \tag{4.59}$$

where $\omega_{k,T}$ refers to the mass fraction *deep* inside the I boundary. The suffix T, thus, represents the *transferred substance* state and $\omega_{k,T}$ must be known. Equation 4.59 is again a flux condition, therefore, it can be treated in the manner of the q_b condition just described. Again, from the converged solution, $\omega_{k,b}$ can be extracted.

When heterogeneous chemical reactions occur at the wall, $\dot{m}_b$ is typically given by the Arrhenius relationship, which yields

$$\dot{m}_b = f(\omega_{k,b}, T_b). \tag{4.60}$$

The exact implementation of the boundary condition for a heterogeneous reaction requires modification of Equation 4.59. This is explained later through an example of carbon burning (see Equation 4.129).

In problems involving evaporation or condensation, the value of $\omega_{k,b}$ itself can be specified from the equilibrium relation (or saturation condition).

$$\omega_{k,b} = f(T_b). \tag{4.61}$$

Thus, in mass transfer problems with or without surface chemical reaction, $\dot{m}_b$ can be known and this knowledge can be used to evaluate ψ_b from Equation 4.54. It is important to remember, however, that the most general problem of mass transfer is usually quite complex and, therefore, several manipulations are typically introduced to simplify the boundary condition treatment [33, 38].

4.6.3 Free Stream

The free-stream boundary condition has relevance only when external[1] boundary layers are considered. The free stream is really a *fictitious* boundary and is identified

---

[1] In internal flows, only wall or symmetry conditions are relevant because in these flows the *flow width* is a priori known. Thus, for developing flow between two parallel plates a distance b apart, for example, the flow width b remains constant with x. However, in 2D plane diffusers or nozzles, b may vary with x but still be known a priori.

with the notion that the variation in Φ in the transverse direction *asymptotically* approaches a value Φ_∞ (say) there. Thus, the fictitious notion of a boundary layer thickness is associated with

$$\frac{\Phi - \Phi_I}{\Phi_\infty - \Phi_I} = A, \tag{4.62}$$

where suffix I refers to the inner boundary (wall or symmetry) and A is typically taken to be 0.99 by convention. Note, however, that this boundary layer thickness will be different for different meanings of Φ and the magnitude of thickness typically depends on the Prandtl number[2] Pr_Φ defined as

$$Pr_\Phi \equiv \frac{\nu}{\Gamma_\Phi}. \tag{4.63}$$

The Prandtl number is a property of the fluid. In fact, in Table 4.1, we may replace k/C_p by μ/Pr_T and ρD_k by μ/Pr_{ω_k}.

There is one further notion associated with the free stream. If we assume the E boundary to be the free boundary (see Figure 4.1), the flow region above the boundary can be taken to be a region in which there is no transverse convection or diffusion and

$$\Phi_b = \Phi_\infty(x), \tag{4.64}$$

where $\Phi_\infty(x)$ is specified. However, the physical location where this boundary condition is to be applied is not a priori known because of the *asymptotic* nature of variation of Φ in the vicinity of this boundary. To circumvent this problem, Patankar and Spalding [50] relied on estimating the *entrainment rate* $(-\dot{m}_E)$ into the boundary layer that occurs from the fluid above the E-boundary.

Thus, as previously mentioned, since there is no net flux of Φ in the transverse direction, from Equation 4.17, it follows that

$$(a + b\omega)_E \left.\frac{\partial \Phi}{\partial \omega}\right|_E = \frac{\partial}{\partial \omega}\left[c\frac{\partial \Phi}{\partial \omega}\right]_E. \tag{4.65}$$

However, at the E boundary, $\omega = 1$. Therefore,

$$(a + b\omega)_E = a + b = -\psi_{EI}^{-1}\frac{\partial \psi_E}{\partial x}. \tag{4.66}$$

Thus, Equation 4.65 can be written as

$$\frac{\partial \psi_E}{\partial x} = -\psi_{EI}\left(\frac{\partial \Phi}{\partial \omega}\right)^{-1}\frac{\partial}{\partial \omega}\left[c\frac{\partial \Phi}{\partial \omega}\right] = -\psi_{EI}\frac{\partial}{\partial \Phi}\left[c\frac{\partial \Phi}{\partial \omega}\right]$$

$$= -\frac{\partial}{\partial \Phi}\left[r\,\Gamma_\Phi\frac{\partial \Phi}{\partial y}\right] = -r_E\,\dot{m}_E. \tag{4.67}$$

---

[2] The term Prandtl number applies to variables T and h. When $\Phi = \omega_k$, the appropriate dimensionless number is called the Schmidt number (Sc). For velocity variables, of course, $Pr_\Phi = 1$. We thus use Pr_Φ generically to cover all Φs.

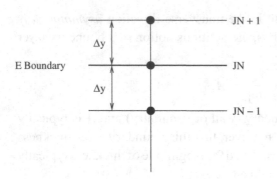

Figure 4.4. The grid construction near the E boundary.

Now, to estimate the required $\dot{m}_E$, we adopt the following special procedure. Since the E boundary is located at $j = JN$ (see Figure 4.4),

$$\frac{\partial}{\partial \Phi}\left[r\,\Gamma_\Phi\,\frac{\partial\Phi}{\partial y}\right]_{JN} = \left(\frac{\partial\Phi}{\partial y}\right)^{-1}\left[\Gamma_\Phi\,\frac{\partial^2\Phi}{\partial y^2} + \frac{\partial\Gamma_\Phi}{\partial y}\,\frac{\partial\Phi}{\partial y}\right]_{JN} r_{JN}. \quad (4.68)$$

However, near the E boundary, $\partial\Gamma_\Phi/\partial y|_{JN}$ can be set to zero. Now, let Δy be the distance between the JN and $JN - 1$ nodes. We next *construct* an imaginary node $JN + 1$ at Δy above the E boundary. Then,

$$\left.\frac{\partial\Phi}{\partial y}\right|_{JN} = \frac{\Phi_{JN+1} - \Phi_{JN-1}}{2\,\Delta y},$$

$$\left.\frac{\partial^2\Phi}{\partial y^2}\right|_{JN} = \frac{\Phi_{JN+1} - 2\,\Phi_{JN} + \Phi_{JN-1}}{\Delta y^2}. \quad (4.69)$$

Noting that $\Phi_{JN+1} = \Phi_{JN} = \Phi_\infty$, we can simplify the derivative expressions further and, therefore, Equation 4.68 can be written as

$$\frac{\partial}{\partial \Phi}\left[r\,\Gamma_\Phi\,\frac{\partial\Phi}{\partial y}\right]_{JN} \simeq 2\,\left.\frac{r\,\Gamma_\Phi}{\Delta y}\right|_{JN} = \frac{2\,r_{JN}\,\Gamma_\Phi}{y_{JN} - y_{JN-1}}. \quad (4.70)$$

Thus, from Equation 4.67, since $r_{JN} = r_E$

$$\dot{m}_{E,std} \simeq -\frac{1}{r_E}\,\frac{\partial\psi_E}{\partial x} \simeq \frac{2\,\Gamma_{\Phi,E}}{y_{JN} - y_{JN-1}}. \quad (4.71)$$

Using the above estimate, it follows that

$$\psi_E(x) \simeq \psi_E(x - \Delta x) - \frac{2\,r_E\,\Gamma_{\Phi,E}\,\Delta x}{y_{JN} - y_{JN-1}}. \quad (4.72)$$

With this estimate, it is now possible to evaluate coefficients in Equation 4.43. This is because, when the E boundary is a free boundary, the I boundary can only be a wall or a symmetry boundary for which $\psi_I(x)$ is already known.

Equation 4.71 is of course an approximate formula for $\dot{m}_E$. To derive an exact formula, we note that Γ_Φ will be different for different Φs and, as already noted, the respective boundary layer thicknesses will also be different. Our interest lies in selecting that Φ for which the thickness is *largest*. Usually, the largest thickness will correspond to the largest Γ_Φ, and for this selected Φ, we evaluate

$$R = \frac{|\Phi_{JN} - \Phi_{JN-1}|}{\Delta \Phi^*}, \qquad \Delta \Phi^* = 10^{-3} \text{ (say)}, \qquad (4.73)$$

where $\Delta \Phi^*$ is a sufficiently small reference quantity. Since Equation 4.43 is iteratively solved, Patankar [52] has suggested the following formula for exact evaluation:

$$\dot{m}_E \text{(exact)} = \dot{m}_{E,std} \times R^n, \qquad (4.74)$$

where, from computational experience, $n \simeq 0.1$ is found to be a convenient value in most cases. Thus, when Equation 4.43 has converged, ψ_E, as evaluated from Equation 4.72, will provide a correct estimate of total mass flow rate $\psi_{EI} = \psi_E - \psi_I$ through the boundary layer at the given x. Once this mass flow rate is known, the y dimension and hence the largest boundary layer thickness among all Φs can be estimated.

4.7 Source Terms

4.7.1 Pressure Gradient

In external boundary layers, the pressure gradient is specified or indirectly evaluated from

$$\frac{dp}{dx} = -\rho U_\infty \frac{d U_\infty}{d x}, \qquad (4.75)$$

where $U_\infty (x)$ is specified. In internal flows, however, a special procedure must be adopted to specify the pressure gradient. The procedure relies on satisfying the overall mass flow rate balance at every streamwise location x. Thus, in a general duct, let $A_d (x)$ represent the duct area between the axis of symmetry (I boundary) and the wall (E boundary). Then

$$A_d = \int_I^E r \, dy = \psi_{EI} \int_0^1 \frac{d\omega}{\rho u}. \qquad (4.76)$$

Therefore,

$$\frac{A_d}{\psi_{EI}} = C \text{ (constant)} = \sum \frac{\Delta \omega_j}{\rho_j u_j}. \qquad (4.77)$$

The task now is to replace u_j in terms of the pressure gradient. To do this, Patankar [52] writes the discretised version of the momentum equation as

$$AP_j u_j = AN_j u_{j+1} + AS_j u_{j-1} + D_j - \Delta V_j p_x, \qquad (4.78)$$

where p_x is the pressure gradient and D_j contains source terms arising from other body forces. To solve this equation by TDMA, let the postulated equation be

$$u_j = A_j u_{j+1} + B_j - R_j p_x, \tag{4.79}$$

where $R_1 = R_{JN} = 0$. Then, the recurrence relations will take the following form:

$$A_j = \frac{AN_j}{DEN}, \quad B_j = \frac{AS_j B_{j-1} + D_j}{DEN}, \quad R_j = \frac{AS_j R_{j-1} + \Delta V_j}{DEN}, \tag{4.80}$$

where $DEN = AP_j - AS_j A_{j-1}$. Note that $A(2)$, $B(2)$, and $R(2)$ can be recovered from Equation 4.78. Therefore, the coefficients in Equation 4.80 can be determined for $j = 3$ to $JN - 1$ by recurrence. Now, let u_j be further postulated as

$$u_j = F_j - G_j p_x, \tag{4.81}$$

where, again by recurrence, F_j and G_j can be determined for $j = JN - 1$ to 2 by

$$F_j = A_j F_{j+1} + B_j, \quad G_j = A_j G_{j+1} + R_j, \tag{4.82}$$

where $A(JN) = G(JN) = 0$. Thus, it is possible to replace u_j in Equation 4.77 by Equation 4.81. The replacement yields a nonlinear equation in p_x:

$$\sum \frac{\Delta \omega_j}{\rho_j (F_j - G_j p_x)} - C = 0. \tag{4.83}$$

This equation can be solved by Newton–Raphson iterative procedure:

$$p_x = p_x^* + \frac{C - S_1}{S_2},$$

$$S_1 = \sum \frac{\Delta \omega_j}{\rho_j (F_j - G_j p_x^*)},$$

$$S_2 = \sum \frac{\Delta \omega_j G_j}{\rho_j (F_j - G_j p_x^*)^2}, \tag{4.84}$$

where p_x^* is the guessed pressure gradient. Iterations are continued until $|C - S_1| < 10^{-4} C$. Usually, about five iterations suffice.

Finally, we note that in free-shear flows, the pressure gradient is zero.

4.7.2 Q''' and R_k

The source terms in the energy and mass transfer equation depend on the problem at hand. In general, however,

$$Q''' = \dot{Q}_{rad} + \dot{Q}_{cr} + \mu \, \Phi_v + \frac{D p}{D t} + \dot{Q}_{md}, \tag{4.85}$$

where $\dot{Q}_{rad} = \partial q_{rad,y} / \partial y$ represents the radiation contribution, $\dot{Q}_{cr}$ represents the generation rate due to endothermic or exothermic chemical reactions, $\mu \, \Phi_v = \mu \, (\partial u / \partial y)^2$ represents the viscous dissipation effect, $D p / D t =$

$u\,\partial p/\partial x$ represents the pressure–work effect in steady flow, and $\dot{Q}_{md} = \partial/\partial y\,\{(\sum_{\text{all } k}\rho\,D_k\,\partial\omega_k/\partial y)\,h_k\}$ represents the contribution of species diffusion mass transfer having specific enthalpy h_k. If h_k equals mixture enthalpy h then $\dot{Q}_{md} = 0$.

When no chemical reaction is present, $R_k = 0$. However, for a reacting boundary layer, R_k will be finite for each species because each may be generated via some reactions and destroyed via some other reactions among the *postulated chemical reactions*. Very often, for gaseous fuels and for highly volatile solid/liquid fuels, an SCR can be assumed [73]. The SCR is specified as

$$1\,\text{kg of fuel} + R_{st}\,\text{kg of oxidant} \rightarrow (1 + R_{st})\,\text{kg of product}, \qquad (4.86)$$

where R_{st} is the stoichiometric ratio for the fuel under consideration. Thus, there are three species and one must specify R_{fu}, R_{ox}, and R_{pr}. However, in an SCR, $R_{fu} = R_{ox}/R_{st} = -R_{pr}/(1 + R_{st})$ so that no net mass is generated or destroyed as a result of chemical reaction. This enables construction of a *conserved scalar* variable $\Psi = \omega_{fu} - \omega_{ox}/R_{st} = \omega_{fu} + \omega_{pr}/(1 + R_{st})$ when mass diffusivities of all species are taken equal. Thus, one may now solve only for ω_{fu} and Ψ with $R_{\Psi} = 0$ instead of three variables. Further, $\dot{Q}_{cr} = |R_{fu}|\Delta H_c$ where ΔH_c is the heat of combustion of the fuel. The value of R_{fu} is obtained from a reaction rate law

$$R_{fu} = R_{fu,kin} = -A\exp\left(-\frac{E}{R_u\,T}\right)\omega_{fu}^m\,\omega_{ox}^n, \qquad (4.87)$$

where, preexponential constant A and constants E, m, and n are specified for the fuel [82] and R_u is the universal gas constant.

If turbulent reacting flow is considered then the *effective* R_{fu} is given by a variant [44] of the eddy-breakup model due to Spalding [74],

$$R_{fu} = -\rho_m\,\frac{\epsilon}{e}\,\min\left\{A\,\omega_{fu},\ A\,\frac{\omega_{ox}}{R_{st}},\ A'\,\frac{\omega_{prod}}{(1 + R_{st})},\ R_{fu,kin}\right\}, \qquad (4.88)$$

where $A = 4$ and $A' = 2$. The postulated arguments in favour of this expression are beyond the scope of this book.

4.8 Treatment of Turbulent Flows

In turbulent flows, Γ_Φ in Table 4.1 will assume an *effective* value. Thus, following Equation 4.63, we have

$$\Gamma_{\Phi,\text{eff}} = \frac{\mu}{Pr_\Phi} + \frac{\mu_t}{Pr_{t,\Phi}}, \qquad (4.89)$$

where suffix t denotes the *turbulent* contribution. The task now is to represent μ_t and $Pr_{t,\Phi}$ via modelled expressions. This exercise, called *turbulence modelling*, implies validity of the Boussinesq approximation for turbulent viscosity. Although there are many variants, all turbulence models of this type stem from a dimensionally correct representation

$$\mu_t \propto \rho \, l \, v', \tag{4.90}$$

where v' is the representative velocity fluctuation scale in the transverse direction y and l is a representative length scale. Two turbulence models used extensively for boundary layer calculations are described in the following.

4.8.1   Mixing Length Model

Since v' is responsible for transverse momentum transfer, it may be written in dimensionally correct form as

$$v' = l_m \left| \frac{\partial u}{\partial y} \right| \tag{4.91}$$

so that

$$\mu_t = \rho \, l_m^2 \left| \frac{\partial u}{\partial y} \right|, \tag{4.92}$$

where l_m is called Prandtl's mixing length. Now, because the velocity gradient can be evaluated from the solution of the momentum equation, l_m must be prescribed to complete evaluation of μ_t. Kays and Crawford [33], after extensive investigations of a variety of wall-boundary-layer flows have prescribed the following formulas:

$$l_m = \begin{cases} \kappa \, y \left[1 - \exp\left(-\frac{y^+}{A^+} \right) \right], & \text{for } \frac{y}{\delta} < 0.2, \tag{4.93} \\[2mm] 0.085 \, \delta & \text{for } \frac{y}{\delta} \geq 0.2, \tag{4.94} \end{cases}$$

where y is the normal distance from the wall, δ is the velocity-boundary-layer thickness and $\kappa = 0.41$. Further,

$$y^+ = \frac{y \, u_\tau}{\nu}, \quad u_\tau = \sqrt{\frac{\tau_w}{\rho}}, \quad \tau_w = \mu \frac{\partial u}{\partial y} |_w. \tag{4.95}$$

Finally, the value of A^+ is sensitised to effects of suction or blowing and local pressure gradient in a generalised manner as

$$A^+ = 25 \left[a \left\{ v_w^+ + b \, p^+ / \left(1 + c \, v_w^+ \right) \right\} + 1 \right]^{-1}, \tag{4.96}$$

where

$$p^+ = \mu \frac{d \, p}{d \, x} \left(\tau_w^3 \rho \right)^{-0.5}, \quad v_w^+ = \frac{v_w}{u_\tau}, \tag{4.97}$$

and $a = 7.1$, $b = 4.25$, and, $c = 10.0$. If $p^+ > 0$ then $b = 2.9$ and $c = 0$.

Laminar-to-Turbulent Transition

To predict laminar-to-turbulent transition, the effective value of Γ_Φ is written as

$$\Gamma_{\Phi,\text{eff}} = \frac{\mu}{Pr_\Phi} + \Upsilon \frac{\mu_t}{Pr_{t,\Phi}}, \tag{4.98}$$

where the *intermittancy* factor Υ is given [1] by

$$\Upsilon = 1 - \exp\left\{-5\left(\frac{x - x_{ts}}{x_{te} - x_{ts}}\right)\right\}. \tag{4.99}$$

In this equation, x_{ts} and x_{te} denote the start and the end of transition, respectively. When $x = x_{te}$, $\Upsilon = 1$ and a fully turbulent state is reached. For $x = x_{ts}$, $\Upsilon = 0$ and the flow is laminar. There are several empirical relations proposed in the literature for estimating x_{ts} and x_{te}; here, two will be given.

Abu-Ghannam and Shaw Model

In the Abu-Ghannam and Shaw [1] model

$$Re_{\delta_2,s} = \frac{U_\infty \delta_{2,s}}{\nu} = 163 + \exp\left[m\left(1 - \frac{Tu}{6.91}\right)\right], \tag{4.100}$$

where $m(K > 0) = 6.91 - 12.75K + 63.64K^2$ and $m(K < 0) = 6.91 - 2.48K - 12.27K^2$ and $K = -\delta_2^2/\nu\,(d\,U_\infty/dx)$. Here, $\delta_{2,s}$ is the boundary layer *momentum* thickness at $x = x_{ts}$. These relations thus identify x_{ts}. The value of x_{te} is identified with

$$x_{te} = x_{ts} + 4.6\frac{\nu_\infty}{u_\infty}\frac{\sigma_o}{B}, \tag{4.101}$$

where $B(K < 0) = 1$, $B(K > 0) = 1 + 1710\,K^{1.4}\exp -(1 + Tu^{3.5})^{0.5}$, and $\sigma_0 = 10^5\,(2.7 - 2.5\,Tu^{3.5})(1 + Tu^{3.5})^{-1}$. Here, Tu is the turbulence intensity in the free stream.

Cebeci Model

In the Cebeci [4] model

$$Re_{\delta_2} = 1.174\left(1 + \frac{22400}{Re_x}\right)Re_x^{0.46}, \tag{4.102}$$

$$x_{te} = x_{ts} + 60\frac{\nu_\infty}{U_\infty}Re_x^{-2/3}, \tag{4.103}$$

where $Re_x = U_\infty x/\nu$.

4.8.2 e–ϵ Model

In this model, the turbulent viscosity is determined from solution of two partial differential equations for scalar quantities e (turbulent kinetic energy) and ϵ

(turbulent energy dissipation[3]). Thus,

$$\mu_t = C_\mu \, \rho \, \frac{e^2}{\epsilon}. \tag{4.104}$$

Fortunately, the modelled equations for e and ϵ can also be cast in the form of Equation 4.1. Thus, we have

Turbulent Kinetic Energy Equation

$$\Phi = e, \qquad \Gamma_e = \mu + \frac{\mu_t}{Pr_{t,e}}, \qquad S_e = G - \rho \epsilon^* \tag{4.105}$$

and

Energy Dissipation Rate Equation

$$\Phi = \epsilon^*, \qquad \Gamma_{\epsilon^*} = \mu + \frac{\mu_t}{Pr_{t,\epsilon^*}},$$

$$S_{\epsilon^*} = \frac{\epsilon^*}{e}[C_1 \, G - C_2 \, \rho \, \epsilon^*] + 2 \, v \, \mu_t \left(\frac{\partial^2 u}{\partial y^2} \right)^2, \tag{4.106}$$

where

$$\epsilon^* = \epsilon - 2v \left(\frac{\partial \sqrt{e}}{\partial y} \right)^2, \tag{4.107}$$

and

$$G = \mu_t \left(\frac{\partial u}{\partial y} \right)^2. \tag{4.108}$$

In these equations, Launder and Spalding [40] specify $Pr_{t,e} = 1$, $Pr_{t,\epsilon^*} = 1.3$, $C_1 = 1.44$,

$$C_\mu = 0.09 \, \exp \left[\frac{-3.4}{(1 + Re_t/50)^2} \right], \tag{4.109}$$

and

$$C_2 = 1.92 \left[1 - 0.3 \, \exp - Re_t^2 \right], \tag{4.110}$$

where the turbulence Reynolds number $Re_t = \mu_t/\mu$. The $e-\epsilon$ model described here, called the *Low Reynolds number* (LRE) turbulence model, permits application of boundary conditions $e = \epsilon^* = 0$ at the wall. Further, the model is equally applicable to prediction of laminar-to-turbulent transition and one need not invoke the intermittency factor required in the mixing length model. In fact, Jones and Launder [30] have successfully applied the model even to the case where a turbulent boundary layer reverts to a laminar boundary layer becuase of strong free-stream acceleration. Several changes to the $e-\epsilon$ model have been proposed by different authors. The more recent among these, for example, are listed in [9].

---

[3] Here $\rho \, \epsilon$ is the turbulent counterpart of the $\mu \, \Phi_v$ term introduced in Equation 4.85.

4.8.3 Free-Shear Flows

In free-shear flows, the mixing length is given by

$$l_m = \beta (y_E - y_I), \tag{4.111}$$

where the E boundary is free and the I boundary is the symmetry axis. The value of constant β depends on the type of flow. According to Spalding [78] $\beta = 0.09$ for a plane jet, $\beta = 0.075$ for a round jet, and $\beta = 0.16$ for a plane wake. In general, however, β must be regarded as an arbitrary constant whose value is determined from experiment.

When the e–ϵ model is used, Equations 4.105 and 4.106 are directly applicable. However, because of the absence of a wall, there will be no region where $Re_t \to 0$. Also, the wall-correction terms $\partial \sqrt{e}/\partial y$ and $2 \nu \mu_t (\partial^2 u/\partial y^2)^2$ vanish. As such, the model will reduce to

$$\Phi = e, \quad \Gamma_e = \mu + \frac{\mu_t}{Pr_{t,e}}, \quad S_e = G - \rho \epsilon, \tag{4.112}$$

$$\Phi = \epsilon, \quad \Gamma_\epsilon = \mu + \frac{\mu_t}{Pr_{t,\epsilon}}, \quad S_\epsilon = \frac{\epsilon}{e} [C_1 G - C_2 \rho \epsilon], \tag{4.113}$$

with $C_1 = 1.44$, $C_2 = 1.92$, $C_\mu = 0.09$, $Pr_{t,e} = 1.0$, and $Pr_{t,\epsilon} = 1.3$. This set is called the *High Reynolds number* (HRE) model.

4.9 Overall Procedure

4.9.1 Calculation Sequence

The previous sections have provided all the essentials to construct the calculation procedure. This is listed in the following.

Evaluations at x_0

1. Choose x_0, where the initial profiles $\Phi (y_j)$ are specified for $j = 1, 2, \ldots, JN$ for the chosen JN.
2. Calculate r_j knowing $\alpha (x_0)$.
3. Set $x_u = x_0$ and evaluate ω_j ($j = 1, 2, \ldots, JN$) from specified u_j for a chosen value of ψ_I^u. This sets ψ_E^u and hence ψ_{EI}^u.

Begin a New Step

4. Choose Δx so that $x_d = x_u + \Delta x$. Calculate ρ_j, μ_j, and C_{p_j} from appropriate known functions of scalar Φ_j^u. Specify or calculate $\dot{m}_I$ or $\dot{m}_E$ as described in Section 4.6.
5. Choose relevant Φ and calculate coefficients and source terms in Equation 4.43 using upstream values. Note that if $\Phi = u$, the pressure gradient for internal and external flows must be appropriately evaluated. Now solve Equation 4.43 using TDMA.

6. Reset y_j, r_j using the u_j just calculated. Also reset ψ_b for a free boundary (b = E or I).
7. Go to step 5 and repeat until convergence of all relevant Φs is reached.
8. Calculate integral quantities δ_1, δ_2, C_{fx}, St_x, etc.
9. Set $x_u = x_d$ and $\Phi_P^u = \Phi_P$ and return to step 3 to execute a new step.
10. Continue untill the domain of interest in the x direction is covered.

4.9.2 Initial Conditions

For internal flows, the flow width at $x = x_0 = 0$ is known and it is easy to specify all $\phi(y_j)$. For external wall boundary layers, the initial profiles by necessity are to be specified at $x = x_0$ to avoid singularity at $x = 0$ where the boundary layer thickness is zero. A suitable choice of x_0 can be made assuming $Re_{x_0} = 10^3$ (say). If and when experimentally measured starting profiles are not available, one may choose the generalised polynomial velocity profile used in the integral method of laminar boundary layer analysis:

$$\left.\frac{u}{u_\infty}\right|_{x_0} = 2\eta - 2\eta^3 + \eta^4 + \frac{\lambda}{6}\left[\eta - 3\eta^2 + 3\eta^3 + \eta^4\right], \qquad (4.114)$$

where $\eta = y / \delta$ and

$$\lambda = \frac{\delta^2}{\nu}\left.\frac{d U_\infty}{d x}\right|_{x_0}. \qquad (4.115)$$

With reference to Figure 4.1, the region $0 < x < x_0$ will typically connote a stagnation flow region for which $\lambda = 7.052$ and $\delta \simeq 2.65 x_0 Re_{x_0}^{-0.5}$. If one is dealing with a flat surface, however, one may set $\lambda = 0$ and evaluate $\delta \simeq 5.83 x_0 Re_{x_0}^{-0.5}$ [65]. Thus, one is now free to choose the y_j distribution and evaluate u_j from equation 4.114.

With these specifications, calculations can continue from the laminar region through the transition region and ending in the turbulent region. If, however, the flow was turbulent from the start of the boundary layer, it is advisable to use an experimentally generated velocity profile. Alternatively, one may use

$$\left.\frac{u}{u_\infty}\right|_{x_0} = \left(\frac{y}{\delta}\right)^{1/7}, \qquad (4.116)$$

where $\delta \simeq 0.37 x_0 Re_{x_0}^{-0.2}$. Similar starting profiles for other Φs can also be prescribed using results from the integral method. For example, for a scalar variable $s = h$ or ω_k, the initial profiles may be specified as follows:

$$\left.\frac{s - s_w}{s_\infty - s_w}\right|_{x_0} = 2\eta_s - 2\eta_s^3 + \eta_s^4 \qquad \text{(Laminar)}$$

$$= \eta_s^{1/7} \qquad \text{(Turbulent)}, \qquad (4.117)$$

where $\eta_s = y/\delta_s$ and $\delta_s = \delta/Pr$ or δ/Sc.

For free-shear flows, again x_0 must be chosen to avoid the *elliptic* flow region very close to where a jet or a wake originates. For advice on the choice of x_0 and the $u(y)$ profile, the reader is referred to Schlichting [65].

4.9.3 Choice of Step Size and Iterations

Iterative calculation is required to deal with nonlinearities arising out of implicitness. In the present procedure, nonlinearities arise from four sources:

1. They can arise from dependence of coefficients and sources in Equation 4.43 on other scalar Φs. Thus, the source term R_k in the equation for ω_k may depend on T, and ρ, and Γ_Φ may depend on ω_k and T.
2. At a downstream station, y_j are not a priori known and therefore the values Δy_n, Δy_s required in several evaluations are not known. These y_js can be evaluated only after the Φ_j^d profile is established.
3. In external boundary layers and free-shear flows, the flow width at a downstream station is not known and we wish to select the largest width among all Φs. This is done via Equation 4.74.
4. In internal flows, the pressure gradient is not known at a downstream station.

By choosing a small enough Δx, one can make the procedure completely noniterative. This can be achieved by evaluating AN, AS, and S in terms of upstream values. We, however, prefer *partial* linearization. Thus, whereas the different Φs required in the evaluation of AN, AS, and S are taken from the upstream station, y_j are established through an iterative solution of equations for all relevant Φs. With this choice, experience shows that we may choose

$$\Delta x \simeq 0.25 \, \delta_2^u. \tag{4.118}$$

This choice ensures both economy and accuracy. However, situations may arise when larger step sizes are also permissible.

4.10 Applications

Flat Plate Boundary Layer

Figure 4.5 shows computed results of friction coefficient Cf_x and Stanton number St_x for a flat plate boundary layer. Computations were begun with a laminar velocity (with $\lambda = 0$ in Equation 4.114) and temperature profiles prescribed at $Re_{x_0} = 10^3$ with $JN = 102$. Such a large number of (nonuniform) grid points are necessary to resolve the profiles in the vicinity of the wall and in the turbulent range. In the mixing length model, transition is sensed by the Cebeci model (Equations 4.102 and 4.103). In the LRE model, the transition is sensed automatically. It is seen that the mixing length model predicts transition at a higher Re_x than the LRE model.

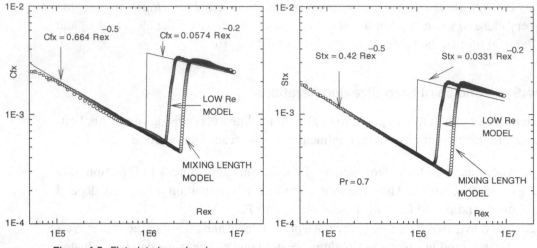

Figure 4.5. Flat plate boundary layer.

The predicted values of Cf_x and St_x are compared with well-known correlations derived from integral analysis. The agreements are satisfactory.

Figure 4.6 shows the velocity and temperature profiles in wall coordinates. The predictions of the mixing length model [Figure 4.6(a)] nearly agree with the two-layer prescriptions of the law of the wall [33] except in the very outer layers. The predictions from the LRE model [Figure 4.6(b)] are somewhat higher than those of the law of the wall. The dimensionless temperature is defined as $T^+ = (T - T_w) \rho C_p u_\tau / q_w$.

Burning of Carbon

We consider burning of carbon in a laminar plane stagnation flow of *dry* air so that the free-stream velocity varies as $U_\infty = Cx$. The surface is held at constant wall temperature T_w. The objective is to predict the burning rate of carbon as a function of T_w. The postulated chemical reactions at the surface are [82] as follows:

Reaction 1

$$C^* + O_2 \rightarrow CO_2, \quad \Delta H_1 = 32.73 \text{ MJ/kg of C},$$

$$k_1 = \begin{cases} 593.83 \, T_g \exp\left(-18{,}000/T_w\right) \text{m/s}, & T_w < 1{,}650 \text{K}, \\ (2.632 \times 10^{-5} \, T_w - 0.03353) \, T_g \, \text{(m/s)}, & T_w > 1{,}650 \text{K}, \end{cases}$$

$$\dot{m}''_{c1w} = \rho_w k_1 \frac{M_C}{M_{O_2}} \omega_{O_2} \quad kg/m^2 - s \tag{4.119}$$

where T_g is the near-wall gas temperature,

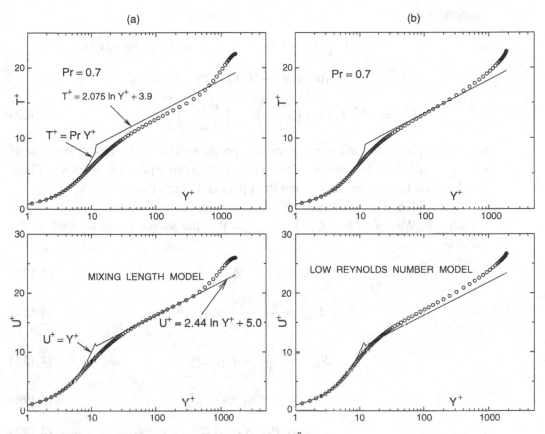

Figure 4.6. Velocity and temperature profiles at $Re_x = 5 \times 10^6$.

Reaction 2

$$C^* + \frac{1}{2}O_2 \rightarrow CO, \quad \Delta H_2 = 9.2 \text{ MJ/kg of C},$$

$$k_2 = 1.5 \times 10^5 \exp\left(-17{,}966/T_w\right) \text{ m/s},$$

$$\dot{m}''_{c2w} = 2\rho_w k_2 \frac{M_C}{M_{O_2}} \omega_{O_2} \quad \text{kg/m}^2\text{-s}, \tag{4.120}$$

and

Reaction 3

$$C^* + CO_2 \rightarrow 2CO, \quad \Delta H_3 = -14.4 \text{ MJ/kg of C},$$

$$k_3 = 4.016 \times 10^8 \exp\left(-29{,}790/T_w\right) \quad \text{m/s},$$

$$\dot{m}''_{c3w} = \rho_w k_3 \frac{M_C}{M_{CO_2}} \omega_{CO_2} \quad \text{kg/m}^2\text{-s}. \tag{4.121}$$

The above 3 reactions are *surface reactions*. In addition, we have the following *gas-phase reaction*:

Reaction 4

$$CO + \frac{1}{2} O_2 \rightarrow CO_2, \quad \Delta H_4 = 10.1 \text{ MJ/kg of CO},$$

$$k_4 = 2.24 \times 10^{12} \exp\left(-20{,}137/T\right) \text{ s}^{-1},$$

$$R_{CO} = \rho^{1.75} k_4 \omega_{CO} \left(\frac{\omega_{O_2}}{M_{O_2}}\right)^{0.25} \left(\frac{\omega_{H_2O}}{M_{H_2O}}\right)^{0.5}, \tag{4.122}$$

where ω_{H_2O} is treated as a parameter of the problem. The steam mass fraction is, of course, small enough so that it does not take part in other possible reactions. These rate laws are taken from Smoot and Pratt [68] and Turns [82].

The problem thus requires solution of equations for $\Phi = u$, ω_{O_2}, ω_{CO_2}, ω_{CO}, and enthalpy h. We define $h = C_p (T - T_{\text{ref}})$ so that the source terms for each of the variables are

$$S_u = \rho C^2 x \Delta V, \tag{4.123}$$

$$S_{\omega_{O_2}} = -\frac{1}{2} \frac{M_{O_2}}{M_{CO}} R_{CO} \Delta V, \tag{4.124}$$

$$S_{\omega_{CO_2}} = \frac{M_{CO_2}}{M_{CO}} R_{CO} \Delta V, \tag{4.125}$$

$$S_{\omega_{CO}} = -R_{CO} \Delta V, \tag{4.126}$$

$$S_h = R_{CO} \Delta H_4 \Delta V. \tag{4.127}$$

The total carbon burn rate is given by

$$\dot{m}_c'' = \dot{m}_{c1w}'' + \dot{m}_{c2w}'' + \dot{m}_{c3w}''. \tag{4.128}$$

To effect the wall boundary condition for mass fractions, we modify Equation 4.59 to account for surface reaction:

$$\dot{m}_c'' = (\omega_{k,w} - \omega_{k,T})^{-1} \left(\rho D_k \left.\frac{\partial \omega_k}{\partial y}\right|_{y=0} + \dot{m}_{\omega_k}''\right), \tag{4.129}$$

where $\dot{m}_{\omega_k}''$ is the surface generation rate of species k and $\omega_{k,T} = 0$ for all species. After discretisation, the wall mass fractions can be deduced from

$$\omega_{O_2,w} = \frac{\rho D/\Delta y \, \omega_{O_2,nw} - (\dot{m}_{c1w}'' + 0.5 \dot{m}_{c2w}'') M_{O_2}/M_C}{\rho D/\Delta y + \dot{m}_c''}, \tag{4.130}$$

$$\omega_{CO_2,w} = \frac{\rho D/\Delta y \, \omega_{CO_2,nw} + (\dot{m}_{c1w}'' - \dot{m}_{c3w}'') M_{CO_2}/M_C}{\rho D/\Delta y + \dot{m}_c''}, \tag{4.131}$$

$$\omega_{CO,w} = \frac{\rho D/\Delta y \, \omega_{CO,nw} + (\dot{m}_{c2w}'' + 2 \dot{m}_{c3w}'') M_{CO}/M_C}{\rho D/\Delta y + \dot{m}_c''}, \tag{4.132}$$

and the enthalpy at the wall boundary is given by

$$h_{\mathrm{w}} = C_p(T_{\mathrm{w}} - T_{\mathrm{ref}}). \tag{4.133}$$

With this enthalpy, we account for the surface heat generation via the source term S_h for the near-wall (suffix nw) control volume. Thus, for $j = 2$

$$S_h = S_h + \left[\dot{m}_c'' C_{p_c}(T_T - T_{\mathrm{ref}}) + \sum_k \dot{m}_{ckw}'' \Delta H_k \right] \Delta x, \tag{4.134}$$

where $T_T = T_{\mathrm{w}}$ and the carbon specific heat is $C_{p_c} = 1{,}300$ J/kg-K. In the free stream at the E boundary, we specify $U_\infty = Cx$, $T_\infty = 298$ K, $\omega_{O_2,\infty} = 0.232$, $\omega_{CO,\infty} = 0.0$, and $\omega_{CO_2,\infty} = 0.0$. The reference temperature is taken as $T_{\mathrm{ref}} = T_\infty$ so that $h_\infty = 0$.

To start the computations, it is assumed that for the starting length x_0 ($Re_{x_0} = 1{,}000$), the surface is inert. So, the inlet profiles for mass fractions and enthalpy are easily specified as uniform, corresponding to the free-stream state. The velocity profile is of course derived from Equation 4.114 with λ and δ corresponding to the stagnation flow condition. Computations are now continued till $Re_x = 10^5$ so that the combustion is well established and the burn rate is constant with x. The density and viscosity are assumed to vary over the width of the boundary layer according to

$$\rho = \frac{p\, M_{\mathrm{mix}}}{R_{\mathrm{u}}\, T}, \tag{4.135}$$

$$\mu = 18.6 \times 10^{-6} \left(\frac{T}{303} \right)^{1.5} \left[\frac{303 + 110}{T + 110} \right] \quad \text{N-s/m}^2, \tag{4.136}$$

where $p = 10^5$ N/m^2 and $R_{\mathrm{u}} = 8{,}314$ J/kmol-K. The molecular weight of the mixture is evaluated from

$$M_{\mathrm{mix}} = \left[\frac{\omega_{O_2}}{M_{O_2}} + \frac{\omega_{CO_2}}{M_{CO_2}} + \frac{\omega_{CO}}{M_{CO}} + \frac{\omega_{N_2}}{M_{N_2}} + \frac{\omega_{H_2O}}{M_{H_2O}} \right]^{-1}, \tag{4.137}$$

where $\omega_{N_2} = 1 - \omega_{O_2} - \omega_{CO_2} - \omega_{CO} - \omega_{H_2O}$. The gas specific heat is, however, assumed constant and is calculated from $C_p = 919.2 + 0.2\, T_{\mathrm{m}}$ J/kg-K and $T_{\mathrm{m}} = 0.5(T_{\mathrm{w}} + T_\infty)$. Computations are carried out for $800 < T_{\mathrm{w}} < 2{,}000$ K and $Pr = 0.72$. The value of the Schmidt number is uncertain in this highly variable property reacting flow. Following Kuo [38], we take the Schmidt number for all species as 0.51. To facilitate evaluation of R_{CO}, the water vapour fraction is taken as $\omega_{H_2O} = 0.001$, but the vapour is assumed chemically inert.

For the purpose of comparison with published [38] experimental data, the predicted burning rate is normalised with respect to the *diffusion controlled* burning rate. Thus we form the ratio

$$\mathrm{BRR} = \frac{\dot{m}_c''\,(\text{predicted})}{\dot{m}_c''\,(\text{dc})}, \tag{4.138}$$

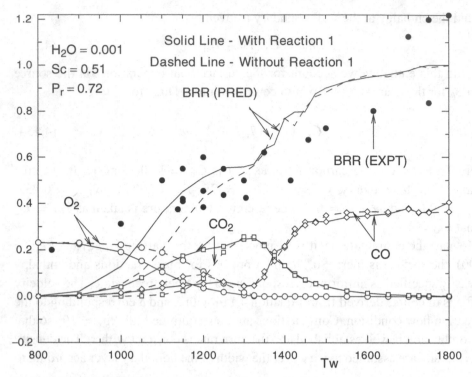

Figure 4.7. Variation of BRR, $\omega_{O_2,w}$, $\omega_{CO_2,w}$, and $\omega_{CO,w}$ with T_w.

where the denominator is estimated[4] for the stagnation flow from [33]

$$\dot{m}_c'' \, (dc) = \frac{0.57}{x} \frac{\mu_\infty}{Pr^{0.6}} \, Re_x^{0.5} \ln(1 + B) \left(\frac{Pr}{Sc}\right)^{0.4} \left(\frac{T_w}{T_\infty}\right)^{0.1} \qquad (4.139)$$

and the driving force $B = 0.174$. Figure 4.7 shows the variation of the ratio BRR with T_w. The experimental data for the burn rate are shown by filled circles. Data are predicted with (solid lines) and without (dashed lines) Reaction 1 to ascertain the influence of this reaction at low temperatures. It is seen that the experimental BRR has considerable scatter and exceeds unity, against expectation. However, this may be due to the normalising factor used by Kuo [38]. Nonetheless, the data show a mild plateau for $1{,}100 < T_w < 1{,}400$. This tendency is nearly predicted by the present computations, particularly when Reaction 1 is included. For $T_w > 1{,}350$, the experimental data show a sudden rise that is again observed in present predictions. The predicted BRR $\rightarrow 1$ at 1,800 K as expected. However, for $T_w < 1{,}000$ K, the present data grossly underpredict the experimental data; the underprediction is greater when Reaction 1 is ignored.

---

[4] Equation 4.139 is derived from Reynolds-flow model developed by Spalding [73] assuming fluid properties in the free-stream state and then corrected for property variations through the boundary layer.

The predicted wall mass fractions for CO, O_2, and CO_2 are also plotted in Figure 4.7. The wall mass fraction $\omega_{O_2,w}$, starting from 0.232 at 800 K, decreases rapidly to zero at $T_w \sim 1,300$ K. Note that $\omega_{O_2,w}$ decreases more rapidly when Reaction 1 is included, as expected. The wall mass fraction $\omega_{CO_2,w}$ gradually increases with temperature, peaks at $T_w = 1,300$ K, and then rapidly falls to zero. In this range where $\omega_{CO_2,w}$ is significant, the BRR indicates a mild plateau after an initial rapid rise with temperature. The $\omega_{CO,w}$, however, increases with wall temperature. At $T_w > 1,300$, CO evolution becomes significant, indicating dominance of Reaction 3. At very high temperatures, this reaction becomes the most dominant and combustion is now diffusion controlled with $\omega_{O_2,w} = \omega_{CO_2,w} = 0$ and $\omega_{CO,w} \rightarrow 0.406$. Overall, Reaction 1 is important at low temperatures and Reaction 3 is important at high temperatures. It must be noted that although the tendencies predicted here are similar to the similarity solution for BRR obtained by Kuo [38], the quality of predictions in combustion calculations greatly depends on the accuracy of the assumed reaction-rate laws.

Entrance Region of a Pipe

We consider simultaneous development of velocity and temperature profiles in the entrance region of a pipe of radius R. The flow is laminar ($Re = 500$) and the fluid Prandtl number $Pr = 0.7$. An axially constant wall temperature boundary condition is assumed. In this axisymmetric flow, the I boundary coincides with the pipe axis and the E boundary with the pipe wall. Computations are performed with a $JN = 25$ nonuniform grid with closer spacings near the wall. The axial locations are determined from $x = L(I - 1/IMAX - 1)^{1.5}$, where $L = 0.2 \times R \times Re$ and I is the axial step number. Figure 4.8 shows the computed variations of $f \times Re$, Nu_x, and velocity u at the pipe axis with $x^+ = (x/R)/Re/Pr$. Also plotted in the figure are previous numerical solutions for Nu_x reported in [33]. It is seen that the present solutions match perfectly with the previous solutions. The $f \times Re$ product also varies as expected with asymptotic approach to 16.0. Similarly, the velocity $u/\overline{u}$ at the pipe axis also reaches 2.0 at large x^+.

Similar computations are now carried out at higher Reynolds numbers ($1,000 < Re < 10,000$) including the transition range. For this purpose, the LRE model is used and computations are performed with a $JN = 47$ nonuniform grid. Here, $IMAX = 1,000$ and $L = 100 \times D$. Figure 4.9 shows variation of f, Nu ($Pr = 0.7$ and 5.0), and $u_{axis}/\overline{u}$ with Reynolds number in the *fully developed* state ($X/D = 100$). It is seen that for $Re < 1,600$, the characteristics correspond to those of a laminar flow ($u_{axis}/\overline{u} = 2.0$). Accoring to the model, transition occurs abruptly and appears to extend up to $Re \sim 2,500$, as evident from the Nu predictions. The $u_{axis}/\overline{u}$ ratio now drops suddenly from its laminar value of 2.0. At $Re = 10,000$, $u_{axis}/\overline{u} = 1.246$. For Nu, the expected trend is again observed. In the laminar range, Nu approaches the analytically derivable fully developed value of 3.667 for $T_w = $ constant boundary condition for both Prandtl numbers. The thermal development

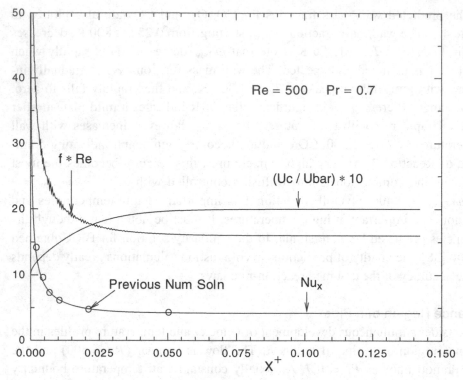

Figure 4.8. Entrance region of a pipe – laminar flow.

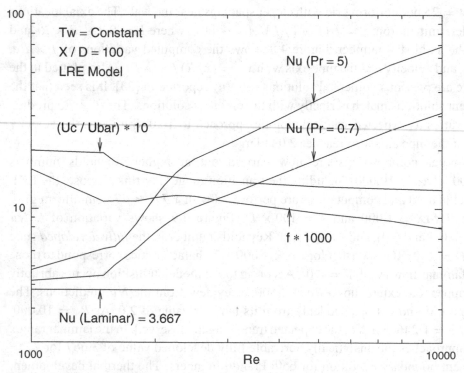

Figure 4.9. Fully developed flow and heat transfer in a pipe.

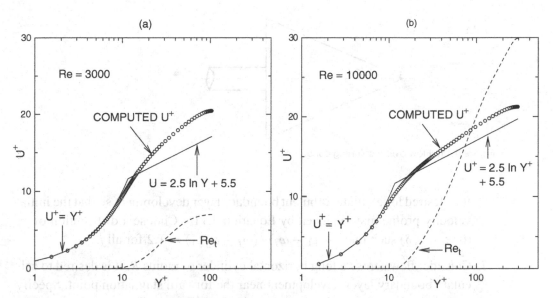

Figure 4.10. Variation of u^+ and Re_t with y^+ – pipe flow.

length is a function of Pr in laminar flow [33]. In turbulent flow, $X/D = 100$ is sufficient for fully developed flow and heat transfer and, therefore, the predicted values of Nu match well with the well-known correlation $Nu = 0.023\,Re^{0.8}\,Pr^{0.4}$. In the turbulent range, the friction factor also corroborates $f = 0.079\,Re^{-0.25}$ well.

Figure 4.10 shows the fully developed velocity profile in wall coordinates at $Re = 3{,}000$ and $10{,}000$. In the transition range, the sublayer is thick. At $Re = 10{,}000$, the predicted profile nearly coincides with the wall law up to $y^+ = 30$ and then departs in the outer layers. The figure also shows variations of turbulence Reynolds number $Re_t = \mu_t/\mu$. At $Re = 3{,}000$, the maximum value of Re_t is lower than that at $Re = 10{,}000$. All these tendencies accord with expectation.

EXERCISES

1. Starting with Equation 4.17, derive Equations 4.22 and 4.26 in their conservative form.

2. Verify Equations 4.37–4.40 through detailed algebra.

3. Derive an equation for $\dot{m}_{I,std}$, similar to Equation 4.71, when the free-stream boundary is located at the I boundary.

4. Derive recurrence relations (4.80) and (4.82).

5. Show that when Re_t is large, the LRE model reduces to the HRE model given in Equations 4.112 and 4.113.

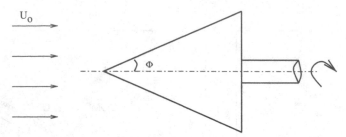

Figure 4.11. Flow over a spinning cone.

6. It is desired to calculate turbulent boundary layer development so that the initial velocity profile may be given by Equation 4.116. Choose a distribution of y_j $(0 < y < \delta)$ such that $(\omega_{j+1} - \omega_j)/(\omega_j - \omega_{j-1}) = 1.2$ for all j.

7. Consider flow across a long horizontal cylinder of radius R. It is desired to calculate boundary layer development near the forward stagnation point. Specify variation of α and r_I with x. Also specify the starting velocity profile.

8. In Exercise 7, it is of interest to calculate the mass transfer of an inert substance in the forward stagnation region. Specify the starting mass fraction profile and select the appropriate boundary conditions for the mass-fraction variable ω and u. (Hint: Use the integral method to specify the ω profile.)

9. It is desired to calculate boundary layer development over a cone spinning with angular velocity Ω (see Figure 4.11). Write the governing equations and the boundary conditions at the I and E boundaries for this problem. Also provide initial conditions. (Hint: Assume that the spinning rate is high so that centrifugal and Coriolis forces must be considered. Also, $\partial p/\partial r$ is not negligible. Hence, dp/dx will vary with y.)

10. Consider an *adiabatic* wall 2 m high, as shown in Figure 4.12. The bottom 1 m is covered with a thick layer of highly volatile solid material having latent heat λ_{fu}. The fuel burns in stagnant dry air under *natural convection* conditions. Assume SCR (4.86) with reaction rate given by (4.87).

    (a) Write all relevant equations governing the phenomenon of burning along with their source terms. (Hint: Use the Boussinesq approximation for the buoyancy term.)

    (b) Write boundary conditions at the I boundary to determine the burning rate. Also write conditions at the E boundary. [Hint: In this problem, the adiabatic condition implies that $T_b = T_T$. Further, the burning surface temperature will equal the evaporation (or boiling) point temperature T_{bp} and is a known property. Further, the SCR assumption implies that $\omega_{fu} = \omega_{ox} = 0$ at the burning surface.]

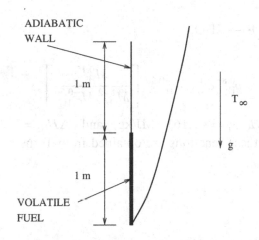

ADIABATIC
WALL

1 m

T_∞

Figure 4.12. Burning from a vertical wall.

g

1 m

VOLATILE
FUEL

(c) Write initial conditions for each variable assuming *pure* natural convection heat transfer between $x = 0$ and $x = x_0$.

11. In the stagnation-flow carbon-burning problem described in the text, the water vapour was treated as inert and its mass fraction was held constant. However, water vapour can react with carbon, resulting in the following two additional surface reactions:

$$C^* + H_2O \rightarrow CO + H_2$$

$$C^* + 2H_2 \rightarrow CH_4.$$

The reaction rate of the first reaction is about twice that of Reaction 3 (i.e., $2k_3$). For the second reaction, $k = 0.035 \exp(-17,900/T_w)$. Assuming $\omega_{H_2O,\infty} = 0.01$, write the equations to be solved along with their source terms and boundary conditions. [Hint: You will need to postulate the following additional gas-phase reactions to approximately account for the presence of H_2, H_2O, and CH_4:

$$CH_4 \rightarrow \frac{1}{2}C_2H_4 + H_2,$$

$$R_{CH_4} = 10^{20.32} \exp\left(-\frac{24,962}{T}\right) \rho_m^{1.97} \omega_{CH_4}^{0.5} \omega_{O_2}^{1.07} \omega_{H_2}^{0.4} \left[\frac{M_{CH_4}^{0.5}}{M_{O_2}^{1.07} M_{C_2H_4}^{0.4}}\right],$$

$$C_2H_4 \rightarrow 2CO + 2H_2,$$

$$R_{C_2H_4} = 10^{17.7} \exp\left(-\frac{25,164}{T}\right) \rho_m^{1.71} \omega_{C_2H_4}^{0.9} \omega_{O_2}^{1.18} \omega_{CH_4}^{-0.37} \left[\frac{M_{C_2H_4}^{0.1}}{M_{O_2}^{1.18} M_{CH_4}^{-0.37}}\right],$$

$$H_2 + \frac{1}{2}O_2 \rightarrow H_2O,$$

$$R_{H_2} = 10^{16.52} \exp\left(-\frac{20,634}{T}\right) \rho_m^{1.71} \omega_{H_2}^{0.85} \omega_{O_2}^{1.42} \omega_{H_2}^{-0.56} \left[\frac{M_{H_2}^{0.15}}{M_{O_2}^{1.42} M_{C_2H_4}^{-0.56}}\right],$$

with $\Delta H_{CH_4} = 50.016$ MJ/kg, $\Delta H_{C_2H_4} = 47.161$ MJ/kg, and $\Delta H_{H_2} = 120.9$ MJ/kg. The reaction rates for these reactions are obtained from Turns [82].]

5 2D Convection – Cartesian Grids

5.1 Introduction

5.1.1 Main Task

In the previous chapter, we considered convective–diffusive transport in long (x direction) and thin (y direction) flows. This implied that although convective fluxes were significant in both x and y directions, significant diffusion fluxes occurred only in the y direction; diffusion fluxes in the x direction are negligible. We now turn our attention to flows in which diffusive fluxes are comparable in both x and y directions. Thus, the general transport Equation (1.25) may be written[1] as

$$\frac{\partial(\rho\,\Phi)}{\partial t} + \frac{1}{r}\frac{\partial(r\,q_j)}{\partial x_j} = S, \qquad j = 1, 2, \tag{5.1}$$

where

$$q_j = \rho\,u_{\mathrm{f}j}\,\Phi - \Gamma_{\mathrm{eff}}\frac{\partial \Phi}{\partial x_j}. \tag{5.2}$$

In Equation 5.2, the first term on the right-hand side represents the convective flux whereas the second term represents the diffusive flux. Note that suffix f is attached to the velocity appearing in the convective flux; the significance of this suffix will become clear in a later section. In Equation 5.1, r stands for radius. This makes the equation applicable to axisymmetric flows governed by equations written in cylindrical polar coordinates. When plane flows are considered, $r = 1$ and Equation 1.25 is readily recovered. By way of reminder, we note that Φ may stand for 1, u_i ($i = 1, 2$), u_3 (velocity in the x_3 direction), ω_k, T or h, and e and ϵ, and Γ_{eff} is the effective exchange coefficient (see Equation 4.89).

Flows with comparable convective–diffusive fluxes in each direction occur routinely in most practical equipment although they are usually three dimensional. Here, only 2D situations are considered for convenience and because the primary

---

[1] Note that ρ_{m} signifying mixture density is now written as ρ for convenience.

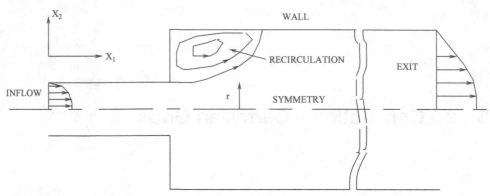

Figure 5.1. 2D flow situation.

objective is to learn the main issues of discretisation. Figure 5.1 shows a practical situation that can be represented by 2D equations (5.1). The figure shows flow at the connection between two pipes of different diameters. The flow is assumed to be axisymmetric. Immediately downstream of the pipe enlargement, the flow will exhibit *recirculation* and thus, in the absence of any predominant flow direction, convective–diffusive fluxes in the x_1 and x_2 directions will be comparable. This implies that property Φ at any x_1 in the recirculation region will be influenced by property values both *upstream* as well as *downstream* of x_1. Similar *two-way* influence is also expected in the x_2 direction. Such two-way influences are called *elliptic* influences [49] and, therefore, Equation 5.1 is an elliptic partial differential equation.[2]

5.1.2 Solution Strategy

Before discretising Equation 5.1, we shall make distinction between the following two problems:

1. the problem of flow prediction and
2. the problem of scalar transport prediction.

Here, scalar transport means transport of all Φs (u_3, ω_k, T, h, e, ϵ, etc.) *other than* velocities ($\Phi = u_1, u_2$) that are vectors. Note that u_3, although a vector, is included in the list of scalars. This is because variations in direction x_3 are absent and, with respect to x_1 and x_2 directions, u_3 may be treated as a scalar. The reason for this distinction between scalars and vectors is twofold.

It is clear from Equation 5.2 that calculation of scalar transport will be facilitated only when the velocity field is established. In fact, if source S and the properties

---

[2] The reader will recall the equation $a\,\Phi_{xx} + 2\,b\,\Phi_{xy} + c\,\Phi_{yy} = S(\Phi_x, \Phi_y, \Phi, x, y)$, where, when the discriminant $b^2 - a\,c = 0$, the equation is parabolic; when $b^2 - a\,c < 0$, the equation is elliptic; and when $b^2 - a\,c > 0$, the equation is hyperbolic.

ρ and Γ were not functions of scalar Φs then the flow equations for $\Phi = u_1, u_2$ will be *independent* of the scalar transport equations. This is the first reason for distinguishing the flow-field equations from other scalar transport equations. To appreciate the second reason, we first set out the equations governing the flow field (the Navier–Stokes equations):

$$\frac{\partial(\rho)}{\partial t} + \frac{1}{r}\frac{\partial}{\partial x_1}\{r\,\rho\,u_{f1}\} + \frac{1}{r}\frac{\partial}{\partial x_2}\{r\,\rho\,u_{f2}\} = 0, \tag{5.3}$$

$$\frac{\partial(\rho\,u_1)}{\partial t} + \frac{1}{r}\frac{\partial}{\partial x_1}\{r\,\rho\,u_{f1}\,u_1\} + \frac{1}{r}\frac{\partial}{\partial x_2}\{r\,\rho\,u_{f2}\,u_1\}$$
$$= -\frac{\partial p}{\partial x_1} + \frac{1}{r}\frac{\partial}{\partial x_1}\left[r\,\mu_{\text{eff}}\frac{\partial u_1}{\partial x_1}\right] + \frac{1}{r}\frac{\partial}{\partial x_2}\left[r\,\mu_{\text{eff}}\frac{\partial u_1}{\partial x_2}\right] + S_{u1}, \tag{5.4}$$

$$\frac{\partial(\rho\,u_2)}{\partial t} + \frac{1}{r}\frac{\partial}{\partial x_1}\{r\,\rho\,u_{f1}\,u_2\} + \frac{1}{r}\frac{\partial}{\partial x_2}\{r\,\rho\,u_{f2}\,u_2\}$$
$$= -\frac{\partial p}{\partial x_2} + \frac{1}{r}\frac{\partial}{\partial x_1}\left[r\,\mu_{\text{eff}}\frac{\partial u_2}{\partial x_1}\right] + \frac{1}{r}\frac{\partial}{\partial x_2}\left[r\,\mu_{\text{eff}}\frac{\partial u_2}{\partial x_2}\right] + S_{u2}. \tag{5.5}$$

A few comments having a bearing on the solution strategy are now in order.

1. In Equations 5.3–5.5, there are three unknowns (u_1, u_2, and p). Therefore, the equation set is solvable.
2. In boundary layer flows, the pressure gradient is specified (external flows) or is evaluated via the overall duct mass flow rate balance (internal flows). In elliptic flows, however, $\partial p/\partial x_1$ and $\partial p/\partial x_2$ are not a priori known.
3. Thus, if we regard Equation 5.4 as the determinant of u_1 field and Equation 5.6 as the determinant of u_2 field, then the pressure field can be established only via the mass conservation equation (5.3). The situation is somewhat similar to the case of internal boundary layer flows but is not as straightforward.
4. The suffix f is attached to velocities satisfying the mass conservation equation. The velocity field without suffix f may or may not satisfy mass conservation directly although, *in a continuum*, it is expected that the u_i and u_{fi} fields are identically overlapping and, therefore, the former must also satisfy mass conservation.
5. The reader may find this distinction between the u_i and u_{fi} fields somewhat unfamiliar. This is because most textbooks a priori assume a fluid continuum. Numerical solutions are, however, developed in a discretised space and the distinction mentioned here becomes relevant. This will become clear in a later section.

These points reveal the fact that there is no explicit differential equation for determination of the pressure field with p (or its variant) as the dependent variable.

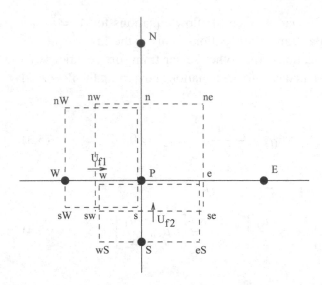

Figure 5.2. The staggered grid.

Such an equation, however, can be *derived* from explicit satisfaction of the mass conservation equation. In the sections to follow, the SIMPLE method for determination of the pressure field is presented. This method was developed by Patankar and Spalding [51]. It is among the most extensively used methods in CFD practice. In fact, most CFD packages employ this method. The acronym SIMPLE stands for *Semi-Implicit Method for Pressure-Linked Equations.*[3]

The original SIMPLE method [51] was derived for Cartesian grids in which the scalar Φs (including pressure p) and the velocity vectors were defined in a *staggered* arrangement (see Figure 5.2). To understand this arrangement, consider typical node P (i, j) with the surrounding control volume whose faces are located at e, w, n, and s. In the staggered arrangement, pressure $p_{i,j}$ is *stored/defined* at the node P. The same holds for other scalars $\Phi_{i,j}$. However, the vector u_{f1} (i, j) is stored at the cell face w and vector u_{f2} (i,j) is stored at cell face s. Thus, the vectors and the scalars are stored in staggered locations. It is easy to identify appropriate control volumes surrounding the cell-face locations as shown in Figure 5.2. Thus, in the (i, j) address system, there are three partially overlapping control volumes.

Now, the SIMPLE method *requires* that to determine the pressure field, the mass conservation equation must be satisfied over the control volume (ne-se-sw-nw) surrounding node P where $p_{i,j}$ is stored. Thus, using the IOCV method, the discretised version of Equation 5.3 is written as

$$[(\rho \, r \, u_{f1})_e - (\rho \, r \, u_{f1})_w] \, \Delta x_2 + [(\rho \, r \, u_{f2})_n - (\rho \, r \, u_{f2})_s] \, \Delta x_1 = - \left(\rho_P - \rho_P^0\right) \frac{\Delta V}{\Delta t},$$

(5.6)

[3] In compressible flows, $p = \rho \, R_g \, T$, where R_g is the gas constant, must be added to the equation set (5.3–5.6). This equation of state is used to determine density ρ.

where $\Delta V = r_{\mathrm{P}}\,\Delta x_1\,\Delta x_2$ and superscript o represents values at the old time. Superscript n is dropped for convenience.

Equation 5.6 indicates that the velocities with suffix f appear at the cell faces of the control volume surrounding node P. Therefore, in SIMPLE-staggered, momentum equations, Equation (5.4) is solved over control volume n-nW-sW-s and Equation 5.6 is solved over the control volume w-wS-eS-e *without* explicit commitment to satisfy mass conservation over these control volumes. The overall strategy for solution of the flow equations is as follows:

1. Guess a p^l field and solve momentum equations (5.4) and (5.6) over control volumes surrounding cell faces to yield u_{f1}^l and u_{f2}^l fields.
2. These fields, in general, will not satisfy the mass conservation equation (5.6).
3. Derive a *mass-conserving pressure-correction* equation to satisfy mass conservation over the control volume *surrounding* node P.
4. Use the pressure correction p' so determined to correct the guessed pressure p^l and velocities u_{f1}^l and u_{f2}^l.

For a complete description of the SIMPLE-staggered method, the reader is referred to [49, 51].

5.2   SIMPLE – Collocated Grids

5.2.1   Main Idea

Although the SIMPLE-staggered grid method enjoyed considerable success particularly when Cartesian grids were employed, the procedure was found to be inconvenient when curvilinear or unstructured grids were to be employed to compute over ever more complex domains. Further, even on Cartesian grids, the process of discretisation required considerable *book keeping* because the dimensions of the control volumes of vector and scalar variables were different.

Since the early 1980s, therefore, researchers began to explore the possibility of implementing the SIMPLE procedure using *collocated* variables.[4] That is, the velocity and the scalar variables were to be stored/defined at the *same* node P (i, j). This, it was felt, would permit attention to be directed to a single transport equation (5.1), thereby reducing the book-keeping requirements considerably.

Although convenient, this departure also brought within its wake a major difficulty with respect to the pressure-field prediction. It was found that if the pressure-correction equation as derived for staggered grids was used to predict pressure on collocated grids, the predicted pressure distribution showed *zigzagness*. Depending on the identified cause of this problem, different researchers (see, for example, [59])

---

[4] In the literature, the procedure with collocated variables is sometimes referred to as a procedure employing nonstaggered or collocated grids.

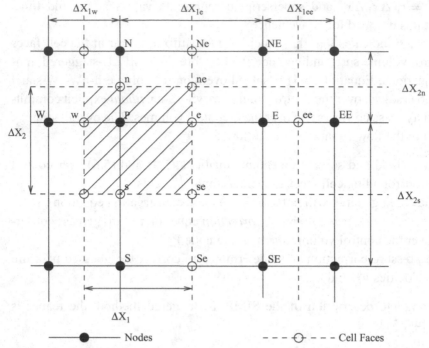

Figure 5.3. The collocated grid.

proposed different cures with differing amounts of complexity. Here, we shall describe the method developed by Date [14] that elegantly eliminates the problem of the zigzag pressure prediction. It will be shown in a later section that this matter is connected with the recognition of the need to modify the normal-stress expression as discussed in Chapter 1.

5.2.2 Discretisation

For collocated variables, we need to consider only one control volume (hatched) surrounding typical node P, as shown in Figure 5.3. Further, the cell faces are assumed to be midway between the adjacent nodes. As usual, using the IOCV method ($dV = r\,dx_1\,dx_2$), we integrate Equation 5.1 so that

$$\int_s^n \int_w^e \frac{1}{r} \left\{ \frac{\partial(r\,q_1)}{\partial x_1} + \frac{\partial(r\,q_2)}{\partial x_2} \right\} dV = \int_s^n \int_w^e \left[S - \frac{\partial(\rho\,\Phi)}{\partial t} \right] dV. \quad (5.7)$$

Now, replacing the qs from Equation 5.2, we can show that

$$[C_e\,\Phi_e - d_e\,(\Phi_E - \Phi_P)] - [C_w\,\Phi_w - d_w\,(\Phi_P - \Phi_W)]$$

$$+ [C_n\,\Phi_n - d_n\,(\Phi_N - \Phi_P)] - [C_w\,\Phi_w - d_w\,(\Phi_P - \Phi_W)]$$

$$= S\,\Delta V - (\rho\,\Phi - \rho^o\,\Phi^o)_P \frac{\Delta V}{\Delta t}, \quad (5.8)$$

where the convective coefficients are given by

$$C_e = \rho_e r_e u_{f1,e} \Delta x_2, \qquad C_w = \rho_w r_w u_{f1,w} \Delta x_2,$$

$$C_n = \rho_n r_n u_{f2,n} \Delta x_1, \qquad C_s = \rho_s r_s u_{f2,s} \Delta x_1, \tag{5.9}$$

and the diffusion coefficients are

$$d_e = \frac{\Gamma_{\text{eff},e} r_e \Delta x_2}{\Delta x_{1e}}, \qquad d_w = \frac{\Gamma_{\text{eff},w} r_w \Delta x_2}{\Delta x_{1w}},$$

$$d_n = \frac{\Gamma_{\text{eff},n} r_n \Delta x_1}{\Delta x_{2n}}, \qquad d_s = \frac{\Gamma_{\text{eff},s} r_s \Delta x_1}{\Delta x_{2s}}. \tag{5.10}$$

Now, in terms of the notation just introduced, the discretised mass conservation equation (5.6) (with $\Phi = 1$) can be written as

$$\left(\rho_P - \rho_P^0\right) \frac{\Delta V}{\Delta t} + C_e - C_w + C_n - C_s = 0. \tag{5.11}$$

Further, the expressions for $C \Phi$ at the cell faces can be generalised to account for any of the convection schemes introduced in Chapter 3. When this is done and Equation 5.11 is employed, it can be shown that Equation 5.8 reduces to

$$AP\, \Phi_P = AE\, \Phi_F + AW\, \Phi_W + AN\, \Phi_N + AS\, \Phi_S + D, \tag{5.12}$$

where

$$AE = d_e \left[A + \max\left(-P_{ce}, 0\right)\right], \qquad P_{ce} = C_e/d_e, \tag{5.13}$$

$$AW = d_w \left[A + \max\left(P_{cw}, 0\right)\right], \qquad P_{cw} = C_w/d_w, \tag{5.14}$$

$$AN = d_n \left[A + \max\left(-P_{cn}, 0\right)\right], \qquad P_{cn} = C_n/d_n, \tag{5.15}$$

$$AS = d_s \left[A + \max\left(P_{cs}, 0\right)\right], \qquad P_{cs} = C_s/d_s, \tag{5.16}$$

$$AP = AE + AW + AN + AS + \frac{\rho_P^0 \Delta V}{\Delta t}, \tag{5.17}$$

$$D = S \Delta V + \frac{\rho_P^0 \Delta V}{\Delta t} \Phi_P^0. \tag{5.18}$$

In these equations

$$A = \begin{cases} 1 & \text{(UDS)} \\ \max\left(0, 1 - 0.5\,|P_c|\right) & \text{(HDS)} \\ \max\left\{0, (1 - 0.1\,|P_c|)^5\right\} & \text{(Power)} \\ 1 - 0.5\,|P_c| & \text{(CDS)}. \end{cases} \tag{5.19}$$

From the point of view of computer coding, the utility of this generalised representation for all variables (scalars as well as vectors) is obvious.

5.2.3 Pressure-Correction Equation

In the collocated-grid SIMPLE algorithm, the nodal velocities are determined using Equations 5.12 written for $\Phi = u_1$ and u_2. The pressure gradients appearing in the source terms of these equations are of course evaluated by central difference [for example, $\partial p/\partial x_1|_P = (p_E^l - p_W^l)/(2\,\Delta x_1)$, where p^l is the guessed pressure field and l is the iteration number]. The task now is to correct the u_i^l and p^l fields such that mass conservation over the control volume surrounding node P is satisfied. To do this, and to remain consistent with the SIMPLE-staggered grid, we *imagine* that the momentum equations are also being solved for the cell-face velocities u_{fi}^l. The discretised versions of these imagined equations with underrelaxation will appear as

$$u_{f1}^{l+1} = \frac{\alpha}{AP^{u_{f1}}}\left[\sum_k A_k u_{f1,k}^{l+1} - \Delta V \frac{\partial p^{l+1}}{\partial x_1} + D_{u_1}^l\right] + (1-\alpha)u_{f1}^l, \quad (5.20)$$

$$u_{f2}^{l+1} = \frac{\alpha}{AP^{u_{f2}}}\left[\sum_k A_k u_{f2,k}^{l+1} - \Delta V \frac{\partial p^{l+1}}{\partial x_2} + D_{u_2}^l\right] + (1-\alpha)u_{f2}^l, \quad (5.21)$$

where $D_{u_1}^l$ and $D_{u_2}^l$ contain source terms (if any) other than the pressure gradient, α is the underrelaxation factor, and the summation symbol indicates summation over all immediate neighbours of the cell-face location under consideration. Thus, when Equation 5.20 is written for cell face e, for example, running counter k refers to locations ee, Ne, w, and Se. Now, at iteration level $l+1$, it is expected that

$$\frac{\partial(\rho^{l+1})}{\partial t} + \frac{1}{r}\frac{\partial}{\partial x_1}\left\{r\,\rho^{l+1}u_{f1}^{l+1}\right\} + \frac{1}{r}\frac{\partial}{\partial x_2}\left\{r\,\rho^{l+1}u_{f2}^{l+1}\right\} = 0. \quad (5.22)$$

Substituting Equations 5.20 and 5.21 in Equation 5.22 we can show that

$$\frac{\partial(\rho^{l+1})}{\partial t} + \frac{1}{r}\frac{\partial}{\partial x_1}\left\{r\,\rho^{l+1}u_{f1}^l\right\} + \frac{1}{r}\frac{\partial}{\partial x_2}\left\{r\,\rho^{l+1}u_{f2}^l\right\}$$

$$= \frac{1}{r}\frac{\partial}{\partial x_1}\left[\frac{r\,\rho^{l+1}\alpha}{AP^{u_{f1}}}\left\{AP^{u_{f1}}u_{f1}^l - \sum_k A_k u_{f1,k}^{l+1} + \Delta V \frac{\partial p^{l+1}}{\partial x_1} - D_{u_1}^l\right\}\right]$$

$$+ \frac{1}{r}\frac{\partial}{\partial x_2}\left[\frac{r\,\rho^{l+1}\alpha}{AP^{u_{f2}}}\left\{AP^{u_{f2}}u_{f2}^l - \sum_k A_k u_{f2,k}^{l+1} + \Delta V \frac{\partial p^{l+1}}{\partial x_2} - D_{u_2}^l\right\}\right].$$

$$(5.23)$$

To develop the pressure-correction equation, we introduce the following substitutions:

$$u_{f1}^{l+1} = u_{f1}^l + u_{f1}', \qquad u_{f2}^{l+1} = u_{f2}^l + u_{f2}', \qquad p^{l+1} = p^l + p_m', \quad (5.24)$$

where, p'_m is the *mass-conserving* pressure correction. Thus, Equation 5.23 will read as[5]

$$\frac{1}{r}\frac{\partial}{\partial x_1}\left\{\frac{\rho^{l+1}\,r\,\alpha\,\Delta V}{AP^{u_{f1}}}\frac{\partial p'_m}{\partial x_1}\right\} + \frac{1}{r}\frac{\partial}{\partial x_2}\left\{\frac{\rho^{l+1}\,r\,\alpha\,\Delta V}{AP^{u_{f2}}}\frac{\partial p'_m}{\partial x_2}\right\}$$

$$= \frac{\partial(\rho^{l+1})}{\partial t} + \frac{1}{r}\frac{\partial}{\partial x_1}\left\{r\,\rho^{l+1}\,u^l_{f1}\right\} + \frac{1}{r}\frac{\partial}{\partial x_2}\left\{r\,\rho^{l+1}\,u^l_{f2}\right\}$$

$$- \left[\frac{1}{r}\frac{\partial}{\partial x_1}\left\{\frac{\rho^{l+1}r\,\alpha\,\Delta V}{AP^{u_{f1}}}R_{uf1}\right\} + \frac{1}{r}\frac{\partial}{\partial x_2}\left\{\frac{\rho^{l+1}r\,\alpha\,\Delta V}{AP^{u_{f2}}}R_{uf2}\right\}\right], \quad (5.25)$$

where residuals per unit volume, R_{uf1} and R_{uf2}, are given by

$$R_{uf1} = \frac{AP^{u_{f1}}\,u^l_{f1} - \sum A_k\,u^l_{f1,k} - D^l_{u_1}}{\Delta V} + \frac{\partial p^l}{\partial x_1}, \quad (5.26)$$

$$R_{uf2} = \frac{AP^{u_{f2}}\,u^l_{f2} - \sum A_k\,u^l_{f2,k} - D^l_{u_2}}{\Delta V} + \frac{\partial p^l}{\partial x_2}. \quad (5.27)$$

The discretised version of the mass-conserving pressure-correction Equation 5.25 will read as

$$AP\,p'_{m,P} = AE\,p'_{m,E} + AW\,p'_{m,W} + AN\,p'_{m,N} + AS\,p'_{m,S} - \dot m_P \mid \dot m_R, \quad (5.28)$$

where

$$AE = \frac{\rho^{l+1}\,r^2\,\alpha\,\Delta x_2^2}{AP^{uf1}}\bigg|_e, \qquad AW = \frac{\rho^{l+1}\,r^2\,\alpha\,\Delta x_2^2}{AP^{uf1}}\bigg|_w,$$

$$AN = \frac{\rho^{l+1}\,r^2\,\alpha\,\Delta x_1^2}{AP^{uf2}}\bigg|_n, \qquad AS = \frac{\rho^{l+1}\,r^2\,\alpha\,\Delta x_1^2}{AP^{uf2}}\bigg|_s.$$

$$AP = AE + AW + AN + AS, \quad (5.29)$$

$$\dot m_P = \left(\rho^{l+1}\,r\,u^l_{f1}\big|_e - \rho^{l+1}\,r\,u^l_{f1}\big|_w\right)\Delta x_2$$

$$+ \left(\rho^{l+1}\,r\,u^l_{f2}\big|_n - \rho^{l+1}\,r\,u^l_{f2}\big|_s\right)\Delta x_1 + \left(\rho^{l+1}_P - \rho^0_P\right)\frac{\Delta V}{\Delta t}, \quad (5.30)$$

$$\dot m_R = AE\,R_{uf1}\,\Delta x_1\big|_e - AW\,R_{uf1}\,\Delta x_1\big|_w + AN\,R_{uf2}\,\Delta x_2\big|_n - AS\,R_{uf2}\,\Delta x_2\big|_s. \quad (5.31)$$

A number of comments with respect to Equations 5.25–5.31 are now in order.

1. On both staggered and collocated grids, the pressure is stored at node P and the mass conservation equation is solved over the control volume surrounding node P. Therefore, Equation 5.25 is applicable to both types of grids.

[5] In deriving Equation 5.25, it is assumed that $\sum_k A_k u'_{f1,k} = \sum_k A_k u'_{f2,k} = 0$. This is consistent with the SIMPLE-staggered grid practice [51].

2. In incompressible flows, density is independent of pressure. Therefore, $\rho^{l+1} = \rho^l = \rho$ (say). Derivation of the pressure-correction equation for compressible flow is left to the reader as an exercise (see Date [15, 17]).

3. On staggered grids, the momentum equations are solved at the cell faces and, therefore, residuals R_{uf1} and R_{uf2} must vanish at full convergence, rendering $\dot{m}_R = 0$. Although this state of affairs will prevail only at convergence, one may ignore $\dot{m}_R$ *even during iterative solution*. Thus, effectively, the pressure-correction equation applicable to computations on staggered grids is

$$
\frac{1}{r} \frac{\partial}{\partial x_1} \left\{ \frac{\rho^{l+1} r \alpha \Delta V}{AP^{u_{f1}}} \frac{\partial p'_m}{\partial x_1} \right\} + \frac{1}{r} \frac{\partial}{\partial x_2} \left\{ \frac{\rho^{l+1} r \alpha \Delta V}{AP^{u_{f2}}} \frac{\partial p'_m}{\partial x_2} \right\}
$$

$$
= \frac{\partial(\rho^{l+1})}{\partial t} + \frac{1}{r} \frac{\partial}{\partial x_1} \left\{ r \rho^{l+1} u^l_{f1} \right\} + \frac{1}{r} \frac{\partial}{\partial x_2} \left\{ r \rho^{l+1} u^l_{f2} \right\}. \qquad (5.32)
$$

This equation is derived in [51] via an alternative route. It is solved with the boundary condition

$$
\left. \frac{\partial p'_m}{\partial n} \right|_b = 0. \qquad (5.33)
$$

The explanation for this boundary condition is given in a later section.

4. On collocated grids, cell-face velocities must be evaluated by interpolation to complete evaluation of $\dot{m}_P$ because only nodal velocities u_i are computed through momentum equations. Thus, $\dot{m}_P$ in Equation 5.30 is evaluated as

$$
\overline{\dot{m}}_P = \left(\rho^{l+1} r \, \overline{u}^l_1 \big|_e - \rho^{l+1} r \, \overline{u}^l_1 \big|_w \right) \Delta x_2
$$

$$
+ \left(\rho^{l+1} r \, \overline{u}^l_2 \big|_n - \rho^{l+1} r \, \overline{u}^l_2 \big|_s \right) \Delta x_1 + \left(\rho^{l+1}_P - \rho^o_P \right) \frac{\Delta V}{\Delta t}. \qquad (5.34)
$$

Now, to evaluate $\overline{u}_i$, we use *multidimensional* averaging rather than simple one-dimensional averaging. Thus, for example,

$$
\overline{u}^l_{1,e} = \frac{1}{2} \left[\frac{1}{2} \left(u^l_{1,P} + u^l_{1,E} \right) + \frac{\Delta x_{2,n} u^l_{1,se} + \Delta x_{2,s} u^l_{1,ne}}{\Delta x_{2,n} + \Delta x_{2,s}} \right],
$$

$$
u^l_{1,se} = \frac{1}{4} \left(u^l_{1,P} + u^l_{1,E} + u^l_{1,S} + u^l_{1,SE} \right),
$$

$$
u^l_{1,ne} = \frac{1}{4} \left(u^l_{1,P} + u^l_{1,E} + u^l_{1,N} + u^l_{1,NE} \right). \qquad (5.35)
$$

Similar expressions can be derived for other interpolated cell-face velocities.

5. On collocated grids, we do not explicitly satisfy momentum equations at the cell-face locations. Therefore, there is no guarantee that $\dot{m}_R$ will vanish *even at*

convergence. We, therefore, write $R_{u\text{fl},e}$ in Equation 5.31, for example, as

$$
R_{u_{\text{fl},e}} = \overline{\left.\frac{A P^{u_{\text{fl}}} u^l_{\text{fl}} - \sum A_k u^l_{\text{fl},k} - D^l_{u_1}}{\Delta V}\right|_e} + \left.\frac{\partial p^l}{\partial x_1}\right|_e . \tag{5.36}
$$

This equation is the same as Equation 5.26 written for location e, but the net momentum transfer terms are again multidimensionally averaged. This averaging is done because, when computing on collocated grids, one does not have the cell-face coefficients A_k.[6] Now, again using Equation 5.26, we get

$$
\left.\frac{A P^{u_{\text{fl}}} u^l_{\text{fl}} - \sum A_k u^l_{\text{fl},k} - D^l_{u_1}}{\Delta V}\right|_e = \overline{R_{u_{\text{fl},e}}} - \overline{\left.\frac{\partial p^l}{\partial x_1}\right|_e} . \tag{5.37}
$$

Thus, effectively,

$$
R_{u_{\text{fl},e}} = \overline{R_{u_{\text{fl},e}}} - \overline{\left.\frac{\partial p^l}{\partial x_1}\right|_e} + \left.\frac{\partial p^l}{\partial x_1}\right|_e . \tag{5.38}
$$

6. Now, $\overline{R_{u_{\text{fl},e}}}$ is again evaluated in the manner of Equation 5.35. Thus, $\overline{R_{u_{\text{fl},e}}}$ will contain residuals only at nodal locations P, E, N, S, NE, and SE. These residuals will of course vanish at full convergence because momentum equations are being solved at the nodal positions. Therefore, $\overline{R_{u_{\text{fl},e}}} = 0$ and

$$
R_{u_{\text{fl},e}} = \left.\frac{\partial p^l}{\partial x_1}\right|_e - \overline{\left.\frac{\partial p^l}{\partial x_1}\right|_e} . \tag{5.39}
$$

The practice followed here is same as that followed on staggered grids (see item 3).

7. Now, to evaluate the multidimensionally averaged pressure-gradient in Equation 5.39, we write

$$
\overline{\left.\frac{\partial p^l}{\partial x_1}\right|_e} = \frac{1}{2}\left[\frac{1}{2}\left(\left.\frac{\partial p^l}{\partial x_1}\right|_P + \left.\frac{\partial p^l}{\partial x_1}\right|_E\right) + \frac{\Delta x_{2,n}\,\partial p^l/\partial x_1|_{se} + \Delta x_{2,s}\,\partial p^l/\partial x_1|_{ne}}{\Delta x_{2,n} + \Delta x_{2,s}}\right]
$$

$$
= \frac{1}{4}\left[\frac{p^l_E - p^l_W}{\Delta x_{1,e} + \Delta x_{1,w}} + \frac{p^l_{EE} - p^l_P}{\Delta x_{1,e} + \Delta x_{1,w}}\right]
$$

$$
+ \frac{1}{4}\frac{\Delta x_{2,s}}{\Delta x_{2,n} + \Delta x_{2,s}}\left[\frac{p^l_E + p^l_{NE} - p^l_P - p^l_N}{\Delta x_{1,e}}\right]
$$

$$
+ \frac{1}{4}\frac{\Delta x_{2,n}}{\Delta x_{2,n} + \Delta x_{2,s}}\left[\frac{p^l_E + p^l_{SE} - p^l_P - p^l_S}{\Delta x_{1,e}}\right] . \tag{5.40}
$$

---

[6] Note that, in principle, evaluation of these coefficients can be carried out. However, the computational effort involved will be prohibitively expensive in multidimensions. For example, in a three-dimensional calculation, one will need to evaluate eighteen extra coefficients at the cell faces in addition to the six coefficients evaluated at the nodal locations.

To simplify the evaluation, we introduce the following definitions:

$$\overline{p}_{x_1,\mathrm{P}} = \frac{\Delta x_{1,\mathrm{w}}\, p_\mathrm{E} + \Delta x_{1,\mathrm{e}}\, p_\mathrm{W}}{\Delta x_{1,\mathrm{w}} + \Delta x_{1,\mathrm{e}}}, \tag{5.41}$$

$$\overline{p}_{x_2,\mathrm{P}} = \frac{\Delta x_{2,\mathrm{s}}\, p_\mathrm{N} + \Delta x_{2,\mathrm{n}}\, p_\mathrm{S}}{\Delta x_{2,\mathrm{s}} + \Delta x_{2,\mathrm{n}}}, \tag{5.42}$$

$$\overline{p}_\mathrm{P} = \frac{1}{2}\left(\overline{p}_{x_1,\mathrm{P}} + \overline{p}_{x_2,\mathrm{P}}\right), \tag{5.43}$$

$$\overline{p}_{x_1,\mathrm{E}} = \frac{\Delta x_{1,\mathrm{e}}\, p_\mathrm{EE} + \Delta x_{1,\mathrm{ee}}\, p_\mathrm{P}}{\Delta x_{1,\mathrm{e}} + \Delta x_{1,\mathrm{ee}}}, \tag{5.44}$$

$$\overline{p}_{x_2,\mathrm{E}} = \frac{\Delta x_{2,\mathrm{s}}\, p_\mathrm{NE} + \Delta x_{2,\mathrm{n}}\, p_\mathrm{SE}}{\Delta x_{2,\mathrm{s}} + \Delta x_{2,\mathrm{n}}}, \tag{5.45}$$

$$\overline{p}_\mathrm{E} = \frac{1}{2}\left(\overline{p}_{x_1,\mathrm{E}} + \overline{p}_{x_2,\mathrm{E}}\right). \tag{5.46}$$

Substituting these definitions in Equation 5.40 and replacing p_EE and p_W in favour of p_E and p_P, we can show that

$$\left.\frac{\overline{\partial p^l}}{\partial x_1}\right|_\mathrm{e} = \frac{1}{2}\left[\frac{p_\mathrm{E}^l - p_\mathrm{P}^l}{\Delta x_{1,\mathrm{e}}} + \frac{\overline{p}_\mathrm{E}^l - \overline{p}_\mathrm{P}^l}{\Delta x_{1,\mathrm{e}}}\right] = \frac{1}{2}\left.\frac{\partial(p^l + \overline{p}^l)}{\partial x_1}\right|_\mathrm{e}, \tag{5.47}$$

and, therefore, from Equation 5.39

$$R_{u_{\mathrm{f1,e}}} = \frac{1}{2}\left.\frac{\partial(p^l - \overline{p}^l)}{\partial x_1}\right|_\mathrm{e} = \left.\frac{\partial p_\mathrm{sm}'}{\partial x_1}\right|_\mathrm{e}, \tag{5.48}$$

where

$$p_\mathrm{sm}' = \frac{1}{2}\left(p^l - \overline{p}^l\right). \tag{5.49}$$

The suffix sm here stands for *smoothing* pressure correction.

8. Repeating items 4, 5, 6, and 7 at other cell faces, we obtain

$$R_{u_{\mathrm{f1,w}}} = \left.\frac{\partial p_\mathrm{sm}'}{\partial x_1}\right|_\mathrm{w}, \qquad R_{u_{\mathrm{f2,n}}} = \left.\frac{\partial p_\mathrm{sm}'}{\partial x_2}\right|_\mathrm{n}, \qquad R_{u_{\mathrm{f2,s}}} = \left.\frac{\partial p_\mathrm{sm}'}{\partial x_2}\right|_\mathrm{s}. \tag{5.50}$$

Thus, substituting these equations in Equation 5.31, it follows that

$$\dot{m}_\mathrm{R} = AE\left.\frac{\partial p_\mathrm{sm}'}{\partial x_1}\Delta x_1\right|_\mathrm{e} - AW\left.\frac{\partial p_\mathrm{sm}'}{\partial x_1}\Delta x_1\right|_\mathrm{w}$$

$$+ AN\left.\frac{\partial p_\mathrm{sm}'}{\partial x_2}\Delta x_2\right|_\mathrm{n} - AS\left.\frac{\partial p_\mathrm{sm}'}{\partial x_2}\Delta x_2\right|_\mathrm{s}. \tag{5.51}$$

9. In evaluating coefficients AE, AW, AN, and AS, we need AP coefficients at the cell faces (see Equation 5.29). However, these can be evaluated by

one-dimensional averaging as

$$AP_e^{uf1} = \frac{1}{2}\left(AP_P^u + AP_E^u\right),$$

$$AP_n^{uf2} = \frac{1}{2}\left(AP_P^u + AP_N^u\right), \tag{5.52}$$

where $AP^u = AP^{u1} = AP^{u2}$ on collocated grids.

These derivations show that Equations 5.30 and 5.31 can be replaced by Equations 5.34 and 5.51, respectively. Thus, the mass-conserving pressure-correction equation (5.25) can be effectively written as

$$
\frac{1}{r}\frac{\partial}{\partial x_1}\left\{\frac{\rho^{l+1}\,r\,\alpha\,\Delta V}{AP^{uf1}}\frac{\partial p_m'}{\partial x_1}\right\} + \frac{1}{r}\frac{\partial}{\partial x_2}\left\{\frac{\rho^{l+1}\,r\,\alpha\,\Delta V}{AP^{uf2}}\frac{\partial p_m'}{\partial x_2}\right\}
$$
$$
= \frac{\partial(\rho)}{\partial t} + \frac{1}{r}\frac{\partial}{\partial x_1}\left\{r\,\rho^{l+1}\,\overline{u_1}^{\,l}\right\} + \frac{1}{r}\frac{\partial}{\partial x_2}\left\{r\,\rho^{l+1}\,\overline{u_2}^{\,l}\right\}
$$
$$
- \left[\frac{1}{r}\frac{\partial}{\partial x_1}\left\{\frac{\rho^{l+1}\,r\,\alpha\,\Delta V}{A\Gamma^{uf1}}\frac{\partial p_{sm}'}{\partial x_1}\right\} + \frac{1}{r}\frac{\partial}{\partial x_2}\left\{\frac{\rho^{l+1}\,r\,\alpha\,\Delta V}{AP^{uf2}}\frac{\partial p_{sm}'}{\partial x_2}\right\}\right]. \tag{5.53}
$$

This equation represents the appropriate form of the mass-conserving pressure-correction equation on collocated grids.

5.2.4 Further Simplification

It is possible to further simplify Equation 5.53. To understand this simplification, consider, for example, the grid disposition near the west boundary as shown in Figure 5.4. When computing at the near-boundary node P $(2, j)$, the pressure gradient $\partial p/\partial x_1|_P$ must be evaluated in the momentum equation for velocity $u_{1,P}$. This will require knowledge of the boundary pressure $p_b = p(1, j)$. On collocated grids, this pressure is not known and, therefore, is evaluated by *linear* extrapolation from interior flow points. Thus,

$$p_b = \frac{L_{bE}}{L_{PE}}\,p_P - \frac{L_{bP}}{L_{PE}}\,p_E, \tag{5.54}$$

where L denotes length. The same procedure is adopted at Nb and Sb. Now, assuming that the pressure variation near a boundary is *locally* linear in both x_1 and x_2 directions, it follows that

$$p_b - \overline{p}_b = p_P - \overline{p}_P \quad \text{or} \quad p_{sm,b}' = p_{sm,P}', \tag{5.55}$$

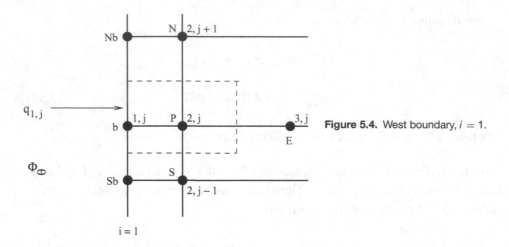

Figure 5.4. West boundary, $i = 1$.

and, therefore,

$$\left.\frac{\partial p'_{sm}}{\partial x_1}\right|_b = \left.\frac{\partial p'_{sm}}{\partial n}\right|_b = 0. \tag{5.56}$$

The same condition is also applicable to p'_m (see Equation 5.33). Now, Equation 5.53 shows that multipliers of gradients of p'_m and p'_{sm} are identical and, since the boundary conditions for these two variables are also identical, we may write the mass-conserving pressure correction equation in the following form:

$$\frac{1}{r}\frac{\partial}{\partial x_1}\left\{\Gamma_1^{p'}\frac{\partial p'}{\partial x_1}\right\} + \frac{1}{r}\frac{\partial}{\partial x_2}\left\{\Gamma_2^{p'}\frac{\partial p'}{\partial x_2}\right\}$$

$$= \frac{\partial(\rho^{l+1})}{\partial t} + \frac{1}{r}\frac{\partial}{\partial x_1}\left\{r\,\rho^{l+1}\,\overline{u}_1^{\,l}\right\} + \frac{1}{r}\frac{\partial}{\partial x_2}\left\{r\,\rho^{l+1}\,\overline{u}_2^{\,l}\right\}, \tag{5.57}$$

where $\Gamma_1^{p'} = \rho^{l+1}\,r\,\alpha\,\Delta V/AP^{ufl}$ and $\Gamma_2^{p'} = \rho^{l+1}\,r\,\alpha\,\Delta V/AP^{uf2}$. Equation 5.57 must be solved with the following boundary condition:

$$\left.\frac{\partial p'}{\partial n}\right|_b = 0, \tag{5.58}$$

where the *total* pressure correction p' is given by

$$p' = p'_m + p'_{sm}, \tag{5.59}$$

and the discretised form of Equation 5.57 is

$$AP\,p'_P = AE\,p'_E + AW\,p'_W + AN\,p'_N + AS\,p'_S - \overline{m}_P, \tag{5.60}$$

where $\overline{m}_P$ is given by Equation 5.34 and the coefficients by Equation 5.29. In passing we note that Equation 5.57 for collocated grids has great resemblance to Equation 5.32, which is applicable to staggered grids, although the dependent variables have different meanings.

5.2.5 Overall Calculation Procedure

The sequence of calculations on collocated grids is as follows.

1. At a given time step, guess the pressure field $p_{i,j}^l$. This may be the pressure field from the previous time step.
2. Solve (see the next section) the momentum equations (5.12) once each for $\Phi = u_1$ and u_2 with problem-dependent boundary conditions. Designate the velocity fields so generated by u_1^l and u_2^l.
3. Form $\overline{m}_{i,j}$ (Equation 5.34) using multidimensional[7] interpolations of cell-face velocity. Now, solve Equation 5.60 with boundary condition (5.58) iteratively to yield the total pressure-correction $p'_{i,j}$ field. The number of iterations may not exceed 5 to 10.
4. Recover the mass-conserving pressure correction via Equation 5.59. Thus,

$$p'_{m,i,j} = p'_{i,j} - p'_{sm,i,j} = p'_{i,j} - \frac{1}{2}\left(p'_{i,j} - \overline{p}'_{i,j}\right), \qquad (5.61)$$

where $\overline{p}'_{i,j}$ is evaluated from Equation 5.43.
5. Correct the pressure and velocity fields according to

$$p_{i,j}^{l+1} = p_{i,j}^l + \beta\, p'_{m,i,j}, \qquad 0 < \beta < 1, \qquad (5.62)$$

$$u_{1,i,j}^{l+1} = u_{1,i,j}^l - \left.\frac{r\,\alpha\,\Delta x_2}{A\,P^{u1}}\right|_{i,j} (p'_{m,i+1/2,j} - p'_{m,i-1/2,j}), \qquad (5.63)$$

$$u_{2,i,j}^{l+1} = u_{2,i,j}^l - \left.\frac{r\,\alpha\,\Delta x_1}{A\,P^{u2}}\right|_{i,j} (p'_{m,i,j+1/2} - p'_{m,i,j-1/2}). \qquad (5.64)$$

Note that $A\,P^{u1} = A\,P^{u2}$.
6. Solve the discretised equations (5.12) for all other scalar $\Phi_{i,j}$ relevant to the problem at hand.
7. Check convergence through evaluation of residuals (see the next section) for momentum and scalar Φ equations. Care is, however, required in calculation of mass residuals as will be discussed shortly.
8. If the convergence criterion is not satisfied, treat $p^{l+1} = p^l$, $\Phi^{l+1} = \Phi^l$ and return to step 2.
9. To execute the next time step, set all $\Phi^0 = \Phi^{l+1}$ and return to step 1.

---

[7] Although multidimensional interpolation is prescribed, in actual computations, one-dimensional interpolations suffice in most applications.

5.3 Method of Solution

5.3.1 Iterative Solvers

Equations 5.12 for any Φ and Equation 5.60 for p' have the same form, which for any node (i, j) can be generalised as

$$(AP_{i,j} + Sp_{i,j})\,\Phi_{i,j}^{l+1} = AE_{i,j}\,\Phi_{i+1,j}^{l+1} + AW_{i,j}\,\Phi_{i-1,j}^{l+1}$$
$$+ AN_{i,j}\,\Phi_{i,j+1}^{l+1} + AS_{i,j}\,\Phi_{i,j-1}^{l+1} + Su_{i,j}, \qquad (5.65)$$

where $Su = D$, $AP = AE + AW + AN + AS$, and $Sp = (\rho^{\circ}\,\Delta V / \Delta t)$. Note that Su and Sp can be further augmented to effect underrelaxation, boundary conditions, and to some extent domain complexity. If there are IN nodes in the i direction and JN nodes in the j direction, Equation 5.65 represents a set of $(IN - 2) \times (JN - 2)$ equations for the *interior* nodes for each Φ. These equations can be solved by matrix-inversion-type direct methods. However, in multidimensional convection, *iterative methods* are usually preferred in which Equation 5.65 is solved *sequentially* for each Φ. There are two extensively used methods of this type: GS and alternating direction integration (ADI).

Gauss–Seidel (GS) Method

In the GS method, for each Φ, coefficients AE, AW, AN, AS, Su, and Sp are evaluated based on Φ values at iteration level l for each node (i, j), $i = 2$ to $IN - 1$ and $j = 2, JN - 1$. Then the nodal value is updated in a double DO loop:

```
     DO 1 J = 2, JN-1
     DO 1 I = 2, IN-1
     ANUM = AE (I, J)*FI(I+1, J) + AW(I, J)*FI(I - 1, J)
         + AN(I, J)*FI(I, J + 1) + AS(I, J)*FI(I, J - 1)
         + SU(I, J)
     ADEN = AP(I, J) + SP(I, J)
     FI(I, J) = ANUM / ADEN
  1     CONTINUE
```

This method is sometimes called a point-by-point method because each node (i, j) is visited in turn. Note that as one progresses from $i = 2$ and $j = 2$, some of the neighbouring Φ values are already updated whereas others still retain their values at iteration level l. Thus, the net evaluation is really a mixed evaluation. Yet, at the end of the DO loop, values at all nodes are treated as having $(l + 1)$-level values. Convergence is declared when the residuals (see the next subsection) fall below a certain low value. This iterative method, though very robust and simple to implement, is very slow to converge.

ADI Method

The ADI method is a line-by-line method in which Equation 5.65 is first solved for all $j = $ constant lines (say). This is called the j-direction sweep. The solution thus obtained may be called the $\Phi^{l+1/2}$ solution. Now, using this solution, Equation 5.65 is again solved for $i = $ constant lines to generate the Φ^{l+1} solution. This is called the i-direction sweep. The implementation details are as follows. For the j sweep, Equation 5.65 is written as

$$(AP_{i,j} + Sp_{i,j})\, \Phi_{i,j}^{l+1/2} = AE_{i,j}\, \Phi_{i+1,j}^{l+1/2} + AW_{i,j}\, \Phi_{i-1,j}^{l+1/2} + SJ_{i,j}, \quad (5.66)$$

where

$$SJ_{i,j} = AN_{i,j}\, \Phi_{i,j+1}^{l} + AS_{i,j}\, \Phi_{i,j-1}^{l} + Su_{i,j}. \quad (5.67)$$

Now, dividing by coefficient of $\Phi_{i,j}$, Equation 5.66 for fixed j can also be written as

$$\Phi_{i}^{l+1/2} = a_i\, \Phi_{i+1}^{l+1/2} + b_i\, \Phi_{i-1}^{l+1/2} + c_i, \quad i = 2, \ldots, IN - 1, \quad (5.68)$$

where $a_i = AE_{i,j}/(AP_{i,j} + Sp_{i,j})$, $b_i = AW_{i,j}/(AP_{i,j} + Sp_{i,j})$, and $c_i = SJ_{i,j}/(AP_{i,j} + Sp_{i,j})$.

It is clear that Equation 5.68 can be solved using TDMA for each $j = 2$ to $JN - 1$ to complete the j sweep. To execute the i sweep, Equation 5.65 is again written as

$$(AP_{i,j} + Sp_{i,j})\, \Phi_{i,j}^{l+1} = AN_{i,j}\, \Phi_{i,j+1}^{l+1} + AS_{i,j}\, \Phi_{i,j-1}^{l+1} + SI_{i,j}, \quad (5.69)$$

where

$$SI_{i,j} = AE_{i,j}\, \Phi_{i+1,j}^{l+1/2} + AW_{i,j}\, \Phi_{i-1,j}^{l+1/2} + Su_{i,j}. \quad (5.70)$$

Equation 5.69 can again be cast in the form of Equation 5.68 and subsequently solved for each $i = $ constant line by TDMA. The two sweeps complete one iteration. Thus, in the ADI method, the domain is swept *twice* per iteration. In spite of this, the procedure proves to be much faster than the GS procedure. In Chapter 9, some additional methods for convergence enhancement are described.

5.3.2 Evaluation of Residuals

The convergence of the iterative procedure is checked by evaluating the imbalance in Equation 5.12. Thus, for each Φ, we evaluate

$$R_{\Phi} = \left[\sum_{\text{all nodes}} \left\{ AP\, \Phi_P - \sum_k A_k\, \Phi_k - D \right\}^2 \right]^{0.5}. \quad (5.71)$$

When the maximum value of R_{Φ} among all Φs is less than the convergence criterion (typically 10^{-5}), the iteration is stopped. Often, R_{Φ} is normalized with a reference quantity specific to a problem having units of $AP\, \Phi$.

Special care is, however, needed for the mass residual. On staggered grids, the mass residual R_m is checked via Equation 5.30 [51]. That is,

$$R_m = \left[\sum_{\text{all nodes}} (\dot{m}_P)^2 \right]^{0.5}. \tag{5.72}$$

However, on collocated grids, one cannot use this equation directly because $\overline{\dot{m}}_{i,j} \neq 0$ even at convergence. Therefore, Equation 5.72 is written as

$$R_m = \left[\sum_{\text{all nodes}} \left(AP\, p'_{m,i,j} - \sum_k A_k\, p'_{m,k} \right)^2 \right]^{0.5}, \tag{5.73}$$

where AP and A_k are coefficients of the pressure-correction equation. It will be recognized that this equation simply represents the discretised version of the left-hand side of Equation 5.32 (or see Equation 5.28 with $\dot{m}_R = 0$). Thus, R_m is evaluated *after* $p'_{m,i,j}$ is recovered in step 4 of the calculation procedure. This is an important departure from the staggered-grid practice that a casual reader may overlook.

5.3.3 Underrelaxation

Global Relaxation

As mentioned in Chapter 2, in steady-state problems ($\Delta t \to \infty$), underrelaxation is effected by augmenting Su and Sp as

$$Su_{i,j} = Su_{i,j} + B\, \Phi^l_{i,j}, \quad Sp_{i,j} = Sp_{i,j} + B, \quad B = \frac{(1-\alpha)}{\alpha}(AP_{i,j} + Sp_{i,j}), \tag{5.74}$$

where α is the underrelaxation factor and l is the iteration level. The value of α is the same for all nodes but it may be different for different Φs. This is called *global*, or constant, underrelaxation.

False Transient

In multidimensional problems, underrelaxation is often effected in another way. Thus, consider a steady-state problem in which $\Delta t = \infty$ and, therefore, the transient term is zero. However, one can imagine that the steady state is achieved following a transient and each time step is likened to a change in iteration level by one. In this case, $\Phi^o_{i,j}$ may be viewed as $\Phi^l_{i,j}$ and the time step Δt as the *false-transient* step. Then, combining Equation 5.65 with Equation 5.74, we can deduce that the resulting equation may be viewed as one in which

$$\alpha_{\text{eff},i,j} = \frac{AP_{i,j} + Sp_{i,j}}{AP_{i,j} + Sp_{i,j} + (\rho^o\, \Delta V / \Delta t)_{i,j}}, \tag{5.75}$$

where the suffix eff is added for two reasons. Firstly, note that this equation arises out of comparison with Equation 5.74; secondly, α_{eff} is not a global constant but will vary for each node (i, j). In fact, this variation also proves to be most appropriate. This can be understood as follows. When $AP_{i,j} + Sp_{i,j}$ is small, the change in Φ from iteration level l to $l + 1$ will be large (see Equation 5.65). It is precisely this large change that is to be controlled by underrelaxation. Equation 5.75 shows that α_{eff} is indeed small when $AP_{i,j} + Sp_{i,j}$ is small. Conversely, when $AP_{i,j} + Sp_{i,j}$ is large, the implied change in Φ is small; therefore, we can afford a larger value of α. Thus, underrelaxation through the false-transient method is proportionate to the requirement. Of course, the smaller the value of the false Δt, the smaller is the value of the estimated α_{eff}.

Although in most nonlinear problems use of constant α suffices, the false-transient method needs to be invoked when couplings between equations for different Φs are strong or when the source terms for a given Φ vary greatly over a domain or when the initial guess of different variables is very poor. Most practitioners invoke the false-transient method when the global underrelaxation method fails.

5.3.4 Boundary Conditions for Φ

In fluid flow and convective transport, five types of boundaries are encountered: *inflow, outflow or exit, symmetry, wall,* and *periodic*. At all these boundaries, mainly three types of conditions are encountered:

1. Φ_b specified,
2. $\partial\Phi/\partial n|_b$ specified, and
3. $\partial^2\Phi/\partial n^2|_b$ specified,

where n is normal to the boundary. We shall discuss each boundary type separately.

Inflow Boundary

At the inflow boundary, values of all variables are specified and are therefore known.[8] Thus, at a west boundary (see Figure 5.4), for example, we can write

$$Su_{2,j} = Su_{2,j} + AW_{2,j}\,\Phi_{1,j}, \qquad Sp_{2,j} = Sp_{2,j} + AW_{2,J}, \qquad AW_{2,j} = 0.$$

$$(5.76)$$

---

[8] Care is needed in specifying inflow conditions for turbulence variables e and ϵ. Typically, $e_{in} = (Tu\,u_{in})^2$, where Tu is the prescribed turbulence intensity. Now, the dissipation is specified through the definition of turbulent viscosity. Thus, $\epsilon_{in} = C_\mu\,\rho\,e^2/(\mu\,\text{VISR})$, where the ratio $\text{VISR} = \mu_t/\mu$ is assumed (typically, of the order of 20 to 40). In practical applications, Tu and VISR are rarely known and, therefore, the analyst must assume their magnitudes.

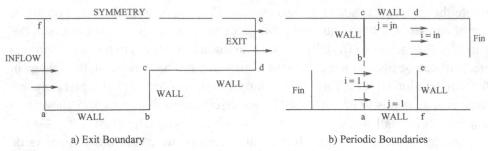

　　　　　a) Exit Boundary　　　　　　　　　　　b) Periodic Boundaries

Figure 5.5. Exit and periodic boundaries.

Wall Boundary

At the wall, either Φ_b or its flux q_b is specified. For the first type, Equation 5.76 applies. If flux is specified, then at the west boundary again,

$$Su_{2,j} = Su_{2,j} + A_{1,j}\, q_{1,j}, \qquad \Phi_{1,j} = \frac{A_{1,j}\, q_{1,j}}{AW_{2,j}} + \Phi_{2,j}, \qquad AW_{2,j} = 0,$$

$$(5.77)$$

where $A_{1,j} = r_j\, \Delta x_{2j}$ is the boundary area.[9]

Symmetry Boundary

At this boundary, there is no flow normal to the boundary and no diffusion either. Thus, with reference to Figure 5.4, for a scalar Φ, $q_{1,j} = 0.0$. For vectors, the normal velocity component $u_1(1, j) = 0$ and $u_2(1, j) = u_2(2, j)$. In all cases, $AW_{2,j} = 0$.

Outflow Boundary

The outflow boundary is one where the fluid leaves the domain of interest. The boundary condition at the outflow or exit plane is most uncertain. To understand the main issues involved, consider Figure 5.5(a) in which de represents the outflow boundary. Now to affect the boundary condition, we may assume that the Peclet number $(u_1\, \Delta x_1 / \Gamma)|_b$ is very large. In this case, the AE coefficient of all near-boundary nodes will be zero and, therefore, no explicit boundary condition Φ_b or $\partial\Phi/\partial n|_b$ is necessary. In many circumstances, this assumption may not be strictly valid. One way to overcome this difficulty is to *shift* boundary de further downstream than required in the original domain specification. Thus, one carries out computations over an extended domain and effect $AE = 0$ at the new location of de. A third alternative is to assume that a *fully developed* state prevails at de so that both the first as well as the second normal derivatives are zero. Most researchers prefer to set the second-order derivative to zero and extract Φ_b by extrapolation while the transport equation is solved with $AE = 0$.

---

[9] In turbulent flows, the wall boundary requires special attention when the HRE form of the e–ϵ model is employed. This matter will be taken up in the next section.

Since none of these alternatives can be relied upon, it is advisable to ensure that the overall mass balance for the domain is maintained throughout the iterative process. This means that the exit-mass flow rate must equal the known inflow rate. Thus, after effecting the boundary condition (marked by superscript *, say) according to any of the alternatives just described, it is important to correct the boundary velocities as

$$u_{1b} = u_{1b}^* F, \qquad u_{2b} = u_{2b}^* F, \qquad F = \dot{m}_{in} \Big/ \sum \dot{m}_{exit}^*, \qquad (5.78)$$

where $\dot{m}_{exit}^*$ is evaluated from the starred velocity boundary condition.

Periodic Boundary

Figure 5.5(b) shows flow between parallel plates with attached fins. In this case, after an initial development length, the flow between two fins will *repeat* exactly. Such a flow is called *periodically* fully developed flow and the periodic boundary condition will imply

$$\Phi_{1,j} = \Phi_{IN,JN+1-j} = 0.5 (\Phi_{2,j} + \Phi_{IN-1,JN+1-j}),$$

$$u_{2(1,j)} - -u_{2(IN,j)} = 0.5 (u_{2(2,j)} - u_{2(IN-1,JN+1-j)}), \qquad (5.79)$$

where IN and JN are the total number of nodes in the ι and j directions, respectively. Note that in this boundary condition specification, the u_2 velocity has *anti-periodicity* whereas all other Φs have *even periodicity*.

5.3.5 Boundary Condition for p_m'

The boundary condition for p_m' is given by Equation 5.33. The reason for this can be understood from step 3 of the calculation procedure. When this step is executed, the u_i^l fields along with their boundary values $u_{i,b}^l$ are already known. Now, when the p' equation is solved, it is *assumed* that these boundary values are correct and, therefore, require no further corrections.

If we now consider Equation 5.12 for $\Phi = u_1^{l+1}$ and $\Phi = u_1^l$ and subtract the latter equation from the former, with $u_1' = u_1^{l+1} - u_1^l$, we have

$$AP\,u_{1,P}' = \sum A_k\,u_{1,k}' - \Delta V\,\frac{\partial p_m'}{\partial x_1}\Big|_P,$$

where ks represent neighbours of P. Also, $\sum A_k u_{1,k}' = 0$ through our assumption introduced in Section 5.2.3. This explains the form of velocity correction introduced in Equation 5.63 for an *interior* node. The same arguments apply to the u_2 velocity corrections given in Equation 5.64.

Now, if the preceding equation is written for the boundary nodes (P $=$ b), clearly $u_{1,b}' = 0$ because no corrections are to be applied to the boundary velocities. Therefore, $\partial p_m'/\partial x_1|_b = 0$. This is boundary condition (5.33). In discretised form,

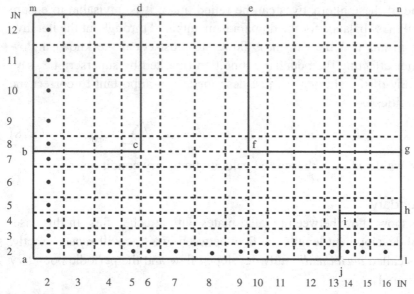

Figure 5.6. Node tagging.

the boundary condition is implemented by setting the boundary coefficient of the pressure-correction equation to zero for the near-boundary node.

Sometimes, we may have a boundary on which pressure is specified and, therefore, remains fixed. For such boundaries, $p'_{m,b} = 0$.

5.3.6 Node Tagging

In Chapter 2, we emphasised that the introduction of Su and Sp can facilitate writing of generalised computer codes by capturing a large variety. In multidimensional codes, further variety can be captured by *tagging* each node of the domain with a number. This is intended to facilitate handling of

1. different types of boundary conditions over different portions of the same *physical* boundary and
2. domains that are not perfect rectangles.

Figure 5.6 shows an arbitrary domain a-b-c-d-e-f-g-h-i-j, which we shall call the domain of interest. However, we regard it as a part of a rectangular domain a-m-n-l with nodes $i = 1$ to IN and $j = 1$ to JN. This will create areas b-c-d-m, f-g-n-e, and j-l-h-i, which are *not* of interest. We term them as *inert* or *blocked* areas. Now, coordinates x_{1i} and x_{2j} are chosen so that the implied cell-face locations exactly coincide with the boundaries of the domain of interest. This ensures that our domain of interest is filled with full (not partial) control volumes as shown in the figure.

Node tagging is now accomplished using the following convention:

1. NTAG (I, J) = 0 identifies all nodes *interior to the domain*. That is, nodes falling on the boundaries a-m, m-n, n-l, and l-a are excluded.
2. NTAG (I, J) = 1 identifies all interior nodes in the *inert* areas.
3. NTAGW (I, J) = 11, 12, 13, 14, 15 identifies nodes adjacent to the WEST boundary with 11 for inflow boundary, 12 for symmetry boundary, 13 for exit boundary, 14 for wall boundary, and 15 for periodic boundary. NTAGW is zero for all other nodes.
4. Similarly, NTAGE (I , J) = 21, 22, 23, 24, 25 identifies nodes adjacent to the EAST boundary, NTAGS (I , J) = 31, 32, 33, 34, 35 identifies nodes adjacent to the SOUTH boundary, and NTAGN (I, J) = 41, 42, 43, 44, 45 identifies nodes adjacent to the NORTH boundary.

Using this convention (which is quite arbitrary), NTAGW will have a finite number for $i = 2$ and $j = 2, 3, \ldots, 7$ (boundary a-b) and for $i = 6$ and $j = 8, 9, \ldots, JN - 1$ (boundary c-d). Similarly, NTAGN will be finite for $j = 7$ and $i = 2, 3, 4, 5$ (boundary b-c), for $j = JN - 1$ and $i = 6, 7, 8, 9$, and again for $j = 7$ and $i = 10, 11, \ldots, IN - 1$ (boundary f-g). NTAGS and NTAGE can be similarly specified.

The choice of numbers 11, 12, 13, etc. in NTAGW is arbitrary but brings one advantage. That is, for near-west boundary nodes, NTAGW/10 = 1 in FORTRAN and, therefore, a WEST boundary is readily identified. Similarly, NTAGN/40 = 1 readily identifies a NORTH boundary. Once this identification is done, the actual numbers (11, 12, etc.) identify the type of boundary condition and therefore $Su_{i,j}$ and $Sp_{i,j}$ for the near-boundary nodes can be set up. This facilitates specification of different boundary conditions at the *same* physical boundary. Thus, if boundary a-b is a wall, a part of it may be insulated and the rest may receive heat flux. Similarly, with respect to mass transfer, a part may be inert but the rest may experience a finite mass transfer flux.

Finally, at the inert or blocked node where NTAG (I, J) = 1, one simply specifies

$$Su_{i,j} = 10^{30} \, \Phi_{\text{desired}}, \qquad Sp_{i,j} = 10^{30}. \qquad (5.80)$$

Examination of Equation 5.65 will show that since $AP_{i,j}$ can never be very large, these settings render $\Phi_{i,j} = \Phi_{\text{desired}}$ at the inert nodes. In Figure 5.6, the inert regions are outside the domain of interest. However, it is easy to appreciate that one can even have inert regions that are enclosed by the overall domain of interest (hence the term blocked region), as shown in Figure 5.7. The figure also shows how a domain with irregular boundaries may be specified by node tagging. Here, the irregular boundary is approximated by a staircase-like zigzag boundary.[10] Such

---

[10] The *accuracy* of the solution will of course depend on the number of steps into which the true boundary is subdivided.

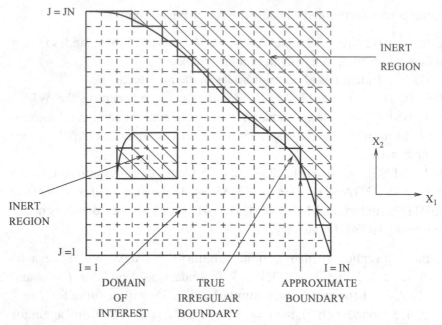

Figure 5.7. Domain with irregular boundary.

an approximation of the true boundary is permissible when the flow is in the x_3 direction (i.e., u_3 is finite but $u_1 = u_2 = 0$ as in the case of laminar fully developed flow in a duct) because the replacement does not imply a *rough* wall.[11] If, however, the velocity components u_1 and u_2 were *finite*, it would be advisable to map the domain by *curvilinear* or *unstructured* grids (see Chapter 6) so that the staircase boundary approximation does not interfere with the expected fluid dynamics (see Exercises 16 and 17).

Finally, note that the exit and wall boundaries may be specified in more than one way, as discussed in the previous subsection. Thus, at a wall one may specify temperature or heat flux. One can introduce further identifying tags for each type.

5.4 Treatment of Turbulent Flows

5.4.1 LRE Model

In multidimensional elliptic flows, the concept of mixing length is not very useful. This is because it is difficult to invent a three-dimensional (3D) algebraic prescription for the mixing length. As was learnt in the previous chapter, however, the LRE e–ϵ model is general and does not require any input that depends on the *distance*

---

[11] The replacement will also be permissible in a pure conduction problem.

from the wall. The 2D elliptic version of this model can be described via Equation 5.1 for $\Phi = e$ and ϵ^* with the following definitions of the source terms [9]:

$$S_e = G - \rho\epsilon^*, \tag{5.81}$$

$$S_{\epsilon^*} = \frac{\epsilon^*}{e} \left[C_1 G - C_2 \rho\epsilon^* \right] + E_{\epsilon^*}, \tag{5.82}$$

where

$$G = \mu_t \left[2 \left(\frac{\partial u_1}{\partial x_1} \right)^2 + 2 \left(\frac{\partial u_2}{\partial x_2} \right)^2 + \left(\frac{\partial u_2}{\partial x_1} + \frac{\partial u_1}{\partial x_2} \right)^2 \right], \tag{5.83}$$

$$\epsilon^* = \epsilon - 2\nu \left[\left(\frac{\partial\sqrt{e}}{\partial x_1} \right)^2 + \left(\frac{\partial\sqrt{e}}{\partial x_2} \right)^2 \right], \tag{5.84}$$

$$E_{\epsilon^*} = 2\nu\nu_t \left[\left(\frac{\partial^2 u_1}{\partial x_1^2} \right)^2 + 2 \left(\frac{\partial^2 u_1}{\partial x_1 \partial x_2} \right)^2 + \left(\frac{\partial^2 u_1}{\partial x_2^2} \right)^2 \right.$$
$$\left. + \left(\frac{\partial^2 u_2}{\partial x_1^2} \right)^2 + 2 \left(\frac{\partial^2 u_2}{\partial x_1 \partial x_2} \right)^2 + \left(\frac{\partial^2 u_2}{\partial x_2^2} \right)^2 \right]. \tag{5.85}$$

The expressions for C_1 and C_2 are the same as those given in Chapter 4. The LRE $e-\epsilon^*$ model permits use of the $e = \epsilon^* = 0$ condition at a wall boundary. Although this is a distinct advantage of the model, accurate predictions require a very large number of nodes, as was learnt through boundary layer predictions. In two dimensions, if more than one boundary is a wall then the number of nodes required becomes very large indeed. This is because, to resolve the *inner* layer near a wall, which typically spans to $y^+ = y u_\tau / \nu = 100$, one may need 60–80 nodes with the first node as close as $y^+ = 1$ whereas the outer layer may require no more than 20–30 nodes. Physically, the inner layer occupies a very thin region near a wall.[12] Thus, computations with the LRE model in 2D and 3D elliptic flows can be quite expensive. In the interest of economy of computations, therefore, it is desirable if an adaptation can be made that restricts calculations only to the outer layers.

5.4.2 HRE Model

In a large majority of flow situations, as is well known, the inner layer exhibits near universality with respect to velocity and temperature profiles – the so-called laws

---

[12] In a fully developed flow in a pipe (radius R), for example, $R^+ = R u_\tau / \nu = (Re/2) \sqrt{f/2}$. Using $f = 0.046 \, Re^{-0.2}$, we estimate that at $Re = 50,000$ (say), $R^+ = 1,285$. This shows that the inner layer is less than 10% of the radius.

of the wall. In the *two-layer* approach, these laws are given by[13]

$$
u^+ = \begin{cases} \dfrac{u}{u_\tau} = y^+, & y^+ < 11.6, \\[2mm] \dfrac{1}{\kappa} \ln \left[E\, y^+ \right], & y^+ > 11.6, \end{cases}
\tag{5.86}
$$

where $\kappa = 0.41$, $E = 9.072$, and wall-friction velocity $u_\tau = \sqrt{\tau_w/\rho}$. Similarly, the temperature law is given by

$$
T^+ = \frac{-(T - T_w)\rho\, C_p\, u_\tau}{q_w} = Pr_t\, (u^+ + PF),
\tag{5.87}
$$

where $Pr_t = 0.9$ and

$$
PF = \begin{cases} \left(\dfrac{Pr}{Pr_t} - 1 \right) u^+, & y^+ < 11.6, \\[3mm] 9.24 \left[\left(\dfrac{Pr}{Pr_t} \right)^{0.75} - 1 \right) \right] \\[3mm] \times \left[1 + 0.28 \exp\left(-0.007\, \dfrac{Pr}{Pr_t} \right) \right], & y^+ > 11.6. \end{cases}
\tag{5.88}
$$

These specifications are empirical but, in the range $30 < y^+ < 100$, they are reasonably accurate. One can thus exploit this near universality to eliminate the inner layer almost completely from the calculations and compute only in the outer layers. In the outer layers, turbulence is vigorous and $Re_t = \mu_t/\mu$ is large (hence the acronym HRE for high Reynolds number model) so that $\epsilon^* \to \epsilon$ and, therefore, the source terms are given by

$$
S_e = G - \rho\epsilon, \qquad S_\epsilon = \frac{\epsilon}{e}\, [C_1\, G - C_2\, \rho\, \epsilon],
\tag{5.89}
$$

where $C_1 = 1.44$ and $C_2 = 1.92$. The task now is to modify our discretised equations for the *near-wall* boundary node P such that the implications of the laws of the wall are embodied in the equations.[14] To achieve this goal, we note the following two characteristics of the $30 < y^+ < 100$ region in which the near-wall node P is assumed to have been placed. These are

$$
u_\tau = C_\mu^{1/4}\, \sqrt{e},
\tag{5.90}
$$

$$
G = \rho\, \epsilon.
\tag{5.91}
$$

Let node P be adjacent to south node b (see Figure 5.8). We shall consider each variable in turn.

---

[13] In all derivations in this subsection, distance y and x_2 are used interchangeably.
[14] In the literature, this is called the *wall function* treatment [39].

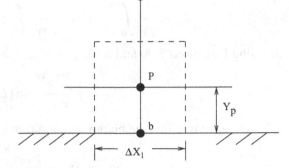

Figure 5.8. Wall function treatment.

$\Phi = u_1$

For an impermeable wall, $C_s = 0$ and, therefore, $AS = \mu_{\text{eff}} \Delta x_1 / y_{\text{P}}$. Also, the no-slip condition requires that $u_{1\text{b}} = 0$ at the *stationary* wall. Thus

$$\tau_{\text{w}} = \mu_{\text{eff}} \left. \frac{\partial u_1}{\partial y} \right|_{y=0} = \frac{\mu_{\text{eff}}}{y_{\text{P}}} (u_{1\text{P}} - u_{1\text{b}}) = \frac{\mu_{\text{eff}}}{y_{\text{P}}} u_{1\text{P}}. \tag{5.92}$$

Now, replacing $u_{1\text{P}}$ from Equation 5.86, we can show that

$$\frac{\mu_{\text{eff}}}{y_{\text{P}}} = \frac{\tau_{\text{w}}}{u_{1\text{P}}} = \frac{\rho \kappa u_\tau}{\ln(E y_{\text{P}}^+)}, \tag{5.93}$$

where $y_{\text{P}}^+ = y_{\text{P}} u_\tau / \nu$. Therefore, using Equation 5.90, we get

$$\frac{\mu_{\text{eff}}}{y_{\text{P}}} = \begin{cases} \dfrac{\mu}{y_{\text{P}}}, & y^+ < 11.6, \\[2mm] \dfrac{\rho \kappa C_\mu^{1/4} \sqrt{e_{\text{P}}}}{\ln(E y_{\text{P}} C_\mu^{1/4} \sqrt{e_{\text{P}}} / \nu)}, & y^+ > 11.6. \end{cases} \tag{5.94}$$

Thus, for variable $\Phi = u_1$, for the near-wall node P, we may set

$$Su = Su + 0, \qquad Sp = Sp + \frac{\mu_{\text{eff}}}{y_{\text{P}}} \Delta x_1, \qquad AS = 0. \tag{5.95}$$

$\Phi = e$

A further characteristic of the inner layer is that the shear stress through the layer is constant and hence equals τ_{w}. Also, experimental data demonstrate that in the $30 < y^+ < 100$ region, $\partial e / \partial y \simeq 0$. Therefore, $AS = 0$. The implications of the law of the wall thus can be absorbed through redefinition of S_e for point P:

$$S_e = G_{\text{P}} - \rho \bar{e}_{\text{P}}, \tag{5.96}$$

where

$$G_{\text{P}} \simeq \mu_{\text{eff}} \left(\frac{\partial u_1}{\partial y} \right)^2 = \mu_{\text{eff}} \left(\frac{u_{1\text{P}}}{y_{\text{P}}} \right)^2 = \tau_{\text{w}} \frac{\partial u_1}{\partial y}. \tag{5.97}$$

and, using Equation 5.91,

$$\bar{\epsilon}_P = \frac{1}{y_P} \int_0^{y_P} \epsilon \, dy = \frac{\tau_w}{\rho \, y_P} \int_0^{y_P} \frac{\partial u_1}{\partial y} \, dy = \frac{u_\tau^2 \, u_{1P}}{y_P} \tag{5.98}$$

or, using Equations 5.90 and 5.93,

$$\bar{\epsilon}_P = \frac{C_\mu^{3/4} \, e_P^{3/2}}{\kappa \, y_P} \ln(E \, y_P^+). \tag{5.99}$$

It is now easy to effect the boundary condition via

$$Su_e = Su_e + \frac{\mu_{\text{eff}} u_{1P}^2 \, \Delta V_P}{y_P^2}, \tag{5.100}$$

$$Sp_e = Sp_e + \frac{\rho \, C_\mu^{3/4} \, e_P^{1/2}}{\kappa \, y_P} \ln(E \, y_P^+) \, \Delta V_P. \tag{5.101}$$

$\Phi = \epsilon$

To evaluate ϵ_P, we combine Equations 5.91 and 5.97 so that

$$\epsilon_P = \frac{\tau_w}{\rho} \frac{\partial u_1}{\partial y} = u_\tau^2 \frac{\partial u_1}{\partial y}. \tag{5.102}$$

But, from Equation 5.86, $\partial u_1 / \partial y = u_\tau / (\kappa y)$. Therefore,

$$\epsilon_P = \frac{u_\tau^3}{\kappa \, y_P} = \frac{C^{3/4} \, e_P^{3/2}}{\kappa \, y_P}. \tag{5.103}$$

To effect this condition, we set

$$Su_\epsilon = 10^{30} \, \epsilon_P, \qquad Sp_\epsilon = 10^{30}. \tag{5.104}$$

$\Phi = T$

In this case, $AS = \Gamma_{\text{eff}} \Delta x_1 / y_P$, where $\Gamma_{\text{eff}} = k_{\text{eff}} / C_p$. Again, we set $AS = 0$ and absorb the boundary condition via an augmented source. Thus

$$Su_T = Su_T + \frac{\Gamma_{\text{eff}} \Delta x_1}{y_P} (T_b - T_P) = Su_T + \frac{q_w}{C_p} \Delta x_1. \tag{5.105}$$

Substituting for $(T_b - T_P)$ from Equation 5.87, it follows that

$$\frac{\Gamma_{\text{eff}}}{y_P} = \frac{\rho \, u_\tau}{Pr_t (u_{1P}^+ + PF)}. \tag{5.106}$$

Thus, if q_w is specified, we set

$$Su_T = Su_T + \frac{q_w}{C_p} \Delta x_1, \qquad Sp_T = Sp_T + 0, \tag{5.107}$$

and recover T_b from Equation 5.105. Similarly, if the wall temperature T_b is specified then

$$Su_T = Su_T + \frac{\Gamma_{\text{eff}} \Delta x_1}{y_P} T_b, \qquad Sp_T = Sp_T + \frac{\Gamma_{\text{eff}} \Delta x_1}{y_P}, \qquad (5.108)$$

and q_w is recovered from Equation 5.105. For further refinements of the wall-function approach, see references [41, 69].

$\Phi = \omega_k$

It is not clear if universal mass transfer laws exist for all mass transfer rates. Following theory developed by Spalding [73], however, it is possible to show that

$$\frac{\Gamma_{\text{eff}}}{y_P} = \frac{\rho u_\tau}{Pr_t(u_{1P}^+ + PF)} \frac{\ln(1+B)}{B}, \qquad (5.109)$$

where the Spalding number B is given by

$$B = \frac{\omega_{k,P} - \omega_{k,b}}{\omega_{k,b} - \omega_{k,T}}, \qquad (5.110)$$

and $\omega_{k,T}$ is the mass fraction deep inside the *wall* from where mass transfer is taking place. Note that as $B \to 0$, $\ln(1+B) \to B$. Further, PF is still given by Equation 5.88 but with Pr replaced by Schmidt number Sc. All other adjustments are the same as those for the temperature variable.

5.5 Notion of Smoothing Pressure Correction

It is important to consider the notion of smoothing pressure correction introduced in our analysis of the collocated-grid calculation procedure. This is because, in the original SIMPLE-staggered grid procedure, such a smoothing correction is not required. However, its introduction is vital if zigzag pressure prediction is to be avoided on collocated grids, particularly when coarse grids are used. To understand the importance of smoothing correction, we consider computation of laminar flow in a square cavity (see Figure 5.9) of side L that is infinitely long in the x_3 direction. The top side (the lid) of this cavity is moving in the positive x_1 direction with velocity U_{lid} (say). Because of the no-slip condition, the linear lid movement sets up fluid circulation in the clockwise direction. In this case, steady-state equations for $\Phi = u_1, u_2$, and p' need to be solved.

Figure 5.10 shows the computed distribution of pressure for $Re = U_{\text{lid}} L / \nu = 100$. In Figure 5.10(a), solutions obtained with a 15×15 grid are shown at the vertical midplane ($x_1/L = 0.5$). The solutions are obtained using both staggered and collocated grids with identical grid dispositions. However, in the latter, smoothing pressure correction is not applied (see step 4 of the calculation procedure). It is clear that whereas the staggered-grid procedure produces a smooth pressure distribution,

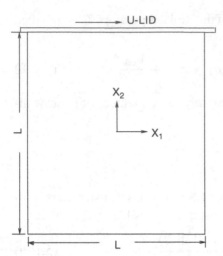

Figure 5.9. Square cavity with a moving lid.

the predicted pressure on the collocated grid is zigzag. Note that the zigzagness is most pronounced in regions where the staggered-grid pressure distribution considerably departs from linearity. Figure 5.10(b) shows the results obtained with a 41×41 grid. Notice that the pressures predicted on both grids are nearly identical and smooth. This suggests that pressure smoothing is in fact not required when fine grids are used. In Figure 5.10(c), the coarse-grid solutions are repeated but now the smoothing pressure correction is applied. It is seen that the predicted pressure distribution on collocated grids is now smooth though not in exact agreement with the staggered-grid pressure distribution because of the coarseness of the grid and also because $\overline{p}$ is evaluated by multidimensional averaging.

Then, what is the role of the smoothing pressure correction? This can be understood from definition (5.49). The smoothing correction represents the difference between the point value of pressure p and the control-volume-averaged pressure $\overline{p}$. The latter is defined by Equation 5.43 as the average of linearly interpolated pressures in the x_1 and x_2 directions. Thus, p'_{sm} can be finite only when spatial variation of pressure p multidimensionally departs from linearity. This is the case at the midplane of the square cavity. On coarse grids, we observe zigzagness if smoothing is not applied. However, when grids are refined, $p'_{sm} \to 0$. That is, as a continuum is approached, no smoothing should be required. The role of smoothing pressure correction is thus simply to predict smooth pressure distribution on *coarse* grids.

We now recall the quantity $\lambda_1 (p - \overline{p})$ introduced in the normal stress expression in Chapter 1. It was stated in that chapter that λ_1 is trivially zero in a continuum but is finite in discretised space. We have recovered $\lambda_1 = 0.5$ in our definition of p'_{sm}. But, as the grid size is refined, one approaches a continuum and, therefore, λ_1 can be set to zero to predict smooth pressure distributions as shown in Figure 5.10(b).

As a corollary, we may now view pressure zigzagness as a *spatial* counterpart of the oscillating compressible sphere of isothermal gas explained by Schlichting [65].

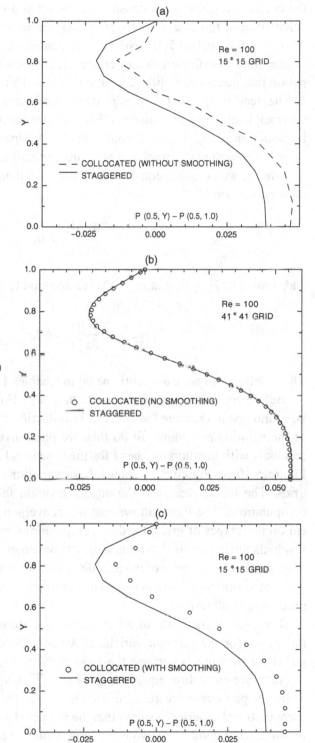

Figure 5.10. Pressure variation with and without smoothing.

On collocated grids, when density is constant and steady state prevails (as in our calculation of the square cavity problem), $\overline{m}_P = \rho \nabla \vec{V}$ and thus $\overline{m}_P \neq 0$, as was recognized in Section 5.3.2. Now, as our control volume is fixed, $\nabla \vec{V} \neq 0$ (which implies rate of volume change) creates dissipation in the system. It is this dissipation that generates $\overline{p}$ different from p. We had anticipated this result in Chapter 1. The need for $p'_{sm} = 0.5(p - \overline{p})$ discovered through our discretisation of equations applicable to a *continuum* is therefore not surprising. In summary, therefore, introduction of p'_{sm} simply accounts for the dissipation introduced in the system. Further discussion of smoothing pressure correction can be found in [16, 17].

Finally, we note that equation 5.41 suggests that $\overline{p}_{x_1,P}$ is a solution to the discretised version of

$$\frac{\partial^2 p}{\partial x_1^2}\bigg|_P = 0, \tag{5.111}$$

and, similarly, $\overline{p}_{x_2,P}$ (Equation 5.42) is a solution to the discretised version of

$$\frac{\partial^2 p}{\partial x_2^2}\bigg|_P = 0. \tag{5.112}$$

These deductions were also anticipated in Chapter 1.

Before considering applications of our SIMPLE-collocated procedure, it would be of interest to examine the effect of introduction of p'_{sm} on the *convergence rate* of the solution procedure. To do this, we plot variation of momentum and mass residuals with iteration number l for the case of 41×41 grid solutions shown in Figure 5.10(b). Figure 5.11 shows these variations for staggered and collocated grids. The initial guess and the underrelaxation factors are identical in the two computations. The figure shows that the convergence histories are almost identical on both types of grids. Further, computations were stopped when momentum residuals fell below 10^{-5}. At this stage of convergence, the mass residuals are seen to be smaller by an order of magnitude. Thus, we may conclude that our SIMPLE-collocated grid procedure is successful in mimicking the SIMPLE-staggered grid procedure in all respects.

The convergence rate of an iterative procedure greatly depends on the initial guess for the relevant variables. Among the different variables, the initial guess for pressure is perhaps the most difficult to provide. Further, in deriving the pressure-correction equation, quantities $\sum A_i u'_i$ and $\sum A_i v'_i$ are set to zero. Thus, the pressure-correction equation is only an approximate one. In spite of this, computational experience shows that the predicted pressure-correction distribution provides very good velocity corrections, which are proportional to the pressure-correction gradient (see Equations 5.63 and 5.64), but a rather poor correction of pressure itself.

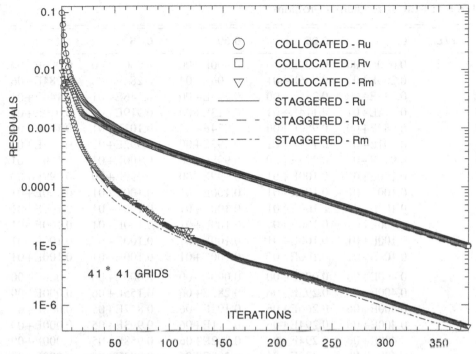

Figure 5.11. Convergence histories.

To appreciate this experience, we consider a 1D flow through a porous medium[15] having porosity ϵ (= volume of fluid/total volume). Then, the governing mass conservation and momentum equations are given by

$$\frac{d\,(\rho^* u)}{d\,x} = 0, \qquad (5.113)$$

$$\frac{d}{d\,x}\,(\rho^* u\,u) = -\frac{d\,p}{d\,x} + 2\,\mu^* \frac{d^2\,u}{d\,x^2} - \mu^* \epsilon\,R\,u, \qquad (5.114)$$

where $\rho^* = \rho/\epsilon^2$, $\mu^* = \mu/\epsilon$, and u is the superficial fluid velocity through the porous medium. The medium resistance parameter $R = 1/K$, where K is the permeability of the medium. If we assume that fluid density is constant then $d\,u/d\,x = 0$ and the momentum equation will reduce to $d\,p/d\,x = -\mu^* \epsilon\,R\,u$. Therefore, taking $\rho = \mu = 1$, $\epsilon = 0.1$, and $R = 4 \times 10^5$ gives the exact solution

$$u = 1, \qquad p = 4 \times 10^5(1 - x/L),$$

where L is the domain length.

We solve this 1D problem using the 2D computer program given in Appendix C[16] in two ways. In Problem 1, the initial guess for pressure is taken from the exact

---

[15] The author is grateful to Prof. D. B. Spalding for recommending this problem for inclusion in this book.

[16] The relevant USER file for this fixed-pressure boundary condition problem is given in Appendix C.

Table 5.1: Porous medium – Problem 2.

l	x/L	0.0	0.25	0.50	0.75	1.0
0		0.000E+00	0.000E+00	0.000E+00	0.000E+00	0.000E+00
1		0.266E+01	0.266E+01	0.106E−02	0.284E−06	0.284E−06
2		0.737E+00	0.737E+00	0.705E+00	0.146E+01	0.146E+01
3		0.114E+01	0.114E+01	0.913E+00	0.919E+00	0.919E+00
4		0.982E+00	0.982E+00	0.974E+00	0.103E+01	0.103E+01
5		0.101E+01	0.101E+01	0.992E+00	0.992E+00	0.992E+00
6	U	0.999E+00	0.999E+00	0.998E+00	0.100E+01	0.100E+01
7		0.100E+01	0.100E+01	0.999E+00	0.999E+00	0.999E+00
8		0.100E+01	0.100E+01	0.100E+01	0.100E+01	0.100E+01
9		0.100E+01	0.100E+01	0.100E+01	0.100E+01	0.100E+01
10		0.100E+01	0.100E+01	0.100E+01	0.100E+01	0.100E+01
11		0.100E+01	0.100E+01	0.100E+01	0.100E+01	0.100E+01
12		0.100E+01	0.100E+01	0.100E+01	0.100E+01	0.100E+01
0		0.400E+06	0.000E+00	0.000E+00	0.000E+00	0.000E+00
1		0.400E+06	0.295E+06	0.283E+06	0.155E+06	0.000E+00
2		0.400E+06	0.266E+06	0.192E+06	0.837E+05	0.000E+00
3		0.400E+06	0.294E+06	0.211E+06	0.993E+05	0.000E+00
4		0.400E+06	0.294E+06	0.202E+06	0.958E+05	0.000E+00
5		0.400E+06	0.298E+06	0.202E+06	0.982E+05	0.000E+00
6	P	0.400E+06	0.299E+06	0.201E+06	0.987E+05	0.000E+00
7		0.400E+06	0.299E+06	0.201E+06	0.993E+05	0.000E+00
8		0.400E+06	0.300E+06	0.200E+06	0.996E+05	0.000E+00
9		0.400E+06	0.300E+06	0.200E+06	0.998E+05	0.000E+00
10		0.400E+06	0.300E+06	0.200E+06	0.999E+05	0.000E+00
11		0.400E+06	0.300E+06	0.200E+06	0.999E+05	0.000E+00
12		0.400E+06	0.300E+06	0.200E+06	0.100E+06	0.000E+00
12	p'_{m}/p	0.000E+00	0.114E−03	−0.150E−03	0.341E−03	0.000E+00
12	p'_{sm}/p	0.000E+00	−0.963E−04	0.188E−03	−0.288E−03	0.000E+00

solution given here, but velocity $u = 0$ at all nodes. In Problem 2, $p(1) = 4 \times 10^5$ and $p(IN) = 0$, but $p = 0$ at all interior nodes of the domain. Again $u = 0$ at all nodes. Thus, in both problems, the guessed velocity is zero and the boundary pressures are held fixed so that $p'(1) = p'(IN) = 0$. Relaxation parameters are taken as $\alpha = \beta = 1$.

For Problem 1, by solving for u and p', the exact solutions (not shown here) for p and u are obtained in just one iteration although the initial guess for u was zero. This is because the initial guess for pressure was itself the exact solution and, therefore, required no correction.

Table 5.1 shows evolutions with iteration number l for Problem 2. Notice that because of the poor initial guess for pressure, the exact velocity solution is obtained in eight iterations whereas the correct pressure prediction requires twelve iterations.

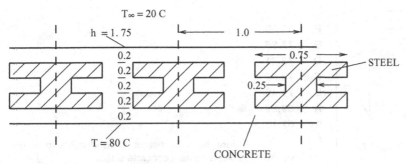

Figure 5.12. Reinforced concrete slab.

Thus, the correct velocity solution is indeed obtained earlier in the iteration process. The last two rows in the table show values of p'_m/p and p'_{sm}/p at convergence. They are indeed small within round-off errors and become even smaller if the iterations are continued.

The general lesson learnt from the example here is that, in a pure flow problem, overall convergence rate is controlled by the evolution of the pressure variable for which there is no exact equation.

5.6 Applications

In this section, a few problems are solved to illustrate the application of the procedure just described. The problems are solved using the generalised computer code given in Appendix C. The reader will find it useful to read the typical USER files given in this appendix to understand the details of implementation.

Conduction Problem

Figure 5.12 shows a concrete slab with I-section steel beams embedded for reinforcement. The conductivities of steel and concrete are 100 and 1 W/m-K, respectively. The lower surface of the slab is at 80°C and the upper surface is exposed to the environment at 20°C with a heat transfer coefficient of 1.75 W/m²-K. It is required to determine the steady-state temperature distribution in the slab.[17]

In this problem, $u_i = 0$; therefore, solution need be obtained for $\Phi = T$ only. The governing differential equation is

$$\frac{\partial}{\partial x_1}\left[K\frac{\partial T}{\partial x_1}\right] + \frac{\partial}{\partial x_2}\left[K\frac{\partial T}{\partial x_2}\right] = 0. \tag{5.115}$$

Equation 5.115 must be solved on the smallest domain, exploiting symmetries. Thus, the chosen domain is $0 \le x_1 \le 0.5$ and $0 \le x_2 \le 1.0$, with $x_1 = 0$ and $x_1 = 0.5$ taken as symmetry boundaries. The boundary conditions at the top and bottom of the slab are shown in the figure.

---

[17] This problem is taken from the book by Patankar [53].

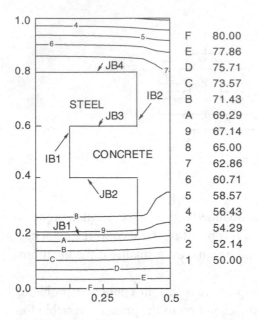

F	80.00
E	77.86
D	75.71
C	73.57
B	71.43
A	69.29
9	67.14
8	65.00
7	62.86
6	60.71
5	58.57
4	56.43
3	54.29
2	52.14
1	50.00

Figure 5.13. Isotherms – conduction in a reinforced cement concrete slab.

Figure 5.13 shows the computed temperature contours. Computations were carried out by employing harmonic-mean conductivities at the cell faces. This is important because conductivities of concrete and steel are different (see interfaces IB1, IB2, JB1, JB2, JB3, and JB4 marked on Figure 5.13). A $13(x_1) \times 22(x_2)$ grid is employed. The figure shows that, in the middle of the slab, the temperature is almost uniform in both steel and concrete. The maximum temperature, 80°C, is prescribed at the lower boundary and the predicted temperature at the top convective boundary is almost uniform at 54°C. The heat loss through the top boundary is thus calculated at 60 W/m^2 and this also equals the heat gain through the bottom boundary since steady-state conditions prevail. Note that if the I-section beams were not present, one would have 1D heat conduction through concrete alone and the heat loss would then be 38.2 W/m^2. The presence of high-conductivity I-section beams has enhanced the rate of heat transfer.

Periodic Laminar Flow and Heat Transfer

Compact heat exchangers often employ an offset-fin configuration to enhance convective heat transfer at the expense of an increased pressure drop. However, when geometric parameters are suitably chosen, the overall thermo-hydraulic performance (i.e., increased heat transfer for the same pumping power or reduced pumping power for the same heat duty) is improved, resulting in a compact heat exchanger design. Figure 5.14 shows an array of interrupted plates or blocks, which may be regarded as a 2D idealisation of the offset-fin heat exchanger; the flow width in the x_3 direction is large. The length and the width of each block are L and t, respectively, and the transverse pitch is H.

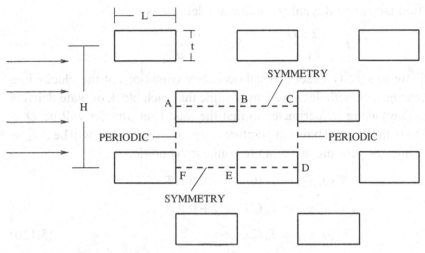

Figure 5.14. Flow in an interrupted passage.

Clearly, under *periodically* fully developed flow and heat transfer past the blocks, suitably defined variables will exhibit distance periodicity $2L$. Thus, for computational purposes, the smallest representative domain (or module) will be A-B-C-D-E-F, as marked in Figure 5.14. Planes A-B-C and D-E-F will experience *symmetry* boundary condition whereas boundaries A-F and C-D will be periodic. Equations for $\Phi = u_1, u_2, T$ and for p' must be solved over this domain.

For the flow variables, the distance periodicity can be accounted for by setting

$$p(x_1, x_2) = -\beta x_1 + p_o(x_1, x_2),\qquad(5.116)$$

where β is the overall pressure gradient (a constant because the flow is fully developed) and p_o is the superposed pressure that is periodic [54]. The same situation also holds for the velocities. Thus, the boundary conditions at planes A-F and C-D are

$$p_o(0, x_2) = p_o(2L, x_2),\qquad u_i(0, x_2) = u_i(2L, x_2).\qquad(5.117)$$

Note that parts of A-F and C-D are solid walls. The symmetry and wall boundary conditions require no elaboration. With the introduction of variable p_o, it will be appreciated that the u_1 and u_2 momentum equations are solved with source terms $\beta - \partial p_o/\partial x_1$ and $-\partial p_o/\partial x_2$, respectively, and the p' equation will provide corrections to pressure p_o. In fact, the equations are solved with an assumed value of β and the average streamwise velocity is evaluated from the resulting predicted velocity field at convergence. The total mass flow through the module can be estimated at any transverse plane but we may evaluate it at plane A-F (say) so that $\dot{m} = \int_0^{H/2} \rho u_1 \, dx_2$ and define u_{av} based on the *frontal area*, as is the practice in heat-exchanger design. Thus,

$$u_{av} = \dot{m}/(\rho H/2).\qquad(5.118)$$

The friction factor and Reynolds number are defined as

$$f = \frac{2\,\beta\,H}{2\,\rho\,u_{\mathrm{av}}^2}, \qquad Re = \frac{\rho\,u_{\mathrm{av}}\,2\,H}{\mu}. \tag{5.119}$$

It is difficult to specify exact thermal boundary conditions at the blocks in a real heat exchanger. Nonetheless, we may assume that each block or plate delivers heat flux q_{w} (say) along its perimeter so that the total heat transfer will be $Q = q_{\mathrm{w}}\,(2\,L + 4\,t/2)$ and the total bulk temperature rise across the module will be $\Delta T_{\mathrm{b}} = Q/(\dot{m}\,C_p)$. Thus, the periodic temperature boundary condition will be

$$T\,(0, x_2) = T_o\,(0, x_2) - 0.5\,\Delta T_{\mathrm{b}},$$

$$T\,(2L, x_2) = T_o\,(2L, x_2) + 0.5\,\Delta T_{\mathrm{b}},$$

$$T_o\,(0, x_2) = T_o\,(2L, x_2). \tag{5.120}$$

In Equations 5.117 and 5.120, all variables must be evaluated at $x_1 = 0\,(I = 1)$ and $x_1 = 2L(I = IN)$. This evaluation is done as follows:

$$\Phi\,(1, J) = \Phi\,(IN, J) = 0.5\,[\Phi\,(2, J) + \Phi\,(IN - 1, J)], \tag{5.121}$$

where $\Phi = p_o, u_i, T$ and it is assumed that the chosen grid disposition is such that $x_1\,(IN) - x_1\,(IN - 1) = x_1\,(2) - x_1\,(1)$. Solution of the temperature equation enables evaluation of the mean bulk temperature $T_{\mathrm{b}} = 0.5\,(T_{\mathrm{b,AF}} + T_{\mathrm{b,CD}})$, where the bulk temperatures at the periodic planes are evaluated from

$$T_{\mathrm{b,AForCD}} = \frac{\int_0^{H/2} \rho\,C_p\,u_1\,T\,dx_2}{\int_0^{H/2} \rho\,C_p\,u_1\,dx_2}. \tag{5.122}$$

Finally, the Stanton number St is evaluated as

$$St = \frac{h_{\mathrm{av}}}{\rho\,C_p\,u_{\mathrm{av}}}, \tag{5.123}$$

where the average heat transfer coefficient is evaluated from

$$h_{\mathrm{av}} = \frac{1}{(2L + 2t)} \int \frac{q_{\mathrm{w}}}{T_{\mathrm{w},s} - T_{\mathrm{b}}}\,ds, \tag{5.124}$$

and s is measured along the heated surfaces. Computations are performed for air $(Pr = 0.7)$ with a $38(x_1) \times 36(x_2)$ grid and the results are shown in Figure 5.15. In all computations, $L/H = 1.0$ and t/H is varied. Also plotted in the figure are experimental data of Kays and London as read from reference [54]. These data have been obtained for $t/H = 0.05$, $L/H = 1.14$ (instead of 1 in the present case), and the $(x_3$-direction width$)/H = 5.9$. Therefore, the geometric data approximate the present 2D computational domain. It is seen from the figure that the predicted friction factor data (solid lines) are in very good agreement with the experimental data (open circles). The predicted $St \times Pr^{2/3}$ (dashed lines) trend, however, deviates from the experimental data (open squares). But, as indicated earlier, it is difficult to approximate the exact boundary conditions of the experiment, which

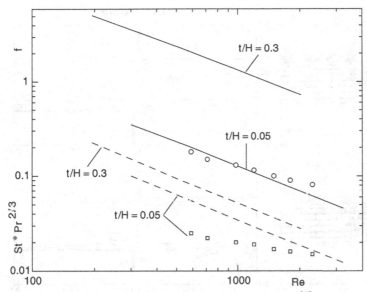

Figure 5.15. Offset Fin $(L/H) = 1$ – variation of f and $St \times Pr^{2/3}$ with Re.

involved condensing steam for heating. This condition implies a nearly uniform temperature at the blocks. However, then, the heat transfer, unlike the flow, will not be periodically fully developed. According to [54], the effect of this deviation from the experimental condition on predicted St may not be greater than 10%. The reader should note that such departures from exact experimental conditions are often made in CFD analysis.

The figure further shows that the effect of t/H on f is more significant than on the Stanton number. An approximate analysis carried out in [33] shows that the effect of a finite thickness fin is to create continuously disrupted laminar boundary layers on the fin surface and thus achieve enhanced heat transfer. Thus, although it is important to include the effect of a finite fin thickness in the analysis, the results show that fin thickness must be optimised in order not to exact a severe penalty in pressure drop.

To demonstrate the effect of Re, velocity vectors and temperature $(T - T_{min})/(T_{max} - T_{min})$ contours at an interval of 0.1 are plotted for $t/H = 0.3$ at three different Reynolds numbers in Figure 5.16. In each case, the core flow is nearly parallel to the x_1 axis but the strength of flow circulation in the fin-wake regions increases with Reynolds number. Similarly, as Re increases, the temperature contours are seen to be closer near the heating surfaces, indicating higher heat transfer rates at higher Re.

Turbulent Flow in a Pipe Expansion

We now consider turbulent flow and heat transfer at a pipe expansion, as shown in Figure 5.1. The radius ratio (R_2/R_1) of the two pipes is 2. For prediction purposes,

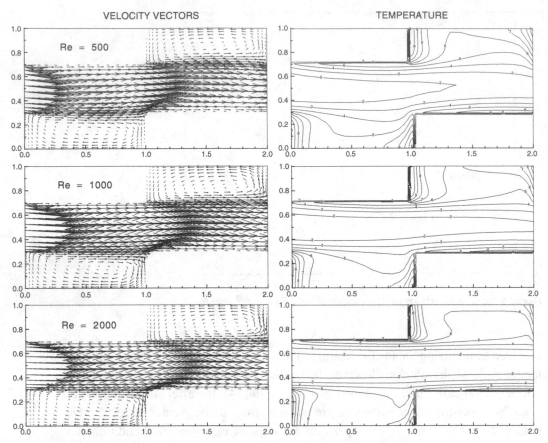

Figure 5.16. Offset Fin ($L/H = 1$, $t/H = 0.3$) – vector & temperature plots.

the HRE e–ϵ model is used. The predictions[18] will be compared with the experimental data of Krall and Sparrow [36] for $Pr = 3.0$ and of Runchal [62] for $Pr = 1,400$. Krall and Sparrow made measurements in a pipe with radius R_2 in which an orifice of radius R_1 is fitted. Downstream of the orifice, a constant wall heat flux is supplied. Runchal employed a converging nozzle (with exit-end radius R_1) fitted in a pipe of radius R_2. He employed an electro-chemical mass transfer technique to measure variation of mass transfer Stanton number downstream of the nozzle. The technique involves use of a NaOH solution whose Schmidt number (>1,000) depends on the solution concentration. The electro-chemical technique measures transfer of ferrocyanide ions to ferricyanide ions at a cathode surface embedded in the pipe wall to estimate the rate of mass transfer. These rates are, however, very low so that the mass transfer measurements can readily simulate the heat transfer situation with $Sc = Pr$. The electro-chemical technique simulates a $T_w = constant$ condition.

---

[18] The USER file for this problem is given in Appendix C.

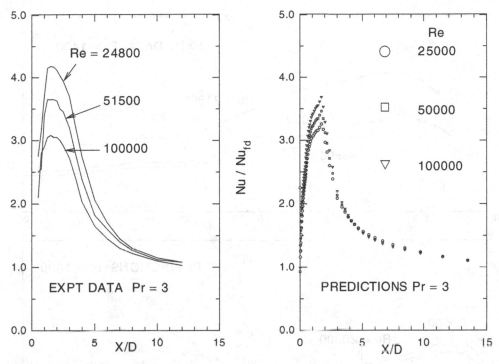

Figure 5.17. Sudden expansion, with $R_2/R_1 = 2$ and $q_w =$ constant

In both cases, the domain downstream of the orifice or nozzle is considered. At the inlet section, the specifications are $u_{in} = 4 \times \overline{u}$, $e_{in} = (0.1 \times u_{in})^2$, and ϵ_{in} is evaluated from the specification $\mu_t/\mu = C_\mu \, \rho \, e_{in}^2/\epsilon_{in} = 0.003 \, Re$ for $0 \leq r < R_1$ and $u_{in} = 0$ (wall) for $R_1 \leq r \leq R_2$. The Reynolds number of the larger pipe is defined as $Re = \rho \, \overline{u} \, 2 \, R_2/\mu$. Computations are carried out with $\rho = 1$ and $\overline{u} = 1$ and $R_2 = 1$. Thus, Re is varied by varying μ. The Nusselt numbers at different axial locations are evaluated from $Nu_x = q_w \, 2 \, R_2/K \, (T_w - T_b)$, where T_b is the bulk temperature and T_w is the wall temperature at each x.

In the computations, 67 (streamwise) × 28 (radial) nodes were used with closer spacings in the recirculation region to accurately predict the point of reattachment. Because of the close near-wall spacings, it was not possible to ensure that the first node away from the wall will have sufficiently large y^+ at all axial stations. Therefore, the *two-layer* wall function is active for velocity (see Equation 5.86). For the temperature equation, PF is given by Equation 5.88.

In Figure 5.17, predicted Nu_x/Nu_{fd} are compared with the experimental data of Krall and Sparrow. Here, as per their recommendation, $Nu_{fd} = 0.0123 \, Re^{0.874} \, Pr^{0.4}$. In these computations, the reattachment point is predicted at $x/(2 \, R_2) \approx 1.84$ at all Reynolds numbers. The predicted Nu_{max} locations (≈ 1.81) thus appear to coincide with the point of flow reattachment. The high values of Nu_{max}/Nu_{fd} indicate that the recirculation region is by no means *dead* with respect

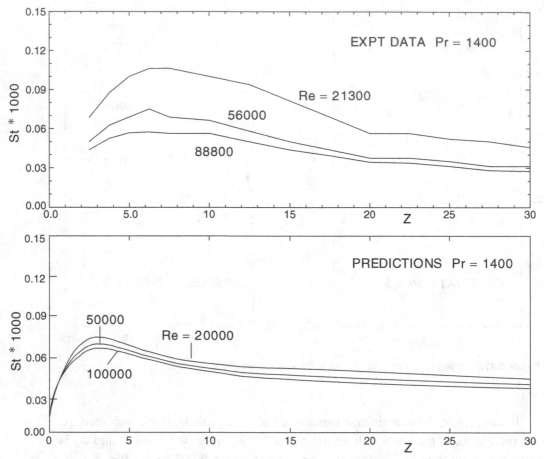

Figure 5.18. Sudden expansion, with $R_2/R_1 = 2$ and $T_w = $ constant.

to heat transfer, although the flow velocities are very low there. This is a special characteristic of recirculating regions in which fluid mixing is enhanced. The predictions also appear to nearly match the trends shown by the experimental data, although the exact magnitudes of Nu_{max} are not well predicted.

A similar comparison with the data of Runchal is shown in Figure 5.18. Here, $Z = x/(R_2 - R_1)$ and $St = Nu_x/(Re\, Pr)$ so that the predicted flow reattachment occurs at $Z = 7.43$. The predictions, however, show that the maximum St occurs at nearly $Z \approx 3.55$. Thus, the point of reattachment and maximum heat transfer *do not* coincide. The experimental data, however, indicate that maximum St occurs at $Z \approx 6.5$. Thus, clearly our wall-function treatment with respect to heat transfer is in need of further refinement for very large Pr. It is possible to do so by invoking a *three-layer* model for heat transfer and setting different limits on the three layers. However, this is not done here to draw the reader's attention to the need for such empirical adjustments. At the same time, it must be noted that the electro-chemical technique really simulates the $T_w = $ *constant* boundary condition only over a *patch* occupied

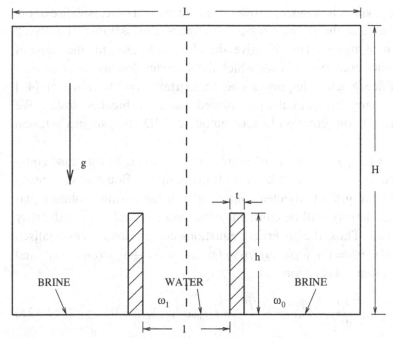

Figure 5.19. Natural convection mass transfer.

by the cathode but remains inert to mass transfer on remaining portions of the wall. This may be an added reason for lack of correspondence between predictions and experiment. Modelling for separated flow regions at high Pr numbers is an area in which basic research is hampered by the extremely sharp variations of temperature in the near-wall region where, although the turbulent viscosity may be negligible, turbulent conductivity may still be significant. Thus, a constant Pr_t assumption may not be justified.

Natural Convection Mass Transfer[19]

Figure 5.19 shows an open channel (width l and height h) placed inside a wider channel of width L and height H. The wider channel is closed at the top. The inner channel wall thickness is t. Both the channels are long in the x_3 direction. The inner channel has water whereas the wider channel has brine at its floor ($x_2 = 0$). The temperatures of water, brine, and the gas (air + water vapour) are the same and equal to the ambient temperature. In this isothermal case, evaporation will ensue because of the difference in vapour pressures at the water (high) and the brine (lower) surface. The vapour pressure at the brine surface can be altered by altering brine concentration. Thus, a mass transfer driving force is established.

The inner channel may be viewed as the well-known Stefan tube in which the evaporation rate of water can be analytically evaluated under the assumption that

[19] The USER file for this problem is given in Appendix C.

the fluid inside the channel is stagnant. However, in the present case, because of the density gradient caused by the vapour-pressure difference, a mass transfer buoyancy force will induce fluid motion. The objective, therefore, is to examine the range of mass transfer Grashof numbers Gr_m for which the stagnant flow assumption may be reasonably justified. Such an inquiry has been undertaken by McBain et al. [47] in which the inner channel is a circular tube placed inside a cubical enclosure. We have modified this 3D configuration to accommodate a 2D analysis in Cartesian coordinates.

We define $L^* = L/l$, $H^* = H/l$, $h^* = h/l$, and $t^* = t/l$. In this case equations for $\Phi = u_1, u_2, \omega$, and p' must be solved. Invoking the Boussinesq approximation, except for the gravity-affected source term in the u_2-momentum equation, we assume the density will be constant. Also viscosity and mass diffusivity are assumed constant. Thus, the governing equations can be nondimensionalised using $u_i^* = u_i/(\nu/l)$, $p^* = (p + \rho g x_2)/\rho (\nu/l)^2$, $\omega^* = (\omega - \omega_0)/(\omega_1 - \omega_0)$, and $x_i^* = x_i/l$. The relevant *source terms* are

$$S_{u_1^*} = -\frac{\partial p^*}{\partial x_1^*}, \quad S_{u_2^*} = -\frac{\partial p^*}{\partial x_2^*} + Gr_m \omega^*, \quad S_{\omega^*} = 0, \qquad (5.125)$$

where $Gr_m = g\,\beta_m\,(\omega_1 - \omega_0)\,l^3/\nu^2$ and $\beta_m = \rho^{-1}\,\partial\rho/\partial\omega^*$.

The *boundary conditions* are

$$u_i^* = 0, \quad \frac{\partial \omega^*}{\partial n^*} = 0 \quad \text{on all walls}, \qquad (5.126)$$

where n is normal to the walls. The $x_1^* = 0$ line is the symmetry boundary and computations are performed over the domain to the right of the symmetry line. The mass transfer boundary conditions on the floor ($x_2^* = 0$) are

$$u_1^* = 0,$$

$$u_2^* = Sc^{-1}(\omega_1^* - \omega_T^*)^{-1} \left.\frac{\partial \omega^*}{\partial x_2^*}\right|_{x_2^*=0}, \quad \omega^* = \omega_1^* \text{ (water)},$$

$$u_2^* = Sc^{-1}(\omega_0^* - \omega_T^*)^{-1} \left.\frac{\partial \omega^*}{\partial x_2^*}\right|_{x_2^*=0} :, \quad \omega^* = \omega_0^* \text{ (brine)}, \qquad (5.127)$$

where $\omega_1^* = 1$ and $\omega_0^* = 0$.

These specifications indicate that in the present mass transfer problem, the momentum equations are coupled with the mass transfer equation in two ways, firstly, through the source term $Gr_m \omega^*$ and, secondly, through the floor boundary condition. The dimensionless total evaporation flux is, therefore, given by

$$F_{\text{conv}} = 2\,Sc^{-1}(1 - \omega_T^*) \int_0^{1/2} \left.\frac{\partial \omega^*}{\partial x_2^*}\right|_{x_2^*=0} dx_1^*. \qquad (5.128)$$

Table 5.2: Normalized evaporation rate R.

Gr_m	1	10	100	500	1,000	2,000	3,000
R	0.7065	0.7086	0.7293	0.756	0.768	0.781	0.792

For a Stefan tube, the pure diffusion mass transfer rate is given by

$$F_{\text{diff}} = \frac{\ln(1+B)}{Sc\,h^*}, \tag{5.129}$$

where the Spalding number $B = -1/(1 - \omega_T^*)$. Therefore, the flux ratio R will be a functional given by

$$R = \frac{F_{\text{conv}}}{F_{\text{diff}}} = f(Gr_m, H^*, L^*, h^*, t^*, Sc, B). \tag{5.130}$$

In the present computations, $h^* = 2$, $L^* = 16$, $H^* = 8$, $t^* = 0.1$, and $Sc = 0.614$ are fixed. Also, in a typical evaporation problem, B is small. We take $\omega_T^* = 50$, giving $B = 0.0204$. Thus, with these specifications, R is a function of Gr_m only. Computations have been performed with 37×37 grid points with closer spacings near the inner channel wall and near the floor. Initially, only the mass transfer equation is solved. This corresponds to a stagnant fluid case. If $\omega^* = 0$ at $x_2^* = h^*$ then the evaporation flux will be given by Equation 5.129. However, in the present configuration, $\omega^* \neq 0$ at $x_2^* = h^*$ because the boundary condition is applied at the brine surface. This results in $R = 0.704$ for this limiting case. Now, the mass transfer equation is solved together with the flow equations for different values of Gr_m. Table 5.2 shows the results of computations. It is seen that the ratio increases with Gr_m. A similar trend has been observed in [47]. To ensure convergence, solutions for lower Gr_m were used to obtain solutions for higher Gr_m.

The trend observed in the $R \sim Gr_m$ relation is further demonstrated in Figure 5.20 through contour and vector plots over the domain $0 < x_1^* < 2.5$ and $0 < x_2^* < 5.5$. The figure shows that the inner channel remains nearly stagnant at $Gr_m = 10$. For higher Gr_m, the region near the top of the inner channel is influenced by the recirculation outside the channel.

False Diffusion in Multidimensions

In Chapter 3, the question of numerical false diffusion was explored through the 1D conduction–convection equation. Here, this matter is again considered for multidimensional flows through a problem devised by Raithby [57] (see Figure 5.21). We consider a square domain of unit dimensions through which a fluid moves with an angle θ with the x axis. The viscosity and conductivity of the fluid are zero so that transport of temperature occurs by pure convection with Peclet number $P = \infty$. At a certain streamline at $y_0 = 0.5(1 - \tan\theta)$, a step discontinuity in temperature is imposed as shown in the figure. Thus, $T = 1$ above the streamline and $T = 0$

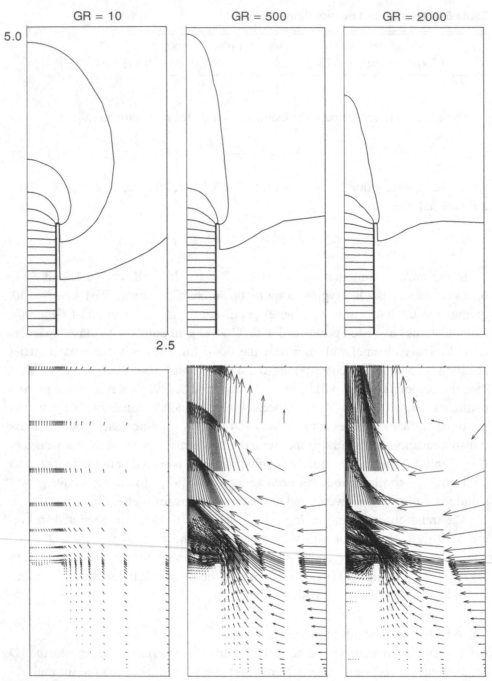

Figure 5.20. Contours of ω^* (at an interval of 0.05) and velocity vectors for natural convection evaporation.

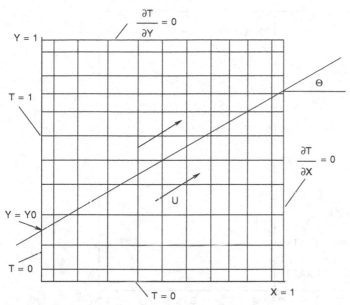

Figure 5.21. Transport of a step discontinuity.

below it. Now, since $P = \infty$, the discontinuity must be preserved in the direction of the flow.

To examine the capability of the UDS for this large Peclet number case, the velocities are prescribed as $u = U \cos \theta$ and $v = U \sin \theta$ at all nodes and the temperature boundary conditions are as shown in Figure 5.21. The equation for T will read as

$$\frac{\partial T}{\partial x} + \tan \theta \frac{\partial T}{\partial y} = 0. \tag{5.131}$$

This equation is solved for different angles θ on a 12×12 grid. Figure 5.22 shows the predicted T profiles at midplane $x = 0.5$. It is seen that the profiles are smeared. The profiles deviate from the exact solution; the deviation increases as θ increases and reaches maximum at $\theta = 45$ degrees. Now, the profiles can be smeared only if *numerical* diffusion is present. This suggests that when the flow inclination with respect to the grid line is large, the numerical diffusion is also large. Conversely, if $\theta = 0$ or 90 degrees, the discontinuity in the temperature profile should be predicted. This is indeed verified by numerical solutions (not shown in the figure). Wolfshtein [89] has devised a method for estimating the false diffusivity (see exercise 12).

What is observed here with UDS remains valid for all convection schemes, although the profile-shape-sensing CONDIF and TVD schemes demonstrate reduced deviations and, therefore, reduced numerical diffusion. However, recognising the angular dependence of false diffusion, some CFD analysts have proposed

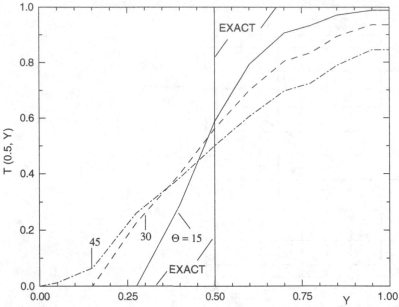

Figure 5.22. Midplane temperature profiles – UDS.

convection schemes that sense the angle θ. In effect, they postulate *flow-oriented* interpolations of cell-face values rather than use the nodal values straddling the cell faces.

EXERCISES

1. Starting with Equation 5.8, validate the generalisations shown in Equation 5.19. Hence, show the correctness of Equation 5.17 for each convection scheme.

2. Derive the value of A in Equation 5.19 for the exponential scheme.

3. Show that if the CONDIF scheme (see Chapter 3, Exercise 10) is used then, for a nonuniform grid, the coefficients AE and AW in Equation 5.12, for example, will read as

$$AE = d_e \left[1 + \frac{|P_{c_e}| - P_{c_e}}{4} \right] + \frac{d_w}{R_x^*} \left[\frac{|P_{c_w}| - P_{c_w}}{4} \right],$$

$$AW = d_w \left[1 + \frac{|P_{c_w}| + P_{c_w}}{4} \right] + d_e R_x^* \left[\frac{|P_{c_e}| + P_{c_e}}{4} \right],$$

where $R_x^* = (\Phi_E - \Phi_P)/(\Phi_P - \Phi_W) \times \Delta x_w/\Delta x_e$.
(Hint: Recognise that CONDIF is essentially a CDS whose coefficients are modified to take account of the shape of the local Φ profile).

4. Using the substitutions shown in Equation 5.24, derive Equation 5.25. Hence, using the IOCV method, derive Equation 5.28.

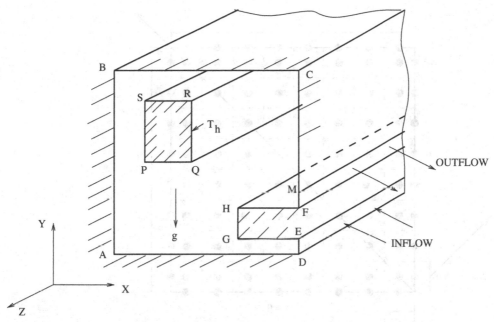

Figure 5.23. Long chamber of Exercise 11.

5. Starting with Equation 5.40, derive Equation 5.47.

6. Show the validity of Equations 5.55 and 5.56.

7. Identify the differences and similarities between Equations 5.57 for collocated grids and Equation 5.32 for staggered grids.

8. Confirm that on collocated grids $A P^{u_1} = A P^{u_2}$.

9. It is of interest to derive a total pressure-correction equation for compressible flows in which $p = \rho R_g T$. To do this, start with Equation 5.57 and write

$$\rho^{l+1} = \rho^l + \rho'_m = \rho^l + \frac{p'_m}{R_g T} = \rho^l + \frac{(p' - p'_{sm})}{R_g T}.$$

With this substitution show that the p'-equation takes the form of a general transport equation for any Φ with appearance of convection–diffusion-like terms. Also, $V_{sound} = \sqrt{\gamma \, \bar{R_g} \bar{T}}$. Hence, show the Mach number dependence in the equation. If CDS is used, can the coefficients in the discretised equation (5.60) turn negative? If yes, suggest a remedy.

10. Explain the need for evaluating the mass residual via Equation 5.73 when computing on collocated grids.

11. Consider the chamber shown in Figure 5.23. The chamber is long in the z-direction so that the flow and heat transfer can be considered 2D. Assume that all relevant dimensions are given. The flow enters the chamber with

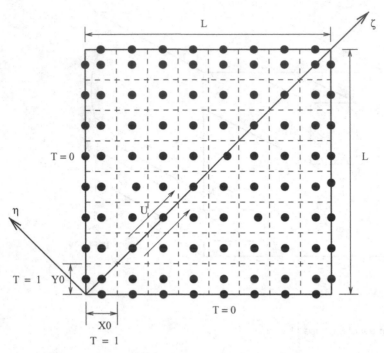

Figure 5.24. Estimating false diffusion.

velocity u_{in} (as shown) and temperature T_{in}. The chamber walls and the lip separating the inflow and outflow are adiabatic. Allow for the presence of a buoyancy effect,

(a) Write the appropriate differential equations and the boundary conditions for all relevant variables.

(b) Carry out any necessary node tagging, defining clearly the convention used. For example, along AB, NTAGW $(2, J) = 14$ (say) to indicate the west adiabatic wall boundary.

12. Solve the problem of false diffusion discussed in the text for the case of $\theta = 45$ degrees in which the boundary conditions are as shown in Figure 5.24. Take $L = 100$ and $y_0 = x_0 = 2\,\Delta S$, where $\Delta S = \Delta X = \Delta Y$. The situation is therefore akin to that of a temperature source convected by U. Now, define orthogonal coordinates ξ and η as shown. Use UDS. Obviously, the maximum temperature T_{max} will occur at $\eta = 0$ for each ξ. Now, locate the value of $\eta_{1/2}$ corresponding to $T/T_{max} = 0.5$. Hence, plot the computed results as T/T_{max} versus $\eta/\eta_{1/2}$ for different values of $\xi/\Delta S > 50$. Show that the profiles collapse on a single curve

$$\frac{T}{T_{max}} = \exp\left\{ -\ln(2)\left(\frac{\eta}{\eta_{1/2}}\right)^2 \right\},$$

where $\eta_{1/2}/\Delta S = (\xi/\Delta S)^{0.5}$. This equation is similar to the solution to the

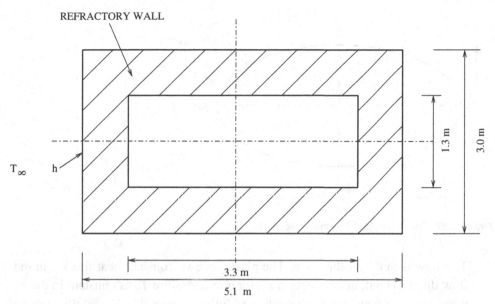

Figure 5.25. Refractory furnace.

equation of a *wake*,

$$U \frac{\partial T}{\partial \xi} - \Gamma_{\text{false}} \frac{\partial^2 T}{\partial \eta^2} = 0, \qquad \frac{T}{T_{\max}} = \exp\left(-\frac{U \eta^2}{4 \Gamma_{\text{false}} \xi}\right).$$

Hence, show that $\Gamma_{\text{false}} \sim 0.361\, U \Delta S$

13. Derive Equation 5.94.

14. Consider a long furnace made from refractory brick ($k = 1.0$ W/m-K), as shown in Figure 5.25. The temperature of the inside surface is 600°C whereas the outside surface is exposed to an environment at 30°C with heat transfer coefficient $h = 10$ W/m²-K. Determine the heat loss from the furnace wall.

15. Consider two parallel plates that are infinitely long in the x_1 and x_3 directions. Fins are attached to the plates in a staggered fashion, as shown in Figure 5.26.

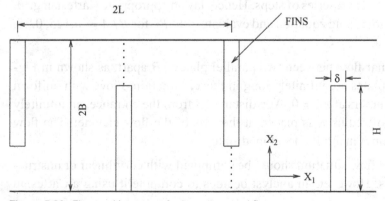

Figure 5.26. Flow and heat transfer in a staggered fin array.

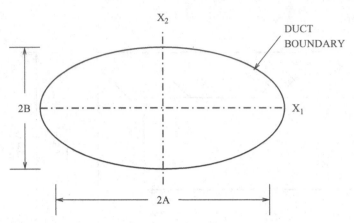

Figure 5.27. Fully developed flow in an ellipse.

The flow is in the x_3 direction. The plates receive constant heat flux q_w in the flow direction but, at any section x_3, their temperature T_w is constant in the x_1 direction. The flow and heat transfer are fully developed.

(a) Assuming laminar flow, identify the equations and the boundary conditions governing the flow and heat transfer

(b) Nondimensionalise the equations and show that

$$f\,Re = F\left(\frac{H}{B}, \frac{L}{B}, \frac{\delta}{B}\right), \qquad Nu = F\left(\frac{H}{B}, \frac{L}{B}, \frac{\delta}{B}, \frac{k_{\text{fin}}}{k_{\text{fluid}}}, \frac{\delta}{H}\right).$$

(c) Compute f and Nu for $B = L = 1$, $H = 1.2$, and $\delta = 0.05$. Take $C_{\text{fin}} = k_{\text{fin}}/k_{\text{fluid}} = 0$, 10, and 100. (Hint: Note that the fin half-width $\delta/2$ must be treated as a blocked region through which 1D heat conduction takes place.)

16. Consider fully developed laminar flow in a duct of elliptic cross section, as shown in Figure 5.27. The flow is in the x_3 direction.

(a) Write the PDE governing distribution of the u_3 velocity. Identify the smallest relevant domain, exploiting the available symmetries.

(b) The duct wall boundary of the domain is curved. This boundary can be approximated by a series of steps. Hence, lay an appropriate Cartesian grid. Solve the governing equation and evaluate $f \times Re$ for $B/A = 0.125, 0.25, 0.5$, and 1.0.

17. Consider laminar flow between two parallel plates $2B$ apart, as shown in Figure 5.28. The plates are infinitely long in the x_3 direction. Flow, with uniform axial velocity, enters at $x_1 = 0$. At a distance S from the entrance, an infinitely long cylinder of radius R is placed at the axis of the flow channel. The flow leaves the channel in a fully developed state.

(a) Ideally, the flow situation should be computed with curvilinear or unstructured grids. However, an analyst decides to compute it using a Cartesian

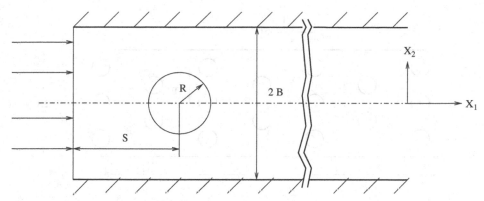

Figure 5.28. Flow in a parallel-plate channel.

mesh. What is the main difficulty that the analyst will face if the drag offered by the cylinder is to be *accurately* determined.

(b) Select the domain length from fluid dynamic considerations. Assume that the Reynolds number based on the channel hydraulic diameter is 40 and $S/R = 3$ and $B/R = 10$.

(c) How should the drag coefficient C_D of the cylinder be determined from the converged solution in discretised form?

18. Consider laminar flow between two parallel plates separated by distance $2b$. Specify the fully developed axial velocity profile at the inflow plane and zero axial velocity gradient at exit. Adapt the 2D computer program in Appendix C for this problem and solve with and without smoothing pressure correction. Observe the predicted velocity and pressure profiles in the two cases. Do you notice any difference? If not, explain why.

19. Engine oil enters a tube (diameter $= 1.25$ cm) at uniform temperature $T_{in} = 160°C$. The oil mass flow rate is 100 kg/hr and the tube wall temperature is maintained at $T_w = 100°C$. If the tube is 3.5 m long, calculate the bulk temperature of oil at exit from the tube and the total pressure drop. The properties of oil are as follows $\rho = 823$ kg/m³, $C_p = 2,351$ J/kg-K, $\nu = 10^{-5}$ m²/s, and $k = 0.134$ W/m-K. Plot the axial variation of Nusselt number Nu_x and the bulk temperature $T_{b,x}$. Assume that the oil enters the tube with uniform velocity. (Hint: You will need to provide close grid spacings near the tube wall to capture steep variations of temperature owing to the high Prandtl number. The grid spacings along the tube axis may expand in the direction of the flow.)

20. Air at 7 bar and 100°C enters a nuclear reactor channel (width $= 3$ mm, length $L = 1.22$ m) at the rate of 7.5 kg/s-m². The heat flux at the channel walls is given by $q_w = 900 + 2,500 \sin(\pi x_1/L)$ W/m². Plot the variation of T_w, T_b, and Nu with axial distance x_1 and find the location of maximum wall temperature. Assume fully developed flow and evaluate properties at 250°C.

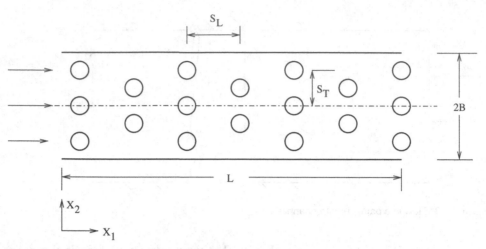

Figure 5.29. Flow in a channel containing rods.

21. Consider fully developed turbulent flow in a pipe of radius R. Assuming that the inner layer extends up to $y^+ = 100$ from the wall, estimate the inner layer thickness as a fraction of R for $Re = 5,000$, $25,000$, $75,000$, and $100,000$.

22. Air at 30°C enters a tube (diameter $D = 5.0$ cm) of a solar air-heater with a uniform velocity of 10 m/s. The tube is 2.1 m long. The tube wall temperature is 90°C. Determine the exit bulk temperature and the pressure drop. Also determine the length-averaged Nusselt number. Use the HRE model.

23. Repeat Exercise 22 assuming that the tube is rough with roughness height $y_r/D = 0.01$. Use the HRE model. For a rough surface, the velocity profile near a wall is given by [65]

$$u^+ = \frac{1}{\kappa} \cdot \ln \left[\frac{y}{y_r} \right] + 8.48.$$

This equation can be cast in the form of Equation 5.86 so that

$$u^+ = \frac{1}{\kappa} \cdot \ln \left[E_r \, y^+ \right], \qquad E_r = \frac{\exp(8.48\,\kappa)}{y_r^+}.$$

Thus, the wall-function treatment remains valid with E replaced by E_r. Similarly, PF (Equation 5.88) must be replaced by $PF_r = 5.19\, Pr^{0.44}\, y_r^{+0.2} - 8.48$ with $Pr_t = 1$ [22]. (Hint: You will need to modify the BOUND subroutine and STAN function in the Library file in Appendix C to account for y_r.)

24. Consider steady turbulent flow in a two-dimensional plane channel (see Figure 5.29) containing an array of rods (of diameter D). Flow enters at $x_1 = 0$ with uniform velocity $u_{1,\text{in}}$. It is of interest to determine the pressure drop over length L. To reduce the computational effort in this densely filled flow situation, model the flow as a *porous-body* flow in which it is assumed that the

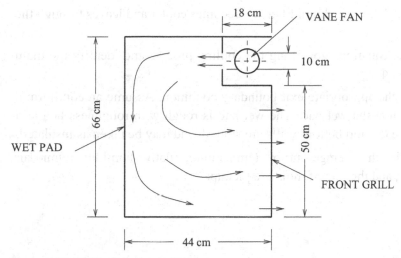

Figure 5.30. Idealised desert cooler.

channel contains no rods but the effect of their presence is captured through two artifacts:

(i) The *effective* fluid density, viscosity, and pressure are taken as $\Phi\,\Psi$, where $\Psi = \rho$, $(\mu + \mu_t)$, and p, respectively, where Ψ is the porosity defined as

$$\Psi = \frac{\text{fluid volume}}{\text{physical volume}}.$$

(ii) The source terms in the u_1 and u_2 momentum equations are augmented by including local flow resistance offered by the rods through *experimentally determined* friction factors f_{u_1} and f_{u_2} defined as

$$f_{u_i} = \frac{\Delta p}{0.5\,\rho\,u_i|V_{\text{tot}}|} = F\left(\frac{S_L}{D}, \frac{S_T}{D}, Re_{D,\text{tot}}\right),$$

where $V_{\text{tot}} = \sqrt{u_1^2 + u_2^2}$ and u_i are *superficial* velocities. Function $F(\ )$ is assumed known but note that S_L and S_T must be re-defined for the u_2 velocity.

(a) Write the equations to be solved and choose an appropriate exit boundary condition assuming $L/(2B) = 10$. Specify the inlet conditions for all variables including the variables characterising turbulence.

(b) Discuss whether the effect of flow resistance terms could be accounted for through source-term linearisation.

25. Figure 5.30 shows an idealised desert cooler in which hot air (40°C and 10% relative humidity) enters the cooler inside through the 10-cm-wide gap with a velocity of 40 m/s. The air picks up moisture at the wet pad, which is supplied

with water at 25°C. The humidified air becomes cooler and leaves through the front grill.

(a) State the equations governing the cooling process and identify the main variables Φ.

(b) Specify the appropriate exit boundary condition. Assume an equilibrium condition at the wet pad. The wet pad is rough with roughness height 5 mm. The top and bottom walls are smooth and may be taken as insulated.

(c) Determine the average outflow temperature, relative humidity, humid-air velocity, and the rate of moisture pickup.

6 2D Convection – Complex Domains

6.1 Introduction

In practical applications of CFD, one often encounters *complex* domains. A domain is called complex when it cannot be elegantly described (or mapped) by a Cartesian grid. By way of illustration, we consider a few examples.

Figure 6.1 shows the smallest symmetry sector of a nuclear rod bundle placed inside a circular channel of radius R. There are nineteen rods: one rod at the channel center, six rods (equally spaced) in the inner rod ring of radius b_1, and twelve rods in the outer ring of radius b_2. The rods are circumferentially equispaced. The radius of each rod is r_o. The fluid (coolant) flow is in the x_3 direction. The flow convects away the heat generated by the rods and the channel wall is insulated. It is obvious that a Cartesian grid will not fit the domain of interest because the lines of constant x_1 or x_2 will intersect the domain boundaries in an arbitrary manner. In such circumstances, it proves advantageous to adopt alternative means for mapping a complex domain. These alternatives are to use

1. curvilinear grids or
2. finite-element-like unstructured grids.

6.1.1 Curvilinear Grids

It is possible to map a complex domain by means of curvilinear grids (ξ_1, ξ_2) in which directions of ξ_1 and ξ_2 may change from point to point. Also, curvilinear lines of constant ξ_1 and constant ξ_2 need not intersect orthogonally either within the domain or at the boundaries. Figure 6.2 shows the nineteen-rod domain of Figure 6.1 mapped by curvilinear grids. The figure shows that curvilinear lines generate clearly identifiable quadrilateral control volumes. When the IOCV method is used, the task is to integrate the transport equations over a typical control volume. To facilitate this, it becomes necessary to first transform the transport equations written in Cartesian

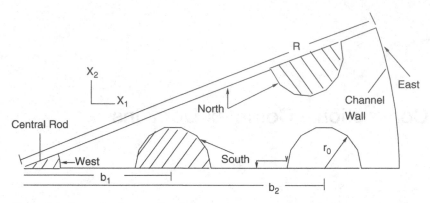

Figure 6.1. Example of a complex domain.

coordinates to curvilinear coordinates via transformation relations

$$x_1 = F_1(\xi_1, \xi_2), \qquad x_2 = F_2(\xi_1, \xi_2). \tag{6.1}$$

In general, these functional relationships must be developed by *numerical* grid generation techniques (see Chapter 8). The grids shown in Figure 6.2 are in fact generated by numerical means. For simpler domains, however, the functional relationships can be specified by algebraic functions. The new set of transport equations in curvilinear coordinates are developed in Section 6.2.

One advantage of mapping domains by curvilinear grids is that one can still retain the familiar (I, J) structure to identify a node (or the corresponding control volume) because, as can be seen from Figure 6.2, along any curvilinear line ξ_1, the total number of intersections with constant-ξ_2 lines remains constant and vice versa. Further advantages of this identifying structure will become clear in Section 6.2.

6.1.2 Unstructured Grids

Another alternative for a complex domain is to map the domain by triangles or any n-sided polygons (including quadrilaterals) or any mix of triangles and polygons. Figure 6.3 shows the mapping of a nineteen-rod bundle by triangles as an example. In this case, the rods are arranged in such a way that the smallest symmetry sector

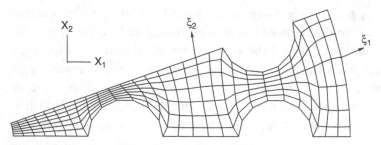

Figure 6.2. Nineteen-rod bundle – curvilinear grids.

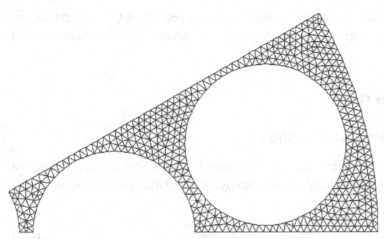

Figure 6.3. Nineteen-rod bundle – unstructured grid.

is a *doubly connected* domain. Such mapping can be generated by commercially available grid generators such as ANSYS. Each triangle may now be viewed as a *control volume* over which the transport equations are to be integrated to arrive at the discretised equations. The process of generating the latter equations is described in Section 6.3.

It will be recognized that a triangle is a very convenient elemental construct because it can map any convex intrusion or concave extrusion at the domain boundaries. More importantly, triangles can also effectively skirt any *blocked* region within the overall domain, as shown in Figure 6.3. Such skirting cannot be elegantly accomplished if curvilinear grids are used for mapping.

The flexibility offered by mapping by triangulation is thus obvious. Further, it is not necessary that all triangles be of the same size or shape. In spite of this flexibility, it becomes necessary to make a significant departure from curvilinear grid practise with respect to *node identification* when unstructured grids are used. It is obvious from Figure 6.3, for example, that one cannot readily identify elements (or nodes) by employing the familiar (I, J) structure as was possible with curvilinear grids. Elements, perforce, must be identified serially with a single identifier N (say). As will be shown in Section 6.3, commercial codes such as ANSYS identify elements in any arbitrary order. Thus, an element having identifier N will interact with elements having arbitrary identifying numbers *without* any generalisable rules. This contrasts with the case of curvilinear grids in which a control volume (I, J) will always interact with control volumes identified by $(I + 1, J)$, $(I - 1, J)$, $(I, J + 1)$, and $(I, J - 1)$.

This serial numbering has consequences for solution of discretised equations evolved on an unstructured grid. This will become clearer in Section 6.3. In passing, we note that there are a variety of methods for triangulation. Automatic triangulation requires detailed considerations from the subject of computational geometry. In

Chapter 8, some simpler approaches will be introduced. Most CFD practitioners, however, employ commercially available packages such as ANSYS for unstructured grid generation.

6.2 Curvilinear Grids

6.2.1 Coordinate Transformation

Our first task is to transform the transport equations in Cartesian coordinates to those in curvilinear coordinates. Thus, employing the chain rule, we can write the first-order derivatives as

$$\frac{\partial}{\partial x_1} = \frac{\partial \xi_1}{\partial x_1}\frac{\partial}{\partial \xi_1} + \frac{\partial \xi_2}{\partial x_1}\frac{\partial}{\partial \xi_2}, \tag{6.2}$$

$$\frac{\partial}{\partial x_2} = \frac{\partial \xi_1}{\partial x_2}\frac{\partial}{\partial \xi_1} + \frac{\partial \xi_2}{\partial x_2}\frac{\partial}{\partial \xi_2}. \tag{6.3}$$

The next task is to determine derivatives of ξ_1 and ξ_2 with respect to x_1 and x_2 knowing functions (6.1). To do this, we note that

$$dx_1 = \frac{\partial x_1}{\partial \xi_1}d\xi_1 + \frac{\partial x_1}{\partial \xi_2}d\xi_2, \tag{6.4}$$

$$dx_2 = \frac{\partial x_2}{\partial \xi_1}d\xi_1 + \frac{\partial x_2}{\partial \xi_2}d\xi_2. \tag{6.5}$$

These relations can be written in matrix form as $|dx| = |A||d\xi|$, or

$$\begin{vmatrix} dx_1 \\ dx_2 \end{vmatrix} = \begin{vmatrix} \partial x_1/\partial \xi_1 & \partial x_1/\partial \xi_2 \\ \partial x_2/\partial \xi_1 & \partial x_2/\partial \xi_2 \end{vmatrix} \begin{vmatrix} d\xi_1 \\ d\xi_2 \end{vmatrix}. \tag{6.6}$$

Now, manipulation of Equations 6.4 and 6.5 will show that

$$d\xi_1 = \frac{1}{\text{Det A}}\left[\text{cof}\left(\frac{\partial x_1}{\partial \xi_1}\right)dx_1 + \text{cof}\left(\frac{\partial x_2}{\partial \xi_1}\right)dx_2\right], \tag{6.7}$$

$$d\xi_2 = \frac{1}{\text{Det A}}\left[\text{cof}\left(\frac{\partial x_1}{\partial \xi_2}\right)dx_1 + \text{cof}\left(\frac{\partial x_2}{\partial \xi_2}\right)dx_2\right], \tag{6.8}$$

where cof denotes *cofactor of* and Det A stands for *determinant of* A. Thus, from the last two equations, it is easy to deduce that

$$\frac{\partial \xi_1}{\partial x_1} = \frac{1}{\text{Det A}}\text{cof}\left(\frac{\partial x_1}{\partial \xi_1}\right) = \frac{1}{\text{Det A}}\left(\frac{\partial x_2}{\partial \xi_2}\right) = \frac{\beta_1^1}{\text{Det A}}, \tag{6.9}$$

$$\frac{\partial \xi_1}{\partial x_2} = \frac{1}{\text{Det A}}\text{cof}\left(\frac{\partial x_2}{\partial \xi_1}\right) = -\frac{1}{\text{Det A}}\left(\frac{\partial x_1}{\partial \xi_2}\right) = \frac{\beta_1^2}{\text{Det A}}, \tag{6.10}$$

$$\frac{\partial \xi_2}{\partial x_1} = \frac{1}{\text{Det A}} \text{cof}\left(\frac{\partial x_1}{\partial \xi_2}\right) = -\frac{1}{\text{Det A}}\left(\frac{\partial x_2}{\partial \xi_1}\right) = \frac{\beta_2^1}{\text{Det A}}, \qquad (6.11)$$

$$\frac{\partial \xi_2}{\partial x_2} = \frac{1}{\text{Det A}} \text{cof}\left(\frac{\partial x_2}{\partial \xi_2}\right) = \frac{1}{\text{Det A}}\left(\frac{\partial x_1}{\partial \xi_1}\right) = \frac{\beta_2^2}{\text{Det A}}, \qquad (6.12)$$

where the βs are called the *geometric coefficients* and are given by

$$\beta_1^1 = \frac{\partial x_2}{\partial \xi_2}, \quad \beta_1^2 = -\frac{\partial x_1}{\partial \xi_2}, \quad \beta_2^1 = -\frac{\partial x_2}{\partial \xi_1}, \quad \beta_2^2 = \frac{\partial x_1}{\partial \xi_1}. \qquad (6.13)$$

Further, it follows that

$$\text{Det A} = \frac{\partial x_1}{\partial \xi_1}\frac{\partial x_2}{\partial \xi_2} - \frac{\partial x_1}{\partial \xi_2}\frac{\partial x_2}{\partial \xi_1} = \beta_1^1 \beta_2^2 - \beta_2^1 \beta_1^2 = J, \qquad (6.14)$$

where symbol J stands for the *Jacobian* of the matrix A. We can now rewrite Equations 6.2 and 6.3 as

$$\frac{\partial}{\partial x_1} = \frac{1}{J}\left[\beta_1^1 \frac{\partial}{\partial \xi_1} + \beta_2^1 \frac{\partial}{\partial \xi_2}\right], \qquad (6.15)$$

$$\frac{\partial}{\partial x_2} = \frac{1}{J}\left[\beta_1^2 \frac{\partial}{\partial \xi_1} + \beta_2^2 \frac{\partial}{\partial \xi_2}\right]. \qquad (6.16)$$

6.2.2 Transport Equation

The first task is to transform the general transport equation (5.1) from the (x_1, x_2) coordinate system to the (ξ_1, ξ_2) coordinate system using relations (6.15) and (6.16). Thus,

$$r\frac{\partial(\rho\,\Phi)}{\partial t} + \frac{1}{J}\left[\beta_1^1 \frac{\partial(r\,q_1)}{\partial \xi_1} + \beta_2^1 \frac{\partial(r\,q_1)}{\partial \xi_2} + \beta_1^2 \frac{\partial(r\,q_2)}{\partial \xi_1} + \beta_2^2 \frac{\partial(r\,q_2)}{\partial \xi_2}\right] = r\,S. \qquad (6.17)$$

This equation can also be written as

$$r\,J\frac{\partial(\rho\,\Phi)}{\partial t} + \frac{\partial\left(\beta_1^1\,r\,q_1\right)}{\partial \xi_1} + \frac{\partial\left(\beta_2^1\,r\,q_1\right)}{\partial \xi_2} + \frac{\partial\left(\beta_1^2\,r\,q_2\right)}{\partial \xi_1} + \frac{\partial\left(\beta_2^2\,r\,q_2\right)}{\partial \xi_2}$$

$$= r\,q_1\left[\frac{\partial\beta_1^1}{\partial \xi_1} + \frac{\partial\beta_2^1}{\partial \xi_2}\right] + r\,q_2\left[\frac{\partial\beta_1^2}{\partial \xi_1} + \frac{\partial\beta_2^2}{\partial \xi_2}\right] + r\,J\,S. \qquad (6.18)$$

Using definitions (6.13), however, we can show that the terms in the square brackets are identically zero. Hence, Equation 6.18 can be written as

$$r\,J\frac{\partial(\rho\,\Phi)}{\partial t} + \frac{\partial}{\partial \xi_1}\left(\beta_1^1\,r\,q_1 + \beta_1^2\,r\,q_2\right) + \frac{\partial}{\partial \xi_2}\left(\beta_2^1\,r\,q_1 + \beta_2^2\,r\,q_2\right) = r\,J\,S. \qquad (6.19)$$

Using Equation 5.2, it is now possible to replace Cartesian fluxes q_1 and q_2. After some algebra, it can be shown that

$$r J \frac{\partial(\rho\, \Phi)}{\partial t} + \frac{\partial}{\partial \xi_1} \left[\rho\, r\, U_{f1}\, \Phi - r \frac{\Gamma_{eff}}{J} dA_1^2 \frac{\partial \Phi}{\partial \xi_1} \right]$$

$$+ \frac{\partial}{\partial \xi_2} \left[\rho\, r\, U_{f2}\, \Phi - r \frac{\Gamma_{eff}}{J} dA_2^2 \frac{\partial \Phi}{\partial \xi_2} \right]$$

$$= \frac{\partial}{\partial \xi_1} \left[r \frac{\Gamma_{eff}}{J} dA_{12} \frac{\partial \Phi}{\partial \xi_2} \right] + \frac{\partial}{\partial \xi_2} \left[r \frac{\Gamma_{eff}}{J} dA_{12} \frac{\partial \Phi}{\partial \xi_1} \right] + r J S,$$

(6.20)

where

$$dA_1^2 = \left(\beta_1^1\right)^2 + \left(\beta_1^2\right)^2,$$

$$dA_2^2 = \left(\beta_2^1\right)^2 + \left(\beta_2^2\right)^2,$$

$$dA_{12} = \beta_1^1 \beta_2^1 + \beta_1^2 \beta_2^2$$

(6.21)

and the contravariant flow velocities are given by

$$U_{f1} = \beta_1^1 u_{f1} + \beta_1^2 u_{f2} = \frac{\partial x_2}{\partial \xi_2} u_{f1} - \frac{\partial x_2}{\partial \xi_1} u_{f2},$$

(6.22)

$$U_{f2} = \beta_2^1 u_{f1} + \beta_2^2 u_{f2} = \frac{\partial x_1}{\partial \xi_1} u_{f2} - \frac{\partial x_1}{\partial \xi_2} u_{f1},$$

(6.23)

where u_{f1} and u_{f2} are the Cartesian velocity components.

6.2.3 Interpretation of Terms

Several new terms appearing in Equation 6.20 can be interpreted using vector mathematics.

Elemental Area

The elemental area dA_i *normal* to the (ξ_j, ξ_k) plane is given by

$$d\vec{A}_i = \left(\frac{\partial \vec{r}}{\partial \xi_j} \times \frac{\partial \vec{r}}{\partial \xi_k} \right) d\xi_j\, d\xi_k,$$

(6.24)

where the position vector $\vec{r} = \vec{i}\, x_1 + \vec{j}\, x_2 + \vec{k}\, x_3$. For our 2D case, if we set $i = 1$, $j = 2$, and $k = 3$ then $\partial \vec{r}/\partial \xi_3 = \partial x_3/\partial \xi_3 = 1$ because the x_3 and ξ_3 directions coincide and are normal to the (ξ_1, ξ_2) plane. Thus, taking unit dimension in the x_3 direction gives

$$dA_1 = \left| \frac{\partial \vec{r}}{\partial \xi_2} \right| d\xi_2 = \left| \vec{i} \frac{\partial x_2}{\partial \xi_2} - \vec{j} \frac{\partial x_1}{\partial \xi_2} \right| d\xi_2 = \left| \vec{i}\, \beta_1^1 + \vec{j}\, \beta_1^2 \right| d\xi_2$$

$$= \sqrt{\left(\beta_1^1\right)^2 + \left(\beta_1^2\right)^2}\, d\xi_2.$$

(6.25)

Similarly, it can be shown that

$$dA_2 = \sqrt{\left(\beta_2^1\right)^2 + \left(\beta_2^2\right)^2}\, d\xi_1.$$ (6.26)

Comparison of the last two equations with Equations 6.21 shows that dA_1 and dA_2 represent areas with $d\xi_1 = d\xi_2 = 1$.

Elemental Volume

The volume element in curvilinear coordinates is given by

$$dV = \frac{\partial \vec{r}}{\partial \xi_i} \cdot \left(\frac{\partial \vec{r}}{\partial \xi_j} \times \frac{\partial \vec{r}}{\partial \xi_k} \right) d\xi_i\, d\xi_j\, d\xi_k.$$ (6.27)

Thus, taking $i = 1$, $j = 2$, and $k = 3$, it follows that

$$dV = \frac{\partial \vec{r}}{\partial \xi_1} \cdot \frac{\partial \vec{r}}{\partial \xi_2}\, d\xi_1\, d\xi_2 = \left(\beta_1^1 \beta_2^2 - \beta_1^2 \beta_2^1 \right) d\xi_1\, d\xi_2.$$ (6.28)

Comparison of Equation 6.28 with Equation 6.14 shows that the Jacobian J is nothing but element volume dV with $d\xi_1 = d\xi_2 = 1$.

The Normal Fluxes

Note that Equation 6.20 can be written in the following form:

$$r\, J \frac{\partial(\rho\, \Phi)}{\partial t} + \frac{\partial}{\partial \xi_1}[r\, q_{\xi_1}] + \frac{\partial}{\partial \xi_2}[r\, q_{\xi_2}] = r\, J\, S,$$ (6.29)

where q_{ξ_1} and q_{ξ_2} are given by

$$q_{\xi_1} = \rho\, U_{\mathrm{f}1}\, \Phi - \frac{\Gamma_{\mathrm{eff}}}{J}\left(dA_1^2 \frac{\partial \Phi}{\partial \xi_1} + dA_{12} \frac{\partial \Phi}{\partial \xi_2}\right),$$ (6.30)

$$q_{\xi_2} = \rho\, U_{\mathrm{f}2}\, \Phi - \frac{\Gamma_{\mathrm{eff}}}{J}\left(dA_2^2 \frac{\partial \Phi}{\partial \xi_2} + dA_{12} \frac{\partial \Phi}{\partial \xi_1}\right).$$ (6.31)

With reference to Figure 6.4, these expressions represent total (convective + diffusive) transport of Φ *normal* to the two curvilinear directions, respectively. The convective transport $\rho\, U_{\mathrm{fi}}\, \Phi$ is thus directed normal to the constant-ξ_i lines. In other words, U_{fi} is directed along the contravariant base vector $\vec{a}^i$. Note that, in general, lines of constant ξ_1 and ξ_2 do not intersect orthogonally. Thus, the total normal diffusive contribution is made up of two components. The first, containing dA_i^2, is due to the property gradient along the covariant base vector direction $\vec{a}_i$, the second, containing dA_{12}, is due to the property gradient along the direction ξ_j, $j \neq i$. If the intersection of coordinate lines were to be orthogonal, $dA_{12} = 0$. Also, from Equations 6.13, it is clear that dA_{12} can be both positive as well as negative.

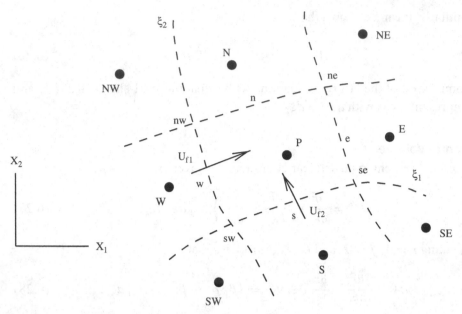

Figure 6.4. Definition of node P and contravariant flow velocities.

6.2.4 Discretisation

Our next task is to discretise Equation 6.20 for the general variable Φ. To do this, we define the typical node P of a curvilinear grid as shown in Figure 6.4. The cell faces (ne-se, se-sw, sw-nw, and nw-ne), as in the case of Cartesian grids, are assumed to be *midway* between the adjacent nodes. In curvilinear coordinates, $\Delta\xi_1 = \Delta\xi_2 = 1$, as already explained. Then, using the IOCV method, integration[1] of Equation 6.20 over the control volume surrounding node P gives

$$\frac{r_P J_P}{\Delta t} \left(\rho_P \Phi_P - \rho_P^0 \Phi_P^0 \right) + [C_e \Phi_e - d_e (\Phi_E - \Phi_P)]$$

$$- [C_w \Phi_w - d_w (\Phi_P - \Phi_W)]$$

$$+ [C_n \Phi_n - d_n (\Phi_N - \Phi_P)]$$

$$- [C_s \Phi_s - d_s (\Phi_P - \Phi_S)]$$

$$= AC_e (\Phi_{ne} - \Phi_{se}) + AC_w (\Phi_{sw} - \Phi_{nw})$$

$$+ AC_n (\Phi_{ne} - \Phi_{nw}) + AC_s (\Phi_{sw} - \Phi_{se})$$

$$+ r_P J_P S, \tag{6.32}$$

---

[1] Each term in Equation 6.20 is integrated as

$$\int_s^n \int_w^e (\text{Term}) d\xi_1 \, d\xi_2.$$

where the convective coefficients are given by

$$C_e = \rho_e r_e U_{f1,e} = \rho_e r_e \left[\beta_{1e}^1 \bar{u}_{1e} + \beta_{1e}^2 \bar{u}_{2e} \right],$$

$$C_w = \rho_w r_w U_{f1,w} = \rho_w r_w \left[\beta_{1w}^1 \bar{u}_{1w} + \beta_{1w}^2 \bar{u}_{2w} \right],$$

$$C_n = \rho_n r_n U_{f2,n} = \rho_n r_n \left[\beta_{1n}^2 \bar{u}_{1n} + \beta_{2n}^2 \bar{u}_{2n} \right],$$

$$C_s = \rho_s r_s U_{f2,s} = \rho_s r_s \left[\beta_{1s}^2 \bar{u}_{1s} + \beta_{2s}^2 \bar{u}_{2s} \right], \tag{6.33}$$

and the diffusion coefficients are

$$d_e = \left. \frac{\left(r\, \Gamma_{\text{eff}}\, d A_1^2 \right)}{J} \right|_e, \qquad d_w = \left. \frac{\left(r\, \Gamma_{\text{eff}}\, d A_1^2 \right)}{J} \right|_w,$$

$$d_n = \left. \frac{\left(r\, \Gamma_{\text{eff}}\, d A_2^2 \right)}{J} \right|_n, \qquad d_s = \left. \frac{\left(r\, \Gamma_{\text{eff}}\, d A_2^2 \right)}{J} \right|_s,$$

$$AC_e = \left. \frac{\left(r\, \Gamma_{\text{eff}}\, d A_{12} \right)}{J} \right|_e, \qquad AC_w = \left. \frac{\left(r\, \Gamma_{\text{eff}}\, d A_{12} \right)}{J} \right|_w,$$

$$AC_n = \left. \frac{\left(r\, \Gamma_{\text{eff}}\, d A_{12} \right)}{J} \right|_n, \qquad AC_s = \left. \frac{\left(r\, \Gamma_{\text{eff}}\, d A_{12} \right)}{J} \right|_s. \tag{6.34}$$

In evaluating the convective coefficients (or the mass fluxes at the cell faces), the $\bar{u}$ at the cell faces are evaluated by linear interpolation from neighbouring nodal velocities. For example, $\bar{u}_{1e} = 0.5(u_{1P} + u_{1E})$. Similarly, the values of Φ at the control-volume corners are also linearly interpolated. For example, $\Phi_{ne} = 0.25(\Phi_P + \Phi_E + \Phi_{NE} + \Phi_N)$. Finally, we note that the diffusion coefficients again have dimensions of *conductance*.

Equation 6.32 applies to $\Phi = u_1, u_2$ and all other scalar variables. When $\Phi = 1$, however, we recover the mass-conservation equation. Thus,

$$\frac{r_P J_P}{\Delta t} \left(\rho_P - \rho_P^o \right) + C_e - C_w + C_n - C_s = 0. \tag{6.35}$$

Now, making use of this equation, we can recast Equation 6.32 again in the following familiar form

$$AP\, \Phi_P = AE\, \Phi_E + AW\, \Phi_W + AN\, \Phi_N + AS\, \Phi_S + D, \tag{6.36}$$

where the convective–diffusive coefficients AE, AW, AN, and AS are given by

$$AE = d_e \left[A + \max\left(-P_{c_e}, 0 \right) \right], \qquad P_{c_e} = C_e/d_e,$$

$$AW = d_w \left[A + \max\left(P_{c_w}, 0 \right) \right], \qquad P_{c_w} = C_w/d_w,$$

$$AN = d_n \left[A + \max\left(-P_{c_n}, 0 \right) \right], \qquad P_{c_n} = C_n/d_n,$$

$$AS = d_s \left[A + \max\left(P_{c_s}, 0 \right) \right], \qquad P_{c_s} = C_s/d_s,$$

$$AP = AE + AW + AN + AS + \left. \frac{r\, \rho^o\, J}{\Delta t} \right|_P. \tag{6.37}$$

In these expressions, A is given by the convection scheme employed (see Chapter 5) and source D is given by

$$D = r_{\mathrm{P}} J_{\mathrm{P}} S + \left. \frac{r \rho^{\circ} J}{\Delta t} \right|_{\mathrm{P}} \Phi_{\mathrm{P}}^{\circ}$$

$$+ AC_{\mathrm{e}} (\Phi_{\mathrm{ne}} - \Phi_{\mathrm{se}}) + AC_{\mathrm{w}} (\Phi_{\mathrm{sw}} - \Phi_{\mathrm{nw}})$$

$$+ AC_{\mathrm{n}} (\Phi_{\mathrm{ne}} - \Phi_{\mathrm{nw}}) + AC_{\mathrm{s}} (\Phi_{\mathrm{sw}} - \Phi_{\mathrm{se}}). \tag{6.38}$$

6.2.5 Pressure-Correction Equation

The appropriate total pressure-correction equation in Cartesian coordinates has already been derived in Chapter 5 (see Equations 5.57 with boundary condition 5.58). Transforming this equation to curvilinear coordinates, we obtain[2]

$$\frac{\partial}{\partial \xi_1} \left[\frac{\rho \, r \, \alpha \, d A_1^2}{A \, P^{uf1}} \frac{\partial p'}{\partial \xi_1} \right] + \frac{\partial}{\partial \xi_2} \left[\frac{\rho \, r \, \alpha \, d A_2^2}{A \, P^{uf2}} \frac{\partial p'}{\partial \xi_2} \right]$$

$$= r J \frac{\partial (\rho)}{\partial t} + \frac{(\rho r \overline{U}_1^l)}{\partial \xi_1} + \frac{(\rho r \overline{U}_2^l)}{\partial \xi_2}. \tag{6.39}$$

When Equation 6.39 is solved, the p' distribution is obtained. The next task is to recover the mass-conserving pressure correction $p'_{\mathrm{m}} = p' - p'_{\mathrm{sm}}$. To evaluate p'_{sm}, we need to calculate $\overline{p} = 0.5(\overline{p}_{x_1} + \overline{p}_{x_2})$ from solution of Equations 5.111 and 5.112. Thus, to calculate $\overline{p}_{x_1}$, for example, we write

$$\left. \frac{\partial^2 p'}{\partial x_1^2} \right|_P = \frac{\partial}{\partial \xi_1} \left[\frac{\beta_1^1 \beta_1^1}{J} \frac{\partial p'}{\partial \xi_1} + \frac{\beta_1^1 \beta_2^1}{J} \frac{\partial p'}{\partial \xi_2} \right]_P$$

$$+ \frac{\partial}{\partial \xi_2} \left[\frac{\beta_2^1 \beta_1^1}{J} \frac{\partial p'}{\partial \xi_1} + \frac{\beta_2^1 \beta_2^1}{J} \frac{\partial p'}{\partial \xi_2} \right]_P = 0. \tag{6.40}$$

With reference to Figure 6.4, the discretised version of Equation 6.40 reads as

$$\left. \frac{\beta_1^1 \beta_1^1}{J} \right|_{\mathrm{e}} (p'_{\mathrm{E}} - p'_{\mathrm{P}}) + \left. \frac{\beta_1^1 \beta_2^1}{J} \right|_{\mathrm{e}} (p'_{\mathrm{ne}} - p'_{\mathrm{se}})$$

$$- \left. \frac{\beta_1^1 \beta_1^1}{J} \right|_{\mathrm{w}} (p'_{\mathrm{P}} - p'_{\mathrm{W}}) - \left. \frac{\beta_1^1 \beta_2^1}{J} \right|_{\mathrm{w}} (p'_{\mathrm{nw}} - p'_{\mathrm{sw}})$$

$$+ \left. \frac{\beta_1^1 \beta_2^1}{J} \right|_{\mathrm{n}} (p'_{\mathrm{ne}} - p'_{\mathrm{nw}}) + \left. \frac{\beta_2^1 \beta_2^1}{J} \right|_{\mathrm{e}} (p'_{\mathrm{N}} - p'_{\mathrm{P}})$$

$$- \left. \frac{\beta_1^1 \beta_2^1}{J} \right|_{\mathrm{s}} (p'_{\mathrm{se}} - p'_{\mathrm{sw}}) - \left. \frac{\beta_2^1 \beta_2^1}{J} \right|_{\mathrm{s}} (p'_{\mathrm{P}} - p'_{\mathrm{S}}) = 0. \tag{6.41}$$

[2] In Equation 6.39, cross-derivative terms containing $d A_{12}$ are dropped. This is because the pressure-correction equation is essentially an estimator of p'_{m} and, therefore, in an iterative procedure the truncated form presented in Equation 6.39 suffices. It is of course possible to recover the effect of the neglected term in a predictor–corrector fashion. $\overline{U}$ are contravariant mean velocities.

Therefore, separating the solution for p_P^l, we get

$$\overline{p}_{x1,P} = p_P^l = \frac{A}{B},$$

$$A = \left\{ \beta_{1,e}^1 \beta_{1,e}^1 \, p_E + \beta_{2,e}^1 \beta_{1,e}^1 \, (p_{ne} - p_{se}) \right\} / J_e$$
$$+ \left\{ \beta_{1,w}^1 \beta_{1,w}^1 \, p_W - \beta_{2,w}^1 \beta_{1,w}^1 \, (p_{nw} - p_{sw}) \right\} / J_w$$
$$+ \left\{ \beta_{2,n}^1 \beta_{2,n}^1 \, p_N + \beta_{2,n}^1 \beta_{1,n}^1 \, (p_{ne} - p_{nw}) \right\} / J_n$$
$$+ \left\{ \beta_{2,s}^1 \beta_{2,s}^1 \, p_S - \beta_{2,s}^1 \beta_{1,s}^1 \, (p_{se} - p_{sw}) \right\} / J_s,$$

$$B = \frac{\beta_{1,e}^1 \beta_{1,e}^1}{J_e} + \frac{\beta_{1,w}^1 \beta_{1,w}^1}{J_w} + \frac{\beta_{2,n}^1 \beta_{2,n}^1}{J_n} + \frac{\beta_{2,s}^1 \beta_{2,s}^1}{J_s}. \tag{6.42}$$

Similarly, evaluation of $\overline{p}_{x_2}$ is accomplished from $\partial^2 p^l / \partial x_2^2 = 0$ and evaluation of $\overline{p}$ is completed.

6.2.6 Overall Calculation Procedure

The overall calculation procedure on curvilinear grids is nearly the same as that on Cartesian grids. Some important features are highlighted in the following.

1. Read coordinates $x_1(i, j)$ and $x_2(i, j)$ for $i = 1, 2, \ldots, IN$ and $j = 1, 2, \ldots, JN$. Hence calculate the geometric coefficients β_j^i and areas and volumes once and for all.
2. At a given time step, guess the pressure field $p_{i,j}^l$. This may be the pressure field from the previous time step.
3. Solve, using ADI, Equation 6.20 for Cartesian velocity components $\Phi = u_1^l$ and u_2^l with appropriate boundary conditions (see next subsection).
4. Evaluate U_{f1} and U_{f2} from Equations 6.22 and 6.23. In these evaluations, the cell-face velocities u_{f1} and u_{f2} are evaluated by arithmetic averaging. Hence, evaluate the source term of the total pressure-correction equation (6.39). Solve Equation 6.39 to obtain the $p_{i,j}'$ field.
5. Evaluate $\overline{p}_{i,j}$ as described in the previous subsection. Hence recover $p_{m,i,j}'$ to correct pressure as $p_{i,j}^{l+1} = p_{i,j}^l + \beta \, p_{m,i,j}'$.
6. Correct Cartesian velocities as

$$u_{1,P}^{l+1} = u_{1,P}^l - \frac{\rho \, r \, \alpha}{AP^{u_1}} \left[(\beta_1^1)_P \, (p_{m,e}' - p_{m,w}') + (\beta_2^1)_P \, (p_{m,n}' - p_{m,s}') \right], \tag{6.43}$$

$$u_{2,P}^{l+1} = u_{2,P}^l - \frac{\rho \, r \, \alpha}{AP^{u_2}} \left[(\beta_1^2)_P \, (p_{m,e}' - p_{m,w}') + (\beta_2^2)_P \, (p_{m,n}' - p_{m,s}') \right]. \tag{6.44}$$

Note that $AP^{u1} = AP^{u2}$.
7. Solve for other relevant scalar Φs.

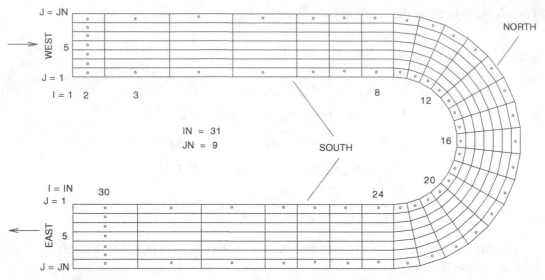

Figure 6.5. Node tagging, for a curvilinear grid with a 180° bend.

8. Check convergence through evaluation of residuals for momentum and scalar Φ
   equations. Evaluate the mass source residual R_m as appropriate for collocated
   grids (see Chapter 5).
9. If the convergence criterion is not satisfied, treat $p^{l+1} = p^l$, $\Phi^{l+1} = \Phi^l$ and
   return to step 3.
10. To execute the next time step, set all $\Phi^o = \Phi$ and return to step 2.

6.2.7 Node Tagging and Boundary Conditions

Because of the applicability of the (i, j) structure on curvilinear grids, there are
many features that are in common with those described for Cartesian grids. Thus,
one can readily use Su and Sp in Equation 6.36 to effect underrelaxation and
boundary conditions. Node tagging too can be done as described in Chapter 5.
Care, however, is needed in identification of the boundary type. To illustrate this,
consider the computational domain for a flow in a duct with a 180° bend shown
in Figure 6.5. The index I increases with ξ_1 and J with ξ_2. The flow enters at
the west boundary. The west boundary is identified with $I = 1$, east with $I = IN$,
south with $J = 1$, and north with $J = JN$. Note that although in the physical
domain (as drawn) the east boundary appears to the west, in the computational
domain it is identified $I = IN$ and the J index is seen to run *downwards*. Thus,
NTAGE $(IN - 1, J)$, $J = 2, 3, \ldots, JN - 1$ will be tagged with 21, 22, 23, or 24
depending on the type of boundary condition. Similarly, the south boundary in the
return flow channel of the bend coincides with $J = 1$ but, in the physical domain,
it is above the north boundary.

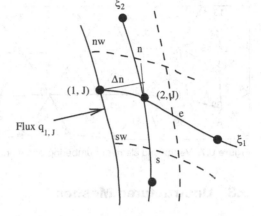

Figure 6.6. Gradient boundary condition.

To illustrate implementation of flux (or normal-gradient) boundary condition, consider the west boundary shown in Figure 6.6. Let q be the specified flux. Then

$$q\, dA_1 = -\Gamma\, dA_1 \left.\frac{\partial \Phi}{\partial n}\right|_{(1,j)} = -\frac{\Gamma}{J}\left[dA_1^2 \frac{\partial \Phi}{\partial \xi_1} + dA_{12} \frac{\partial \Phi}{\partial \xi_2}\right]_{(1,j)}$$

$$= AW_{2,j}\,(\Phi_{1,j} - \Phi_{2,j}) + (AC_w)_{2,j}\,(\Phi_{sw} - \Phi_{nw}). \qquad (6.15)$$

However, this representation involves Φ_{sw} and Φ_{nw}, which are again boundary locations. Therefore, it is advisable to represent the normal flux directly as

$$q\, dA_1 = -\Gamma\, dA_1 \left.\frac{\partial \Phi}{\partial n}\right|_{(1,j)} = -\frac{\Gamma\, dA_1}{\Delta n}\,(\Phi_{2,j} - \Phi_{1,j}), \qquad (6.46)$$

where the normal distance is given by

$$\Delta n = \left(\beta_i^1 \frac{\partial x_1}{\partial \xi_i} + \beta_i^2 \frac{\partial x_2}{\partial \xi_i}\right)\Big/ dA_i. \qquad (6.47)$$

It is now possible to extract an expression for $\Phi_{1,j}$ and implement the boundary condition using Su and Sp in the manner described in the previous chapter. The exit boundary condition where the second derivative of a scalar variable is set to zero can also be derived from this condition. Specification of the exit boundary condition for velocity, however, requires care. This is because the boundary conditions are known only in terms of boundary-normal and tangential velocity components. The Cartesian velocity components are then extracted from this specification. More discussion of this matter is presented in the next section. Boundaries at which Φ is specified require no elaboration. Finally, the *wall-function* treatment for the HRE turbulence model requires special care because the wall shear stress must be evaluated from the wall-normal gradient of velocity parallel (tangential) to the wall. Details of these and other issues of discretisation can be found in Ray and Date [58].

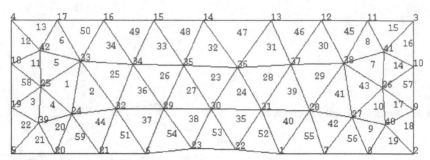

Figure 6.7. Vertex and element numbering on an unstructured grid.

6.3 Unstructured Meshes

6.3.1 Main Task

As mentioned in Section 6.1.2, a typical domain may be mapped by triangular, quadrilateral, and/or *n*-polygonal elements. Here, we again consider a relatively simple domain shown in Figure 6.7. The domain is mapped by triangles using ANSYS. The domain consists of two horizontal parallel plates in which a circular arc bump is provided at the bottom plate. Flow enters the left vertical boundary and leaves through the right vertical boundary.

When a domain is mapped in this way, ANSYS generates two *data files*:

1. a vertex file and
2. an element file.

The entries of these two files are shown in Table 6.1. They correspond to Figure 6.7. In this figure, there are 42 vertices and 59 elements. Note that the vertex numbering is completely arbitrary. The vertex file provides serial numbers of vertices along with their x_1, x_2, and x_3 coordinates. Since the domain is two dimensional, all x_3 are zero. The element file, in contrast, provides serially numbered elements (shown inside triangles) along with the identification numbers of three vertices (since triangular elements are generated) that form the element. Like vertex numbering, element numbers are also assigned arbitrarily.

There are a variety of ways in which transport equations can be discretised on an unstructured grid. The two principal ones are [83] (a) a vertex-centred approach and (b) an element-centred approach.

Vertex-Centred Approach

In the vertex-centred approach, the collocated variables Φ are defined at the vertices. Thus, vertices are treated as nodes. When the transport equations are discretised, a variable at node P (say) is related to variables at vertices in the immediate neighbourhood of P with which node P is connected by a line. The vertex and element files contain sufficient information to identify vertex or node numbers of vertices with which node P is connected. Such a data structure needs to be generated by

Table 6.1: Vertex and element files.

Vertex file				Element file			
NV	x_1	x_2	x_3	NE	NV1	NV2	NV3
1	0.5	0.0	0.0	1	24	33	25
2	1.5	0.0	0.0	2	24	32	33
3	1.5	1.0	0.0	3	19	39	25
4	−1.5	1.0	0.0	4	39	24	25
5	−1.5	0.0	0.0	5	25	33	42
6	−0.5	0.0	0.0	6	42	33	17
⋮	⋮	⋮	⋮	⋮	⋮	⋮	⋮
11	1.3229	1.0	0.0	12	18	42	4
⋮	⋮	⋮	⋮	⋮	⋮	⋮	⋮
24	−1.2708	0.2978	0.0	26	29	35	34
⋮	⋮	⋮	⋮	⋮	⋮	⋮	⋮
31	0.509	0.3404	0.0	33	34	35	15
⋮	⋮	⋮	⋮	⋮	⋮	⋮	⋮
39	−1.3958	0.2127	0.0	56	7	8	27
40	1.357	0.2127	0.0	57	9	10	26
41	1.357	0.7659	0.0	58	19	25	18
42	1.3958	0.7659	0.0	59	20	21	24

writing a separate computer program. It is clear from Figure 6.7 that different vertices will have different numbers of neighbouring vertices. In this approach, to adopt an IOCV method for discretisation, one needs to *construct* a control volume surrounding node P. Figure 6.8(a) shows a typical vertex P along with its neighbours. Different approaches are possible for the control-volume construction, but the one adopted here is as follows:

1. Identify elements having a common vertex at P.
2. Locate *centroids* of each element. This can be done by using known coordinates of vertices of each element.
3. Connect the successive centroids by straight lines (shown dotted in Figure 6.8).

The dotted lines will enclose P and thus form a control volume surrounding P. Such a construction at all vertices will yield a non-overlapping set of control volumes. Discretisation can now be carried out for a typical control volume.

One disadvantage of this approach concerns application of boundary conditions. Thus, consider a vertex (or a node) at the *junction* of two boundaries as shown in Figure 6.8(b). Now, if the boundary conditions at the two boundaries of the junction are different, the boundary condition at the junction node cannot be uniquely defined. It is possible to overcome this difficulty but only at the expense of additional bookkeeping.

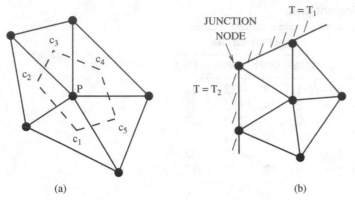

Figure 6.8. Vertex-centred unstructured grid.

Element-Centred Approach

In contrast to the vertex-centred approach, the element-centred approach regards each triangular (or polygonal) element itself as the control volume [see Figure 6.9(a)]. Then, node P is defined at the centroid of the element such that

$$x_{i,P} = \frac{1}{3}(x_{i,1} + x_{i,2} + x_{i,3}), \qquad i = 1, 2, \tag{6.48}$$

and the coordinates of vertices 1, 2, and 3 are known from the vertex file. Note that node P will be identified by the identifier of the element to which it belongs because node P will always remain enclosed within its surrounding control volume.

In this case, node P will have only three neighbours since triangular elements are considered. The identification numbers of neighbouring elements are, however, not a priori known. However, these can be determined from the element file because two neighbouring elements must share the same two vertices. To establish this connectivity between elements, a separate computer program must be written.

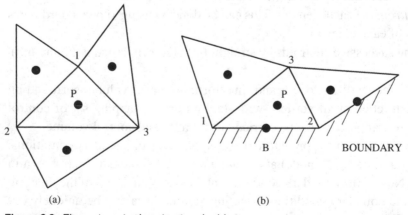

Figure 6.9. Element-centred unstructured grid.

The lines joining vertices will henceforth be called control volumes or cell faces and elements will be referred to as cells. Thus, a triangular element will have three cell faces. The same logic extends to polygonal cells. Now, it is easy to recognize that when nodes are defined at the centroids of cells, there is no node at the boundary to facilitate implementation of the boundary conditions. Therefore, a boundary node must be *defined*. We adopt the convention that the boundary node shall be at the *center of the cell face* coinciding with the domain boundary. This is shown in Figure 6.9(b) by point B. It will be recognised that even if there is a change in boundary condition on either side of a vertex, the boundary condition can now be effected without any ambiguity.

Practitioners of CFD familiar with control-volume discretisation on structured grids prefer the element-centred approach [5, 46, 20] rather than the vertex-centred approach. In the discussion to follow, therefore, the element-centred approach is further developed.

6.3.2 Gauss's Divergence Theorem

The transport equation (5.1) in Cartesian coordinates is again considered here but without the presence of r for brevity.[3] The equation is rewritten as

$$\frac{\partial(\rho\,\Phi)}{\partial t} + \frac{\partial\,q_i}{\partial x_i} = \frac{\partial(\rho\,\Phi)}{\partial t} + \operatorname{div}(\vec{q}) = S, \qquad (6.49)$$

where the vector $\vec{q} = \vec{i}\,q_1 + \vec{j}\,q_2$ and $\vec{i}$ and $\vec{j}$ are unit vectors along Cartesian coordinates x_1 and x_2, respectively.

To implement the IOCV method, Equation 6.49 is now integrated over the elemental control volume shown in Figure 6.9. Thus, with the usual approximations, we have

$$\left(\rho_P\,\Phi_P - \rho_P^o\,\Phi_P^o\right)\frac{\Delta V}{\Delta t} + \int_{\Delta V} \operatorname{div}(\vec{q})dV = S\,\Delta V, \qquad (6.50)$$

where ΔV is the volume (i.e., the area in the 2D domain with unit dimension in the x_3 direction) of the cell surrounding P. This cell volume can be calculated knowing the coordinates of the vertices.

The second term on the left-hand side will now be evaluated by invoking Gauss's divergence theorem [70] applicable to a singly connected region. Thus,

$$\int_{\Delta V} \operatorname{div}(\vec{q})dV = \int_C \vec{q}\cdot\vec{A}, \qquad (6.51)$$

where $\int_C$ is a line integral along the bounding surfaces (or lines in two dimensions) of the control volume and $\vec{A}$ is the local area vector normal (pointing outwards) to

---

[3] This neglect in no way disqualifies the developments to follow.

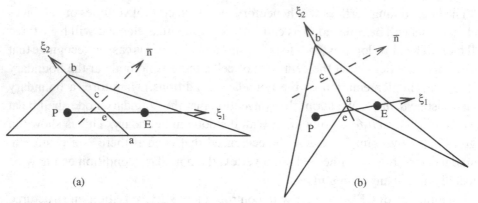

Figure 6.10. Typical cell face ab.

the bounding surface (line). The direction of C is anticlockwise. To make further progress, the line integral is replaced by summation. Thus,

$$\int_C \vec{q} \cdot \vec{A} = \sum_{k=1}^{NK} (\vec{q} \cdot \vec{A})_k, \tag{6.52}$$

where $NK = 3$ for a triangular element and k stands for the kth face of the control volume. Thus, the line integral is discretized into NK segments.

To evaluate the dot product $\vec{q} \cdot \vec{A}$ at each cell face k, consider Figure 6.10, where evaluation at face ab (say) shared by neighbouring cells P and E is to be carried out. Let line PE be along the ξ_1 direction and line ab be along the ξ_2 direction, where the latter direction is chosen such that Jacobian J (see Equation 6.14) is positive. Let lines PE and ab intersect at e. Now, depending on the shapes of cells P and E, e may lie within ab [Figure 6.10(a)] or on an extension of ab [Figure 6.10(b)]. Further, let $\vec{n}$ be the unit normal vector to ab pointing outwards with respect to cell P as shown in the figure. Then, using Equation 6.25, we get

$$\vec{A} = A_{ab} \cdot \vec{n} = \vec{i}\,\frac{\partial x_2}{\partial \xi_2} - \vec{j}\,\frac{\partial x_1}{\partial \xi_2} = \vec{i}\,\beta_1^1 + \vec{j}\,\beta_1^2, \tag{6.53}$$

where

$$\beta_1^1 = x_{2b} - x_{2a}, \qquad \beta_1^2 = -(x_{1b} - x_{1a}), \tag{6.54}$$

$$A_{ab} = A_{ck} = \sqrt{(\beta_1^1)^2 + (\beta_1^2)^2} = \text{area of face ab}, \tag{6.55}$$

and c is the midpoint of ab. The coordinates of c are

$$x_{i,c} = \frac{1}{2}(x_{i,a} + x_{i,b}). \tag{6.56}$$

Substituting Equation 6.53 in Equation 6.52, we have

$$(\vec{q} \cdot \vec{A})_{ck} = (\vec{q} \cdot \vec{n})_{ck}\, A_{ck} = (\beta_1^1 q_1 + \beta_1^2 q_2)_k = \sum_{i=1}^{2} (\beta_1^i q_i)_k = (q_n A_c)_k. \tag{6.57}$$

We now recall that

$$q_i = \rho u_i \Phi - \Gamma \frac{\partial \Phi}{\partial x_i}, \qquad i = 1, 2. \tag{6.58}$$

Therefore,

$$(q_n A_c)_k = \rho_{ck} \Phi_{ck} \sum_{i=1}^{2} \left(\beta_1^i u_i \right)_{ck} - \Gamma_{ck} \sum_{i=1}^{2} \left(\beta_1^i \frac{\partial \Phi}{\partial x_i} \right)_{ck}. \tag{6.59}$$

Now, for brevity, we introduce following notation:

$$C_{ck} = \rho_{ck} \sum_{i=1}^{2} \left(\beta_1^i u_i \right)_{ck} \quad \text{(cell-face mass flow)} \tag{6.60}$$

and

$$-\Gamma_{ck} A_{ck} \left. \frac{\partial \Phi}{\partial n} \right|_{ck} = -\Gamma_{ck} \sum_{i=1}^{2} \left(\beta_1^i \frac{\partial \Phi}{\partial x_i} \right)_{ck} \quad \text{(normal diffusion).} \tag{6.61}$$

Thus, the total transport across the kth cell face is given by

$$(\vec{q} \cdot \vec{A})_{ck} = C_{ck} \Phi_{ck} - \Gamma_{ck} A_{ck} \left. \frac{\partial \Phi}{\partial n} \right|_{ck}. \tag{6.62}$$

Note that the normal diffusion is evaluated directly in terms of a normal gradient rather than in terms of resolved components in ξ_1 and ξ_2 directions as was done on curvilinear grids (see Equations 6.30 and 6.31). It is this feature that makes our diffusion transport evaluation equally applicable to 3D polyhedra.

The convective and diffusive contributions to total transport across each cell face k must now be evaluated. In the literature [19, 46, 20], these contributions are evaluated in a variety of ways, but without invoking any *line structure*. The approach adopted here recognises the importance of a line structure analogous to the one available at the cell face of a structured grid. The existence of such a line structure at the cell face of an unstructured grid, however, is not obvious because the line joining cell centroids P and E intersects cell face ab in an arbitrary manner, as shown in Figure 6.10. Therefore, a line structure must be *deliberately constructed*. This matter is considered in the next subsection.

6.3.3 Construction of a Line Structure

Our interest is to evaluate total transport (Equation 6.62) normal to the kth cell face. To carry out this evaluation, consider the more general face construction shown in Figure 6.10(b). This figure is again drawn more elaborately in Figure 6.11 to carry out the necessary construction of a line structure.

The construction begins by drawing two normals (shown by dotted lines) to ab passing through e and c. Now, two lines parallel to ab are drawn passing through nodes P and E. Let the line through P intersect the face normal through e at P_1 and

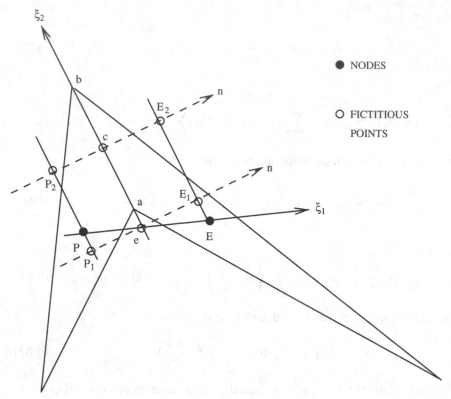

Figure 6.11. Construction of a line structure at cell face ab.

that through c at P_2. Note that these intersections at P_1 and P_2 will be orthogonal. Similarly, let the face-parallel line through E intersect the two normals at E_1 and E_2, respectively.

With this construction, it is clear that Equation 6.62 must be evaluated along the line P_2–c–E_2. These evaluations, it will be appreciated, will now be similar to the evaluations carried out at the cell face of a structured grid control volume. In the next two subsections, the convective and diffusive contributions are evaluated separately.

6.3.4 Convective Transport

Following the usual methodology, the convective transport term in Equation 6.62 is evaluated as

$$C_{ck}\,\Phi_{ck} = C_{ck}[f_{ck}\,\Phi_{P_2} + (1 - f_{ck})\Phi_{E_2}]_k, \qquad (6.63)$$

where f_{ck} are weighting factors that depend on the convection scheme used. If, for example, the UDS is used, then

$$f_{ck}\,(\text{UDS}) = 0.5\left(1 + \frac{|C_{ck}|}{C_{ck}}\right), \qquad (6.64)$$

where, following Equation 6.60,

$$C_{ck} = \rho_{ck} \left[\beta_1^1 u_1 + \beta_1^2 u_2 \right]_{ck}. \tag{6.65}$$

Now, $\rho_{ck}, u_{1,ck}$, and $u_{2,ck}$ are linearly *interpolated* according to the following general formula:[4]

$$\Psi_{ck} = [f_{m,c} \Psi_{E_2} + (1 - f_{m,c}) \Psi_{P_2}]. \tag{6.66}$$

In this evaluation the weighting factor can be deduced from the geometry of construction shown in Figure 6.11 as

$$f_{m,c} = \frac{l_{P_2 c}}{l_{P_2 E_2}} = \frac{l_{P_1 e}}{l_{P_1 E_1}} = \frac{l_{P e}}{l_{P E}}, \tag{6.67}$$

where $l_{p e}$ and $l_{P E}$ can be evaluated from known coordinates of points P, e, and E.

6.3.5 Diffusion Transport

For evaluation of diffusion transport in Equation 6.62, the face area A_{ck} is known from Equation 6.55 and Γ_{ck} can be evaluated from the general formula (6.66) or by harmonic mean. It remains now to evaluate the face-normal gradient of Φ. To do this, it is first recognised that point c, in general, will not be midway between points P_2 and E_2. Therefore, to retain second-order accuracy in the evaluation of this gradient, we employ a Taylor series expansion.

$$\Phi_{P_2} = \Phi_c - l_{P_2 c} \left. \frac{\partial \Phi}{\partial n} \right|_c + \frac{l_{P_2 c}^2}{2} \left. \frac{\partial^2 \Phi}{\partial^2 n} \right|_c + \cdots, \tag{6.68}$$

$$\Phi_{E_2} = \Phi_c + l_{E_2 c} \left. \frac{\partial \Phi}{\partial n} \right|_c + \frac{l_{E_2 c}^2}{2} \left. \frac{\partial^2 \Phi}{\partial^2 n} \right|_c + \cdots. \tag{6.69}$$

Eliminating the second derivative from these two equations and using Equation 6.67, we can show that

$$\left. \frac{\partial \Phi}{\partial n} \right|_c = \frac{\Phi_{E_2} - \Phi_{P_2}}{l_{P_2 E_2}} - \frac{1 - 2 f_{m,c}}{f_{m,c}(1 - f_{m,c})} \left[\frac{f_{m,c} \Phi_{E_2} - \overline{\Phi}_c + (1 - f_{m,c})\Phi_{P_2}}{l_{P_2 E_2}} \right], \tag{6.70}$$

where, from our construction,

$$l_{P_2 E_2} = l_{P_1 E_1} = \vec{l}_{PE} \cdot \vec{n} = \left| \sum_{i=1}^{2} \beta_1^i (x_{i,E} - x_{i,P}) \right| \Big/ A_c. \tag{6.71}$$

---

[4] Note that this interpolation can also be performed *multidimensionally* as stated in Chapter 5. Thus, one may write

$$\Psi_{ck} = \frac{1}{2}[f_{m,c} \Psi_{E_2} + (1 - f_{m,c})\Psi_{P_2}] + \frac{1}{4} (\Psi_a + \Psi_b).$$

In Equation 6.70, the first term on the right-hand side represents first-order-accurate evaluation of the normal gradient whereas the second term imparts second-order accuracy. In this latter term, if $\overline{\Phi}_c$ is evaluated from general formula (6.66) then the term will simply vanish. To retain second-order accuracy, therefore, $\overline{\Phi}_c$ must be interpolated along direction ab. Now, since point c (see Figure 6.11) is midway between a and b,

$$\overline{\Phi}_{ck} = 0.5\,(\Phi_{ak} + \Phi_{bk}). \tag{6.72}$$

Using Equations 6.70 and 6.72, therefore, we can express the total diffusion transport as

$$-\left(\Gamma A \frac{\partial \Phi}{\partial n}\right)_{ck} = -d_{ck}\,(\Phi_{E_2} - \Phi_{P_2})_k + d_{ck}B_{ck}\left[f_{m,c}\Phi_{E_2} - \overline{\Phi}_c + (1 - f_{m,c})\Phi_{P_2}\right]_k, \tag{6.73}$$

where

$$d_{ck} = \frac{(\Gamma A)_{ck}}{l_{P_2 E_2}} \tag{6.74}$$

and

$$B_{ck} = \frac{1 - 2 f_{m,c}}{f_{m,c}\,(1 - f_{m,c})}. \tag{6.75}$$

It will be recognised that d_{ck} is nothing but the familiar diffusion coefficient having significance of a conductance. The symbol B_{ck} is introduced for brevity.

6.3.6   Interim Discretised Equation

At this stage of development, it will be instructive to recapitulate derivations following Equation 6.50. Thus, the volume integral in this equation is replaced by a summation of face-normal contributions in Equation 6.52. The total (convective + diffusive) face-normal contribution at any face is then represented in Equation 6.62. The convective component of the total face-normal contribution is given by Equation 6.63 and the diffusive component by Equation 6.73. Therefore, Equation 6.50 may now be written as

$$
\begin{aligned}
&\left(\rho_P\,\Phi_P - \rho_P^o\,\Phi_P^o\right)\frac{\Delta V}{\Delta t} \\
&+ \sum_{k=1}^{NK} C_{ck}\left[f_c\,\Phi_{P_2} + (1 - f_c)\Phi_{E_2}\right]_k \\
&- \sum_{k=1}^{NK} d_{ck}\,(\Phi_{E_2} - \Phi_{P_2})_k \\
&+ \sum_{k=1}^{NK} d_{ck}\,B_{ck}\left[f_{m,c}\,\Phi_{E_2} - \overline{\Phi}_c + (1 - f_{m,c})\Phi_{P_2}\right]_k = S\,\Delta V. \tag{6.76}
\end{aligned}
$$

This discretised equation, however, is of little use because the values of variables at fictitious points P_2 and E_2 and at vertices a and b are not known. We must therefore relate values at these fictitious points to the values at nodes P and E. This matter is developed in the next subsection.

6.3.7 Interpolation of Φ at P_2, E_2, a, and b

If it is assumed that the Φ variation between P and P_2 is linear then to first-order accuracy

$$\Phi_{P_2} = \Phi_P + \Delta\Phi_P = \Phi_P + \vec{l}_{PP_2} \cdot \nabla\Phi_P, \tag{6.77}$$

where

$$\vec{l}_{PP_2} = \vec{i}\,(x_{1,P_2} - x_{1,P}) + \vec{j}\,(x_{2,P_2} - x_{2,P}), \tag{6.78}$$

and

$$\nabla\Phi_P = \vec{i}\,\frac{\partial\Phi}{\partial x_1}\bigg|_P + \vec{j}\,\frac{\partial\Phi}{\partial x_2}\bigg|_P. \tag{6.79}$$

Taking the dot product in Equation 6.77 therefore gives

$$\Delta\Phi_P = \sum_{i=1}^{2}(x_{i,P_2} - x_{i,P})\frac{\partial\Phi}{\partial x_i}\bigg|_P, \tag{6.80}$$

where $x_{i,P_2} - x_{i,P}$ must be evaluated in terms of points whose coordinates are known. Thus

$$x_{i,P_2} - x_{i,P} = x_{i,P_2} - x_{i,c} + x_{i,c} - x_{i,P}. \tag{6.81}$$

However, from the construction shown in Figure 6.11,

$$x_{i,P_2} - x_{i,c} = x_{i,P_1} - x_{i,e}. \tag{6.82}$$

Therefore, Equation 6.81 is further reformulated as

$$x_{i,P_2} - x_{i,P} = [x_{i,P_1} - x_{i,e} + x_{i,e} - x_{i,P}] + x_{i,c} - x_{i,e}. \tag{6.83}$$

Now, the equation to the face-normal passing through e is given by

$$\vec{n} = \frac{\vec{i}\,(x_{1,e} - x_{1,P_1}) + \vec{j}\,(x_{2,e} - x_{2,P_1})}{l_{P_1e}} = \frac{\vec{i}\,\beta_1^1 + \vec{j}\,\beta_1^2}{A_c}, \tag{6.84}$$

therefore

$$x_{i,P_1} - x_{i,e} = -\frac{l_{P_1e}}{A_c}\,\beta_1^i \tag{6.85}$$

and Equation 6.83 can be written as

$$x_{i,P_2} - x_{i,P} = l_{xi} + d_{xi}, \tag{6.86}$$

where

$$l_{xi} = x_{i,e} - x_{i,P} - \frac{l_{P_1e}}{A_c} \beta_1^i \tag{6.87}$$

$$d_{xi} = x_{i,c} - x_{i,e} = \frac{1}{2}(x_{ia} + x_{ib}) - x_{ie}, \tag{6.88}$$

$$l_{P_1e} = \vec{l}_{Pe} \cdot \vec{n} = \left| \sum_{i=1}^{2} (x_{i,e} - x_{i,P})\beta_1^i \right| \bigg/ A_c. \tag{6.89}$$

Now, since coordinates of e, P, a, and b are known, using Equations 6.86 and 6.80, we can write Equation 6.77 as

$$\Phi_{P_2} = \Phi_P + \Delta\Phi_P = \Phi_P + \sum_{i=1}^{2}(l_{xi} + d_{xi}) \left.\frac{\partial\Phi}{\partial x_i}\right|_P. \tag{6.90}$$

Invoking similar arguments, it can be shown that

$$\Phi_{E_2} = \Phi_E + \Delta\Phi_E = \Phi_E + \sum_{i=1}^{2}\left[d_{xi} - \frac{(1 - f_{m,c})}{f_{m,c}} l_{xi}\right]\left.\frac{\partial\Phi}{\partial x_i}\right|_E. \tag{6.91}$$

Now, Φ_a and Φ_b are evaluated as the average of two estimates in the following manner:

$$\Phi_a = 0.5\left[\Phi_P + \bar{l}_{Pa}\nabla\Phi_P + \Phi_E + \bar{l}_{Ea}\nabla\Phi_E\right], \tag{6.92}$$

$$\Phi_b = 0.5\left[\Phi_P + \bar{l}_{Pb}\nabla\Phi_P + \Phi_E + \bar{l}_{Eb}\nabla\Phi_E\right]. \tag{6.93}$$

6.3.8   Final Discretised Equation

Substituting Equations 6.90 to 6.93 in Equation 6.76 and performing some algebra, we can write the resulting discretised equation as

$$\left(\rho_P\,\Phi_P - \rho_P^o\,\Phi_P^o\right)\frac{\Delta V}{\Delta t} + \sum_{k=1}^{NK} C_{ck}\left[f_c\,\Phi_P + (1 - f_c)\Phi_E\right]_k$$

$$- \sum_{k=1}^{NK} d_{ck}\left(\Phi_E - \Phi_P\right)_k$$

$$= S\,\Delta V + \sum_{k=1}^{NK} D_k, \tag{6.94}$$

where

$$D_k = -d_{ck}\,B_{ck}[f_{m,c}\,\Phi_{E_2} - 0.5\,(\Phi_a + \Phi_b) + (1 - f_{m,c})\Phi_{P_2}]_k$$

$$+ d_{ck}\,(\Delta\Phi_E - \Delta\Phi_P)_k - C_{ck}\,[f_c\,\Delta\Phi_P + (1 - f_c)\Delta\Phi_E]_k\,. \tag{6.95}$$

Further Simplification

Grouping terms in Φ_P and $\Phi_{E,k}$ together, we can write Equation 6.94 as

$$\left[\rho_P \frac{\Delta V}{\Delta t} + \sum_{k=1}^{NK} (C_{ck} f_{ck} + d_{ck}) \right] \Phi_P = \sum_{k=1}^{NK} \{d_{ck} - (1 - f_{ck})C_{ck}\} \, \Phi_{E,k}$$

$$+ S \Delta V + \rho_P^o \frac{\Delta V}{\Delta t} \Phi_P^o + \sum_{k=1}^{NK} D_k. \quad (6.96)$$

It is possible to simplify this equation further. Thus, let coefficient of Φ_{Ek} be AE_k. Then,

$$AE_k = d_{ck} - (1 - f_{ck})C_{ck}. \quad (6.97)$$

Now, for $\Phi = 1$ (i.e., the mass conservation equation), Equation 6.76 gives

$$(\rho_P - \rho_P^o) \frac{\Delta V}{\Delta t} + \sum_{k=1}^{NK} C_{ck} = 0, \quad (6.98)$$

or

$$\rho_P \frac{\Delta V}{\Delta t} = \rho_P^o \frac{\Delta V}{\Delta t} - \sum_{k=1}^{NK} C_{ck}. \quad (6.99)$$

Now, let AP be the multiplier of Φ_P in Equation 6.96. Then using Equations 6.97 and 6.99, it follows that[5]

$$AP = \rho_P \frac{\Delta V}{\Delta t} + \sum_{k=1}^{NK} \{d_{ck} + f_{ck} C_{ck}\} \quad (6.100)$$

$$= \rho_P^o \frac{\Delta V}{\Delta t} + \sum_{k=1}^{NK} \{d_{ck} - (1 - f_{ck}) C_{ck}\} \quad (6.101)$$

$$= \rho_P^o \frac{\Delta V}{\Delta t} + \sum_{k=1}^{NK} AE_k. \quad (6.102)$$

Thus, Equation 6.96 can be compactly written as

$$AP \, \Phi_P^{l+1} = \sum_{k=1}^{NK} AE_k \, \Phi_{Ek}^{l+1} + S \Delta V + \rho_P^o \frac{\Delta V}{\Delta t} \Phi_P^o + \sum_{k=1}^{NK} D_k^l. \quad (6.103)$$

---

[5] Note the similarity of Equation 6.102 with Equation 6.37 derived for curvilinear grids.

The following comments are now in order:

1. Equation 6.103 has the familiar form in which the value of Φ_P is related to its neighbors Φ_{Ek}.
2. Superscripts l and $l+1$ are now added to indicate that terms D_k containing Cartesian derivatives of Φ are treated as sources and therefore lag behind by one iteration. The same applies to the source term S. A method for evaluating nodal Cartesian derivatives is developed in the next subsection.
3. Equation 6.103 applies to an interior node. When the control volume adjoins a boundary, one of the cell faces will coincide with the boundary. In this case, Φ_{Ek} for the boundary face will take the value of Φ_B, where B is shown in Figure 6.9(b). For different types of boundaries, boundary conditions are different for different variables. Therefore, Equation 6.103 must be appropriately modified to take account of boundary conditions. This matter will be discussed in Section 6.3.10.

6.3.9 Evaluation of Nodal Gradients

To evaluate the D_k terms in Equation 6.103, Cartesian gradients of Φ must be evaluated (see Equations 6.90 to 6.93). This evaluation is carried out as follows:

$$\left.\frac{\partial \Phi}{\partial x_i}\right|_P = \overline{\left.\frac{\partial \Phi}{\partial x_i}\right|_P} = \frac{1}{\Delta V} \int_{\Delta V} \left.\frac{\partial \Phi}{\partial x_i}\right|_P dV. \tag{6.104}$$

The volume integral here can again be replaced by a line integral and subsequently by summation. Thus

$$\left.\frac{\partial \Phi}{\partial x_i}\right|_P = \frac{1}{\Delta V} \int_C \left(\beta_1^i\, \Phi\right)_c = \frac{1}{\Delta V} \sum_{k=1}^{NK} \left(\beta_1^i\, \Phi\right)_{ck}, \tag{6.105}$$

where

$$\Phi_{ck} = [f_{mc}\, \Phi_{E_2} + (1 - f_{mc})\Phi_{P_2}]_k$$
$$= [f_{mc}\,(\Phi_E + \Delta\Phi_E) + (1 - f_{mc})(\Phi_P + \Delta\Phi_P)]_k. \tag{6.106}$$

The appearance of $\Delta\Phi_P$ in Equation 6.106 suggests that Equation 6.105 is implicit in $\partial\Phi/\partial x_i$ (see Equation 6.90). However, since the overall calculation procedure is iterative, such implicitness is acceptable.

6.3.10 Boundary Conditions

To describe application of boundary conditions, consider a cell near a boundary (Figure 6.12) with face ab coinciding with the domain boundary. Note that

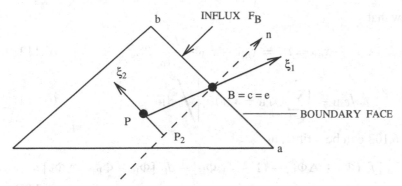

Figure 6.12. Line structure for a near-boundary cell.

a boundary node B has already been defined [see Figure 6.9(b)] such that

$$x_{i,B} = \frac{1}{2}(x_{i,a} + x_{i,b}). \tag{6.107}$$

Thus, since the boundary node is midway between a and b, from the construction shown in Figure 6.11, it is easy to deduce that points B, c, and e will coincide on the boundary face. Therefore, to represent transport at the boundary, an outward normal (shown by a dotted line) is drawn through B. Now, let line PP_2 be orthogonal to this normal and therefore parallel to ab. With this construction, the total outward transport through ab can be written as (see Equation 6.62)

$$(\vec{q} \cdot \vec{A})_B = C_B \, \Phi_B - (\Gamma \, A)_B \left. \frac{\partial \Phi}{\partial n} \right|_B, \tag{6.108}$$

where

$$C_B = \rho_B \left(\beta_1^1 \, u_1 + \beta_1^2 \, u_2 \right)_B, \tag{6.109}$$

$$C_B \, \Phi_B = C_B \left[f_B \, \Phi_{P_2} + (1 - f_B) \Phi_B \right]. \tag{6.110}$$

Now the cell-face normal gradient is represented by the first-order backward-difference formula

$$\left. \frac{\partial \Phi}{\partial n} \right|_B = \frac{(\Phi_B - \Phi_{P_2})}{l_{P_2 B}}. \tag{6.111}$$

In both Equations 6.110 and 6.111,

$$\Phi_{P_2} = \Phi_P + \Delta\Phi_P = \Phi_P + \vec{l}_{PP_2} \cdot \nabla \, \Phi_P = \Phi_P + \sum_{i=1}^{2} (x_{i,P_2} - x_{i,P}) \left. \frac{\partial \Phi}{\partial x_i} \right|_P. \tag{6.112}$$

It is easy to show that

$$x_{i,P_2} - x_{i,P} = l_{xi} = x_{i,B} - x_{i,P} - \frac{l_{P_2 B}}{A_B}\beta_1^i, \tag{6.113}$$

$$l_{P_2 B} = \left| \sum_{i=1}^{2}(x_{i,B} - x_{i,P})\beta_1^i \right| \bigg/ A_B. \tag{6.114}$$

Thus, Equation 6.108 can be written as

$$(\vec{q} \cdot \vec{A})_B = C_B \left[f_B(\Phi_P + \Delta\Phi_P) + (1 - f_B)\Phi_B \right] - d_B \left[\Phi_B - \Phi_P - \Delta\Phi_P \right], \tag{6.115}$$

where the diffusion coefficient is given by

$$d_B = \frac{\Gamma_B A_B}{l_{P_2 B}}. \tag{6.116}$$

Using Equation 6.115, implementation of boundary conditions for scalar and vector variables will be discussed separately.

Scalar Variables: For the near-boundary cell, Equation 6.103 is first rewritten as

$$\left[\rho_P^o \frac{\Delta V}{\Delta t} + \sum_{k=1}^{NK-B} AE_k \right]\Phi_P^{l+1} = \sum_{k=1}^{NK-B} AE_k\, \Phi_{Ek}^{l+1} + \rho_P^o \frac{\Delta V}{\Delta t}\Phi_P^o$$

$$+ S\,\Delta V + \sum_{k=1}^{NK-B} D_k^l - (\vec{q} \cdot \vec{A})_B \tag{6.117}$$

where $NK - B$ implies that the boundary face contribution is *excluded* from the summation and accounted for through the $-(\vec{q} \cdot \vec{A})_B$ term. This accounting can now also be done via Su and Sp as

$$Su - Sp\,\Phi_P = -(\vec{q} \cdot \vec{A})_B$$

$$= -C_B \left[f_B(\Phi_P + \Delta\Phi_P) + (1 - f_B)\Phi_B \right]$$

$$+ d_B \left[\Phi_B - \Phi_P - \Delta\Phi_P \right]. \tag{6.118}$$

Thus, when Φ_B is specified, it is possible to write

$$Su = -C_B \left[f_B\,\Delta\Phi_P + (1 - f_B)\Phi_B \right] + d_B \left[\Phi_B - \Delta\Phi_P \right],$$

$$Sp = C_B\, f_B + d_B. \tag{6.119}$$

Sometimes, boundary *influx* $F_B = \Gamma_B\, \partial\Phi/\partial n\,|_B$ is specified. Then, it can be shown that

$$Su = -C_B \left[f_B\,\Delta\Phi_P + (1 - f_B)\Phi_B \right] + F_B\, A_B,$$

$$Sp = C_B\, f_B. \tag{6.120}$$

These two types of scalar boundary conditions typically suffice to affect physical conditions at inflow, wall, exit, and symmetry boundaries of the domain.

Vector Variables: At inflow and wall boundaries, the velocities $u_{i,B}$ are known and, therefore, Equations 6.119 readily apply. Care is, however, needed when exit and symmetry boundary conditions are considered. Thus, at the *symmetry* boundary, the known conditions are

$$C_B = \rho_B \sum_{i=1}^{2} \beta_1^i u_{i,B} = 0, \qquad (6.121)$$

$$\left. \frac{\partial V_t}{\partial n} \right|_B = 0 \quad \text{or} \quad V_{t,B} = V_{t,P_2}, \qquad (6.122)$$

where V_t is the velocity tangential to face ab, which is therefore directed along ξ_2 (see Figure 6.12). Therefore, the unit tangent vector $\vec{t}$ can be written as

$$\vec{t} = \vec{i}\, l_{x_1} + \vec{j}\, l_{x_2}, \qquad (6.123)$$

where, l_{x_i} are given by Equation 6.87. Thus, the tangential velocity is given by $V_t = \vec{V} \cdot \vec{t} = \sum_{i=1}^{2} l_{x_i} u_i$ and Equation 6.122 can be written as

$$\sum_{i=1}^{2} l_{x_i} u_{i,B} = \sum_{i=1}^{2} l_{x_i} u_{i,P_2} = \sum_{i=1}^{2} l_{x_i} (u_{i,P} + \Delta u_{i,P}). \qquad (6.124)$$

Individual values of $u_{i,B}$ can now be determined from simultaneous solution of Equations 6.121 and 6.124.

At the *exit* boundary, boundary-normal gradients of both normal and tangential velocities are zero. Thus

$$\left. \frac{\partial V_t}{\partial n} \right|_B = 0 \quad \text{or} \quad V_{t,B} = V_{t,P_2}, \qquad (6.125)$$

$$\left. \frac{\partial V_n}{\partial n} \right|_B = 0 \quad \text{or} \quad V_{n,B} = V_{n,P_2}. \qquad (6.126)$$

Equation 6.125 is the same as Equation 6.122 and, therefore, Equation 6.124 readily applies. The normal velocity component, however, is $V_n = \vec{V} \cdot \vec{n}$ and Equation 6.126 will read as

$$\sum_{i=1}^{2} \beta_1^i u_{i,B} = \sum_{i=1}^{2} \beta_1^i u_{i,P_2} = \sum_{i=1}^{2} \beta_1^i (u_{i,P} + \Delta u_{i,P}). \qquad (6.127)$$

Again, the individual components $u_{i,B}$ can be determined from simultaneous solution of Equations 6.124 and 6.127.

6.3.11 Pressure-Correction Equation

In Chapter 5, the total pressure-correction equation in Cartesian coordinates was derived to read as

$$\frac{\partial}{\partial x_i}\left[\Gamma_i^{p'}\frac{\partial p'}{\partial x_i}\right]=\frac{\partial\left(\rho\,\overline{u}_i^l\right)}{\partial x_i}+\frac{\partial\rho}{\partial t}, \tag{6.128}$$

where

$$\Gamma_i^{p'}=\frac{\rho\,\alpha\,\Delta V}{AP^{u_i}}. \tag{6.129}$$

In this definition of $\Gamma^{p'}$, α and AP^{u_i} are, respectively, the underrelaxation factor and the AP coefficient used in the momentum equations. Invoking the Gauss theorem again, the discretised version of Equation 6.128 will read as

$$AP\,p'_{\mathrm{P}}=\sum_{k=1}^{NK}AE_k\,p'_{\mathrm{E}k}-\sum_{k=1}^{NK}C_{ck}-\left(\rho_{\mathrm{P}}-\rho_{\mathrm{P}}^{0}\right)\frac{\Delta V}{\Delta t}+\sum_{k=1}^{NK}D_k^{p'}, \tag{6.130}$$

where $AP=\sum_{k=1}^{NK}AE_k$ and

$$AE_k=d_{ck}^{p'}=\frac{(\Gamma^{p'}A)_{ck}}{l_{\mathrm{P_2E_2}}}. \tag{6.131}$$

Two comments are now important:

1. The $D_k^{p'}$ term in Equation 6.130 will contain Cartesian gradients of p'. However, during iterative calculation, since the pressure-correction equation is treated only as an estimator of p', $D_k^{p'}$ is set to zero.

2. Evaluation of $\Gamma_{ck}^{p'}$ in Equation 6.131 will require evaluation of ΔV and AP^{u_i} at the *cell face* (see Equation 6.129). The evaluation of cell-face volume can be accomplished via a fresh construction at the cell face as shown in Figure 6.13. The construction involves drawing lines parallel to ab passing through P_2 and E_2. Then, two lines parallel to normal $\vec{n}$ (and, hence, parallel to line P_2E_2) are drawn through a and b. The resulting rectangle c_1–c_2–c_3–c_4 will have volume

$$\Delta V_{ck}=l_{\mathrm{ab}}\times l_{\mathrm{P_2E_2}}\times 1=A_{ck}\,l_{\mathrm{P_2E_2}}. \tag{6.132}$$

Using this equation therefore gives

$$AE_k=\frac{\alpha\,(\rho\,A^2)_{ck}}{AP_{ck}^{u}}, \tag{6.133}$$

where $AP_{ck}^{u}=AP_{ck}^{u_1}=AP_{ck}^{u_2}$ can be evaluated from formula (6.66).[6]

---

[6] Alternatively, one may evaluate AP_{ck}^{u} exactly by carrying out a structured-grid-like discretisation over the control volume c_1–c_2–c_3–c_4. This is left as an exercise.

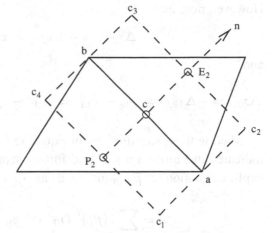

Figure 6.13. Construction of a cell-face control volume.

Thus, the final discretised pressure correction equation is

$$AP\, p_P' = \sum_{k=1}^{NK} AE_k\, p_{Ek}' - \sum_{k=1}^{NK} C_{ck} - (\rho_P - \rho_P^0)\frac{\Delta V}{\Delta t}, \qquad (6.134)$$

where AF_k is given by Equation 6.133. Equation 6.134 must be solved with $\partial p'/\partial n\,|_B = 0$, which can be accomplished simply by setting $AE_k = 0$ for the boundary face. After solving Equation 6.134, the mass-conserving pressure correction is recovered as $p_m' = p' - p_{sm}' = p' - 0.5(p' - \bar{p}')$.

Evaluation of $\bar{p}$

Recall that $\bar{p}_P = 0.5(\bar{p}_{x_1} + \bar{p}_{x_2})$, where $\bar{p}_{x_i}$ is determined from solution of $\partial^2 p/\partial x_i^2\,|_P = 0$. Thus $\bar{p}_{x_1}$, for example, is evaluated from

$$\frac{1}{\Delta V}\int \frac{\partial^2 p}{\partial x_1^2}\bigg|_P\, dV = \frac{1}{\Delta V}\int_C \beta_1^1\frac{\partial p}{\partial x_1}\bigg|_{ck} = \frac{1}{\Delta V}\sum_{k=1}^{NK}\beta_1^1\frac{\partial p}{\partial x_1}\bigg|_{ck} = 0. \quad (6.135)$$

Now, the pressure gradient at the cell face is evaluated by applying Gauss's theorem over the volume c_1–c_2–c_3–c_4. Then, it can be shown that

$$\frac{\partial p}{\partial x_1}\bigg|_{ck} = \frac{\Delta x_{2,E_2}\,p_{E_2} + \Delta x_{2,b}\,p_b + \Delta x_{2,P_2}\,p_{P_2} + \Delta x_{2,a}\,p_a}{\Delta V_{ck}}, \quad (6.136)$$

where

$$\Delta x_{2,E_2} = (x_{2,c_3} - x_{2,c_2}),$$

$$\Delta x_{2,b} = (x_{2,c_4} - x_{2,c_3}),$$

$$\Delta x_{2,P_2} = (x_{2,c_1} - x_{2,c_4}),$$

$$\Delta x_{2,a} = (x_{2,c_2} - x_{2,c_1}).$$

However, note that

$$\Delta x_{2,E_2} = -\Delta x_{2,P_2} = x_{2,b} - x_{2,a} = \beta_1^1$$

and

$$\Delta x_{2,a} = -\Delta x_{2,b} = x_{2,E_2} - x_{2,P_2} = x_{2,E_1} - x_{2,P_1} = \beta_1^2 \left| \sum_{i=1}^{2} \beta_1^i (x_{i,E} - x_{i,P}) \right| \bigg/ A_c^2.$$

Making these substitutions in Equation 6.136 and carrying out the summation indicated in Equation 6.135, and further separating out $\overline{p}_{x1,P} = p_P$, we obtain an explicit equation for $\overline{p}_{x1,P}$ that reads as $\overline{p}_{x_1,P} = A/B$, where

$$A = \sum_{k=1}^{NK} \left[\left(\beta_1^1 \right)^2 \left(p_E + \Delta p_E - \Delta p_P \right) \right] \bigg/ \Delta V_{ck}$$

$$- \sum_{k=1}^{NK} \left[\beta_1^1 \left(x_{2,E_2} - x_{2,P_2} \right) \left(p_b - p_a \right) \right] \bigg/ \Delta V_{ck}, \tag{6.137}$$

and

$$B = \sum_{k=1}^{NK} \left(\beta_1^1 \right)^2 \bigg/ \Delta V_{ck}, \tag{6.138}$$

where p_b and p_a are evaluated using Equations 6.92 and 6.93. Similarly, we obtain an equation for $\overline{p}_{x_2,P} = A/B$, where

$$A = \sum_{k=1}^{NK} \left[\left(\beta_1^2 \right)^2 \left(p_E + \Delta p_E - \Delta p_P \right) \right] \bigg/ \Delta V_{ck}$$

$$- \sum_{k=1}^{NK} \left[\beta_1^2 \left(x_{1,E_2} - x_{1,P_2} \right) \left(p_b - p_a \right) \right] \bigg/ \Delta V_{ck} \tag{6.139}$$

and

$$B = \sum_{k=1}^{NK} \left(\beta_1^2 \right)^2 \bigg/ \Delta V_{ck}. \tag{6.140}$$

6.3.12 Method of Solution

Our interest is in solving the set of equations (6.103) for all interior nodes P. Thus, if there are NE elements, there are NE equations for each variable. Again, equations for each variable are solved sequentially (see the next subsection). It has been noted that the AP coefficients will dominate over the neighbouring coefficients AE_k. But, the positions of AE_k in the coefficient matrix [A] will be arbitrary because of the manner in which neighbouring nodes are numbered during grid generation using ANSYS. This is unlike the case of structured grids (both Cartesian and curvilinear) where AP occupies the diagonal positions and the neighbouring coefficients occupy the off-diagonal positions, forming a pentadiagonal matrix (in the 2D case). It is this special feature of the structured grids that permitted employment of the ADI solution method.

The arbitrary [A] matrix formed on unstructured grids is called a *sparse* matrix. For such matrices, rapidly convergent methods such as conjugate-gradient (CG) and generalised minimal residual (GMRES) are available [3]. These methods are particularly attractive when the number of elements and, hence, the number of equations requiring simultaneous solutions are large. Description of these methods is considered beyond the scope of the present book. However, the diagonally dominant position occupied by the AP coefficient in our equations still permits employment of the simple point-by-point GS procedure. Thus, the equations can be solved by a simple routine as follows:

```
        DO 1 N = 1, NE
        SUM = SU(N)
        DO 2 K = 1, NK(N)
        NEBOR = NHERE(N, K)
2       SUM = SUM + AE(N, K) * FI(NEBOR)
        FI(N) = SUM / (AP(N) + SP(N))
1       CONTINUE
```

where NK(N) stores the number of neighbours of node N, NHERE (N, K) stores the element number of the *k*th neighbouring node of N, and source term SU (N) and AP (N) and SP (N) have already been calculated.

6.3.13 Overall Calculation Procedure

The important features of the overall calculation are described through the procedural steps that follow.

Preliminaries
1. Read element and vertex files. Determine neighbouring elements of each node N to form NHERE (N , K). This is done by searching the shared vertices between neighbouring elements. Note that there will be no neighbouring elements when a boundary face is encountered. At such a face, a boundary node is created and such nodes are identified with numbers NE + 1, NE + 2, etc., where NE are the total number of elements read from the element file. The coordinates of interior nodes are calculated using Equation 6.48 and of boundary nodes using Equation 6.107.
2. Tag the boundary nodes with identification numbers for inflow, symmetry, wall, and exit boundaries. Note that here boundary nodes rather than near-boundary cells are tagged. This is unlike the practice on structured grids.

3. Knowing coordinates of nodes and vertices, calculate $\beta_1^i, l_{x_i}, d_{x_i}$ for $i = 1, 2$ and $f_{m,c}$ and A_c for each face of every node. This is a once-and-for-all calculation and all these quantities are stored in two-dimensional arrays (N, K). In addition, ΔV is calculated for each cell.

Solution Begins

4. At a given time step, guess the pressure field p^l.

5. Solve Equation 6.103 for $\Phi = u_1^l$ and u_2^l. The solution is preceded by evaluation of AE (N, K) and AP (N), SP (N), and the entire source term SU (N) in Equation 6.103. It is assumed that SU and SP are appropriately modified to account for boundary conditions.

6. Perform a maximum of ten iterations on Equation 6.134 for p'. Here AE (N, K) are evaluated from Equation 6.133 and the source term containing mass fluxes is evaluated from Equation 6.65.

7. Recover the p'_m distribution from $p'_m = p' - 0.5\,(p^l - \overline{p}^l)$, where $\overline{p}^l$ is evaluated from Equations 6.137 to 6.140.

8. Apply pressure and velocity corrections at each node. Thus

$$p_P^{l+1} = p_P^l + \beta\,p'_{m,P}, \quad 0 < \beta < 1, \tag{6.141}$$

$$u_{i,P}^{l+1} = u_{i,P}^l - \frac{\alpha\,\Delta V}{AP^{u_i}}\left.\frac{\partial p'_m}{\partial x_i}\right|_P, \tag{6.142}$$

where the pressure gradient is evaluated using Equations 6.105 and 6.106. The mass-source residual R_m is evaluated from Equation 5.73, where AP and A_k coefficients are the same as in Equation 6.134.

9. Solve Equation 6.103 for all other relevant scalar Φs.

10. Check convergence by evaluating residual R_Φ via the imbalance in Equation 6.103 for each Φ as explained in Chapter 5. Special care is again needed in evaluation of the mass residual R_m. This is evaluated from the imbalance in Equation 6.134 in which p' is replaced by p'_m.

11. If the convergence criterion is not satisfied, treat $p^{l+1} = p^l$ and $\Phi^{l+1} = \Phi^l$ and return to step 5.

12. To execute the next time step, set all $\Phi^o = \Phi^{l+1}$ and return to step 4.

6.4 Applications

Flow over Banks of Tubes

In shell-and-tube heat exchangers, the flow on the shell side takes place over a bank of tubes several rows deep. The flow is aligned at various angles to the axis of the tubes. However, for preliminary design work, the flow may be assumed to be transverse to the axis (i.e., a cross flow). This configuration has been extensively researched and experimentally determined data are available [91] for different values

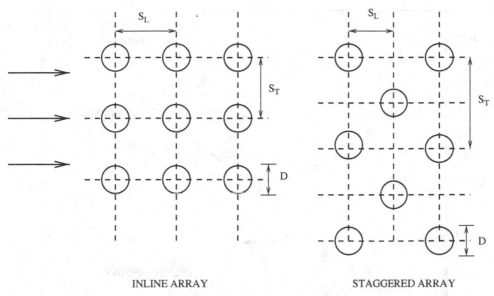

Figure 6.14. Flow across banks of tubes.

of *aligned* and *staggered* arrangement of tubes. The important geometric parameters are (see Figure 6.14) longitudinal pitch S_L, transverse pitch S_T, and tube
diameter D. Here, we consider cases of $S_L/D = S_T/D = 2$ for the inline array and
$S_L/D = S_T/D = 1.5$ for the staggered array.

For the purposes of computations, however, the smallest symmetric domain
must be considered. Such domains are mapped by curvilinear grids as shown in
Figure 6.15. In these domains, the north and south boundaries are partly symmetric
and partly occupied by tube wall but the west and east boundaries are *periodic*. Note,
however, that in the inline array, the periodicity is *even* whereas a *cross-periodicity*
occurs in the staggered array with respect to the u_2 velocity. Computations have
been performed using 45×15 grids for the inline array and 41×15 grids for the
staggered array. For turbulent flow, the standard HRE model with two-layer wall

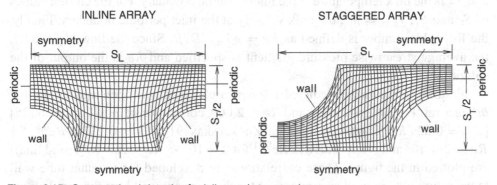

Figure 6.15. Computational domains for inline and staggered arrays.

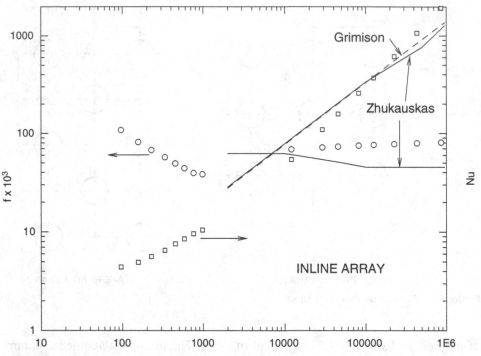

Figure 6.16. Variation of f and Nu with Re for $S_T/D = S_L/D = 2$.

functions has been used with one modification. Thus, in Equation 5.87, $(u^+ + PF)$ is replaced by $[\kappa^{-1} \ln(E\,y^+) + PF]$. All predictions are performed for $Pr = 0.7$ and a constant wall heat flux (q_w) boundary condition is assumed at the tube walls. For laminar flow, global underrelaxation is used to procure convergence whereas for turbulent flow, a *false transient* technique is used. The friction factor and Nusselt number are evaluated as

$$f = 0.5 \frac{dp}{dx} \frac{S_L}{\rho V_{max}^2}, \qquad Nu = \frac{h\,D}{K} = \frac{q_w\,D}{K\,(\overline{T}_w - T_{in})}, \qquad (6.143)$$

respectively, where $\overline{T}_w$ is the average wall temperature over forward and rear tubes and T_{in} is the bulk temperature at the inlet periodic boundary. For the chosen values of S_L and S_T, $V_{max} = \overline{u}_{in}$, the bulk velocity at the inlet periodic boundary. Finally, the Reynolds number is defined as $Re = \rho\,V_{max}\,D/\mu$. Since the flow is periodic, the average streamwise pressure gradient is specified and Re is the output of the solution.

Figure 6.16 shows the predicted f (open circles) and Nu (open squares) for the *inline* array. For the 2×2 array and $Re > 2,000$, correlations due to Grimison [25] [$Nu = 0.229\,Re^{0.632}$ (dotted line)] and Zhukauskas [91] [$Nu = 0.23746\,Re^{0.63}$ for $Re < 2 \times 10^5$ and $Nu = 0.01842\,Re^{0.81}$ for $2 \times 10^5 < Re < 2 \times 10^6$ (solid line)] are plotted in the figure. These correlations are developed for constant tube-wall temperature but are used as a reference for the constant wall heat flux predictions

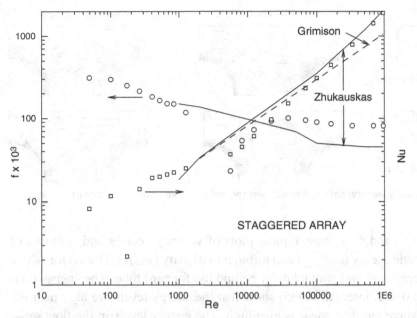

Figure 6.17. Variation of f and Nu with Re for $S_T/D = S_L/D = 1.5$.

considered here. It is well known that for near-unity Prandtl numbers, turbulent flow correlations are typically insensitive to the type of boundary condition. The figure shows that the present turbulent-flow Nu predictions are in good agreement with the correlations. Similar agreement is also obtained by Antonopoulos [2]. The friction-factor data of Zhukauskas are read from an available graph and are shown by a solid line. The presently predicted turbulent-flow friction-factor data are seen to be substantially above the experimental data. Unfortunately, predicted friction-factor data are not reported in [2]. In the laminar range, however, the friction-factor data show the expected steeper slope with Re but no correlations are available for comparison.

Figure 6.17 shows a similar comparison for the *staggered array*. Here again, the turbulent-flow friction-factor data show gross overprediction but Nu data are in excellent agreement with the correlation due to Zhukauskas. The laminar-flow Nu shows a peculiar decline at $Re \sim 120$. This is because of the change in the flow structure at this Reynolds number, which in turn alters the temperature distribution. For $Re < 120$, the maximum temperature occurs at the rear tube, whereas for $Re > 120$, the maximum temperature occurs at the forward tube.

In summary, we may state that for both inline and staggered arrays, the predicted turbulent Nu data are in good agreement with the experimental correlations but the predicted turbulent f data are in poor agreement with the Zhukauskas correlations. Although the latter correlations are taken as standard, it may be noted that there are other researchers whose experimental correlations for f are in much closer agreement with the present predictions.

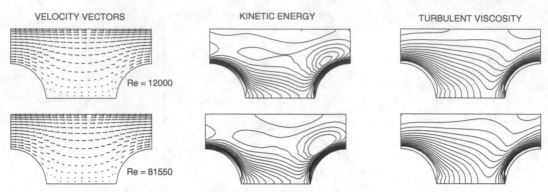

Figure 6.18. Velocity vectors, turbulent kinetic energy, and turbulent viscosity for an inline array.

Figures 6.18 and 6.19 show typical plots of velocity vectors and contours of turbulent kinetic energy ($e/V_{\max}^2$) and turbulent viscosity (μ_t/μ). The vectors show regions of separation and reattachment behind the forward tube. The energy contours (range: 0–0.1, interval: 0.005) show that the energy levels are high near the solid walls where the flow shear is also high. The energy levels in the flow separation region are not insignificant. For the inline array, the viscosity contours for $Re = 12,000$, (range: 0–400, interval: 20) and for $Re = 81,500$, (range: 0–3,000, interval: 150) show that turbulent viscosity is high near the walls, where kinetic energy is high. The levels of viscosity, however, increase with increase in Reynolds number as expected. The viscosity contours for a staggered array show similar trends. However, notice that at similar Reynolds numbers (for $Re = 12,417$, range: 0–200, interval: 10; for $Re = 10^5$, range: 0–2,000, interval: 100) the viscosity levels are lower than those found for the inline array.

Gas-Turbine Combustion Chamber
Flow in a gas-turbine combustion chamber represents a challenging situation in CFD. This is because the flow is three dimensional, elliptic, and turbulent and

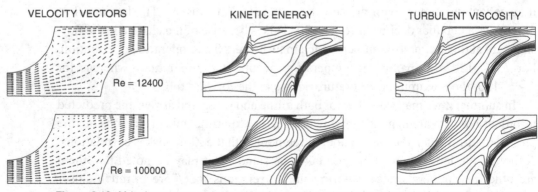

Figure 6.19. Velocity vectors, turbulent kinetic energy, and turbulent viscosity for a staggered array.

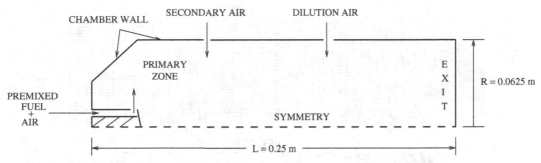

Figure 6.20. Idealised gas-turbine combustion chamber.

involves chemical reaction and the effects of radiation. In addition, the fluid properties are functions of both temperature and the composition of combustion products
and the true geometry of the chamber (a compromise among several factors) is
always very complex.

Figure 6.20 shows an *idealised* chamber geometry. The chamber is taken to
be axisymmetric of exit radius $R = 0.0625$ m and length $L = 0.25$ m. In actual
combustion, aviation fuel (kerosene) is used but we assume that fuel is vaporised
and enters the chamber with air in *stoichiometric* proportion. That is, 1 kg of
fuel is premixed with 17.16 kg of air. Thus, the stoichiometric air/fuel ratio is
$R_{\text{stoic}} = 17.16$. The fuel air mixture enters radially through a circumferential slot
(width $= 3.75$ mm located at $0.105\ L$) with a velocity of 111 m/s and a temperature
of $500°C$ (773 K). Additional air is injected radially through a cylindrical portion
(called casing) of the chamber through two circumferential slots.[7] The first slot
(width $= 2.25$ mm located at $0.335\ L$) injects air (called secondary air) to sustain a
chemical reaction in the primary zone; the second slot (width $= 2.25$ mm located
at $0.665\ L$) provides additional air (called dilution air) to dilute the hot combustion
products before they leave the chamber. The secondary air is injected with a velocity
of 48 m/s and a temperature of $500°C$. The dilution air is injected at 42.7 m/s and
$500°C$. The mean pressure in the chamber is 8 bar and the molecular weights of
fuel, air, and combustion products are taken as 16.0, 29.045, and 28.0, respectively.
The heat of combustion H_c of fuel is 49 MJ/kg.

With these specifications, we have a domain that captures the main features of
a typical gas-turbine combustion chamber. The top panel of Figure 6.21 shows the
curvilinear grid generated to fit the domain. In actual computations, the domain
is extended to $L = 0.8$ m to effect exit boundary conditions. A 50 (axial) $\times$ 32
(radial) grid is used. In this problem, inflow (at three locations), wall (west, north,
and part of south), symmetry, and exit boundaries are encountered. Equations for
$\Phi = u_1, u_2, p', e, \epsilon$, and T must be solved in an axisymmetric mode. In addition,

---

[7] In actual practise, radial injection is carried out through discrete holes. However, because accounting for this type of injection will make the flow three dimensional, we use the idealisation of a
circumferential slot.

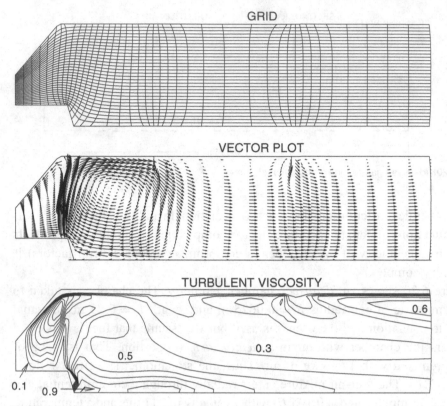

Figure 6.21. Grid and flow variables for a gas-turbine combustion chamber.

equations for scalar variables ω_{fu} and a composite variable $\Psi = \omega_{fu} - \omega_{air}/R_{stoic}$ must also be solved. The latter variable is admissible because a simple one-step chemical reaction,

$$(1) \text{ kg of fuel} + (R_{st}) \text{ kg of air} \rightarrow (1 + R_{st}) \text{ kg of products,}$$

is assumed to take place. Thus, there are eight variables to be solved simultaneously. The source terms of flow variables remain unaltered from those introduced in Chapter 5, but those of T, ω_{fu}, and Ψ are as follows:

$$S_{\omega_{fu}} = -R_{fu}, \quad S_{\Psi} = 0, \quad S_T = \frac{R_{fu} H_c}{C_p}, \quad R_{fu} = C \rho \omega_{fu} \frac{\epsilon}{e}, \quad (6.144)$$

where the volumetric fuel burn rate R_{fu} kg/m³-s is specified following Spalding [74] with $C = 1$. This model is chosen because it is assumed that the fuel-burning reaction is kinetically controlled[8] rather than diffusion controlled. Note that in the specification of S_T, the radiation contribution is ignored.

---

[8] Ideally, R_{fu} should be taken as the *minimum* of that given by expression (6.144) and the laminar Arrhenius expression for the fuel under consideration. Here, Equation 6.144 is used throughout the domain so that the burn rate is governed solely by the turbulent time scale ϵ/e. For further variations on Spalding's model, see [44, 24].

The combustion chamber walls are assumed adiabatic. The inflow boundary specifications, however, require explanation. At the primary slot, $u_1 = 0$, $u_2 = 100$, $e = (0.005 \times u_2)^2$, $\epsilon = C_\mu \rho e^2/(\mu R_\mu)$, where viscosity ratio $R_\mu = \mu_t/\mu = 10$, $T = 773$, $\omega_{\text{fu}} = (1 + R_{\text{stoic}})^{-1}$, and $\Psi = 0$. At secondary and dilution slots, $u_2 = -48$ and -42.5, respectively, and $e = (0.0085 \times u_2)^2$, $R_\mu = 29$, $T = 773$, $\omega_{\text{fu}} = 0$, and $\Psi = -1/R_{\text{stoic}}$ are specified. Finally, fluid viscosity is taken as $\mu = 3.6 \times 10^{-4}$ N-s/m^2 and specific heats of all species are assumed constant at $C_p = 1,500$ J/kg-K. The density is calculated from $\rho = 8 \times 10^5\, M_{\text{mix}}/(R_u T)$, where R_u is the universal gas constant, $M_{\text{mix}}^{-1} = \omega_{\text{fu}}/M_{\text{fu}} + \omega_{\text{air}}/M_{\text{air}} + \omega_{\text{pr}}/M_{\text{pr}}$, and the product mass fraction is $\omega_{\text{pr}} = 1 - \omega_{\text{fu}} - \omega_{\text{air}}$.

In this problem, the equations are strongly coupled and an initial guess for variables is difficult to determine a priori. To ensure convergence, therefore, the *false-transient* technique is used with $\Delta t = 10^{-5}$. Convergence is declared when residuals for all variables (except e and ϵ) are less than 10^{-3}. Further, it is ensured that the exit mass flow rate equals (within 0.1%) the sum of the three flow rates specified at the slots. A total of 12,500 iterations are required.

In the middle panel of Figure 6.21, the vector plot is shown. The plot clearly shows the strong circulation in the primary zone with a reverse flow near the axis necessary to sustain combustion. All scalar variables are now plotted as $(\Phi - \Phi_{\text{min}})/(\Phi_{\text{max}} - \Phi_{\text{min}})$ in the range 0–1 at a contour interval of 0.1. For turbulent viscosity, $\mu_{t,\text{min}} = 0$ and $\mu_{t,\text{max}} = 0.029$; for temperature, $T_{\text{min}} = 773$ K and $T_{\text{max}} = 2,456$ K (adiabatic temperature = 2,572 K); for fuel mass fraction, $\omega_{\text{fu,min}} = 0$ and $\omega_{\text{fu,max}} = 0.055066$, and for composite variable, $\Psi_{\text{min}} = -0.058275$ and $\Psi_{\text{max}} = 0$. The bottom of Figure 6.21 shows that high turbulent viscosity levels occur immediately downstream of the fuel injection slot and secondary and dilution air slots because of high levels of mixing.

The top panel of Figure 6.22 shows that the fuel is completely consumed in the primary zone. Sometimes, it is of interest to know the values of mixture fraction $f = f_{\text{stoic}} + \Psi(1 - f_{\text{stoic}})$, where $f_{\text{stoic}} = (1 + R_{\text{stoic}})^{-1}$. From the contours of Ψ shown in the middle panel of Figure 6.22, therefore, values of f and concentrations of air and products can be deciphered. The temperature contours shown on the bottom panel of Figure 6.22 are similar to those of Ψ. This is not surprising because although T is not a conserved property, enthalpy $h = C_p T + \omega_{\text{fu}} H_c$, like Ψ, is conserved and $\omega_{\text{fu}} \simeq 0$ over a greater part of the domain. The temperatures, as expected, are high in the primary zone and in the region behind the fuel injection slot, but the temperature profile is not at all uniform in the exit section. Combustion chamber designers desire a high uniformity of temperature in the exit section to safeguard the operation of the turbine downstream. Such a uniformity is often achieved by nonaxisymmetric narrowing of the exit section. However, accounting for this feature will make the flow three dimensional and hence is not considered here.

It must be mentioned that combustion chamber flows are extensively investigated through CFD for achieving better profiling of the casing, for determining

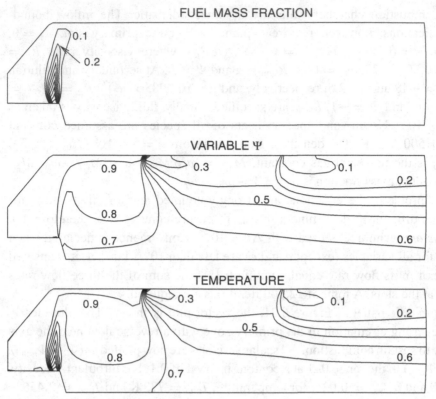

Figure 6.22. Scalar variables for a gas-turbine combustion chamber.

geometry of injection holes to achieve high levels of mixing, for determining exact location of injection ports to minimize NO_x formation, to achieve uniformity of exit temperatures, and to take account of liquid-fuel injection from burners and consequent fuel breakup into droplets.

Laminar Natural Convection in an Eccentric Annulus

Kuehn and Goldstein [37] measured heat transfer in horizontal eccentric cylinders (radius ratio $R_o/R_i = 2$) containing nitrogen ($Pr = 0.706$). The inner cylinder is maintained hot at temperature T_h and the outer cylinder is maintained at colder temperature T_c. The positive vertical eccentricity $\epsilon/L = 0.652$, where $L = R_o - R_i$. This problem has been computed by employing curvilinear grids by Karki and Patankar [32] and Ray and Date [58] among many others. Here, the problem is computed employing triangular (1,340 cells) as well as quadrilateral (1,320 cells) meshes as shown in Figure 6.23. The symmetry about the vertical axis is exploited. Corresponding to experimental conditions, the Rayleigh number $Ra = g\beta(T_h - T_c)L^3/(\nu\alpha) = 4.8 \times 10^4$ is chosen. At this value of Ra, the flow remains laminar in all regions of the cavity between the cylinders.

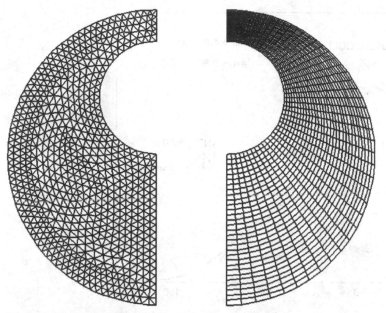

Figure 6.23. Unstructured meshes for natural convection in an eccentric annulus.

In [37], the experimental data are plotted in the form of a *local* conductivity ratio K_{eq}, which is defined as

$$K_{eq,i}(\theta) = \frac{q_{w,i}(\theta)R_i}{K(T_h - T_c)} \ln \frac{R_o}{R_i}, \qquad K_{eq,o}(\theta) = \frac{q_{w,o}(\theta)R_o}{K(T_h - T_c)} \ln \frac{R_o}{R_i}, \quad (6.145)$$

where $\theta = 0$ corresponds to the top of the cylinders and $\theta = 180$ refers to the bottom. The heat fluxes at the inner ($q_{w,i}$) and outer ($q_{w,o}$) cylinders are the output of the computed solution. Figure 6.24 shows a comparison of predicted and experimental (open symbols) data. At the inner hot cylinder, the computed data from the triangular mesh (solid lines) are in superior agreement with the experimental data than those obtained from the quadrilateral mesh (dotted lines). The reverse, however, is the case at the outer cold cylinder. The prediction of peak $K_{eq,o}$ at small angles (i.e., near the top) is in poor agreement with experimental data on both meshes. The cause of this discrepancy between predictions on the two meshes can be attributed to the small difference in the predicted recirculating flow structure (see Figure 6.25) near the top. This difference arises because, compared to the quadrilateral mesh, there are very few cells in the triangular mesh in the top region (see Figure 6.23). Also, the orientations of cell faces with respect to the local direction of the total velocity vector on the two meshes are different. Thus, although the UDS is employed in the calculations on both meshes, false-diffusion errors can be different. The effect of *flow angle* in causing false diffusion was discussed in Chapter 5. The disagreement with experimental data may be due to inadequate correspondence between experimentally and numerically realised boundary conditions

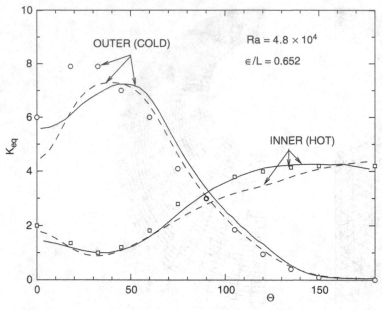

Figure 6.24. K_{eq} versus θ for natural convection in an eccentric annulus.

in this region. It must be mentioned, however, that the results with quadrilateral meshes compare extremely favourably with previous curvilinear grid predictions [32, 58]. It is for this reason that many CFD analysts prefer to use quadrilateral elements near curved surfaces while still employing triangular elements away from such surfaces. Thus, they prefer to use *mixed* elements for the domain as a whole.

Figure 6.25 shows the vector plots on the two meshes. It is seen that there is a strong upward flow near the hot inner cylinder where density is lower. Mass conservation, however, requires that circulation be set up with a downwards flow near the outer cylinder. There is, however, a region of weak contrarotating circulation

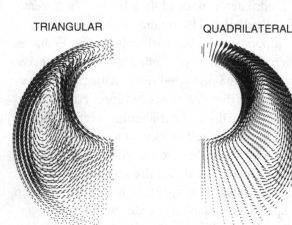

TRIANGULAR   QUADRILATERAL

Figure 6.25. Vector plots for natural convection in an eccentric annulus.

Figure 6.26. Temperature contours (range: 0–1; interval: 0.05) for natural convection in an eccentric annulus.

near the top of the cylinders and the region near the bottom is seen to be almost stagnant. Figure 6.26 shows the predicted isotherms on the two meshes. They are nearly identical. These isotherms corroborate the interferograms measured by Kuehn and Goldstein [37]. Finally, the angularly integrated average value of $\overline{K}_{eq}$ must be identical (so that overall heat balanced is checked) at both inner and outer surfaces of the cylinders. This value was computed at 2.68 on the quadrilateral mesh and at 2.79 on the triangular mesh.

2D Plane Convergent–Divergent Nozzle

Figure 6.27 shows a convergent–divergent plane nozzle whose width in the x_3 direction is large so that the flow may be considered 2D. The bottom boundary represents the axis (centerline) of the nozzle whereas the top boundary is a wall. The flow enters the left boundary and leaves through the right boundary. The total length L of the nozzle is 11.56 cm and the *throat* is midway. The half-heights of the nozzle at entry, throat, and exit are 3.52 cm, 1.37 cm, and 2.46 cm,

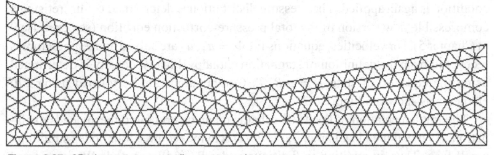

Figure 6.27. 2D plane convergent–divergent nozzle.

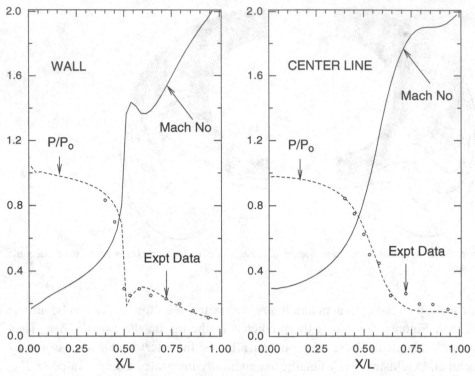

Figure 6.28. Variation of pressure and mach number in the nozzle.

respectively. The inlet Mach number is $M_{in} = 0.232$ and the exit *static* pressure is $p / p_0 = 0.1135$, where p_0 is the stagnation pressure. The stagnation enthalpy is assumed constant. For these specifications, experimental data are available [45]. This flow has been computed by Karki and Patankar [31] using curvilinear grids and the UDS scheme with $\mu = 0$ (i.e., Euler equations are solved). Here, the flow is computed using an unstructured mesh and the TVD scheme (Lin–Lin scheme, see Chapter 3) again with $\mu = 0$. At the inflow plane, since M_{in} is known, u_{in}, T_{in}, and p_{in} are specified using standard isentropic relationships [28]. At the exit plane, except for pressure (which is fixed), all other variables are extrapolated from the near-boundary node values. At the upper wall, a *tangency* condition is applied. This condition is the same as the symmetry condition. At the axis, the symmetry condition is again applied. The pressure distribution is determined by discretising a compressible flow version of the total pressure-correction equation (see exercise 9 in Chapter 5). For velocities, equations for $\Phi = u_1$, u_2 are solved and temperature is recovered from the definition of stagnation enthalpy. Finally, density is determined using the equation of state $p = \rho \, R_g \, T$. Computations are performed using 570 elements as shown in Figure 6.27.

The implementation of the TVD scheme on an unstructured mesh needs explanation. As mentioned in Chapter 3, the TVD scheme requires four nodes straddling a cell face. Thus, in addition to fictitious nodes P_2 and E_2, a node W_2 is selected

Figure 6.29. Mach number contours (range: 0.2–2.0, interval: 0.1) for a plane nozzle.

to the left of P_2 and a node EE_2 is selected to the right of node E_2. The locations of these nodes are such that $l_{c-P_2} = l_{P_2-W_2}$ and $l_{c-E_2} = l_{E_2-EE_2}$ where l is the length measured along the normal to the cell face (see Figure 6.11). Now, it is easy to work out the algebra of the TVD scheme in which $\Phi_{W_2} = \Phi_P + \bar{l}_{P-W_2} \nabla \Phi_P$ and $\Phi_{EE_2} = \Phi_E + \bar{l}_{E-EE_2} \nabla \Phi_E$.

Figure 6.28 shows the predicted variations of pressure (dashed line) and Mach numbers (solid line) at the upper wall and the centerline. The experimental data (open circles) for pressure have been read from a figure in [31]. It is seen that the agreement between experiment and predictions is satisfactory. Note that the predicted Mach number at the upper wall passes through $M = 1$ exactly at the throat ($X/L = 0.5$) and reaches a supersonic state $M = 2.01$ at exit. At the centerline, however, the $M = 1$ location is *downstream* of the throat. Computations of this type can be used to design a convergent–divergent nozzle to obtain a desired exit Mach number. Finally, Figure 6.29 shows the iso-Mach contours. Notice that the iso-Mach lines are slanted.

6.5 Closure

In this chapter, procedures for solution of transport equations on curvilinear and unstructured meshes have been described. By way of a closure, it will be useful to note a few important points.

1. Both procedures require special effort to generate curvilinear or unstructured grids. Some methods for grid generation are introduced in Chapter 8.
2. On curvilinear grids, the familiar (I, J) structure of Cartesian grids remains available. This permits adoption of the fast converging ADI method (as well as some others discussed in Chapter 9) for solution of discretised equations.
3. On unstructured grids, owing to lack of a regular node-addressing structure, a simple point-by-point GS method must be adopted for solution. It is well known that this method is slow to converge, but the convergence rate can be enhanced by adopting fast matrix-inversion techniques such as CG or GMRES.

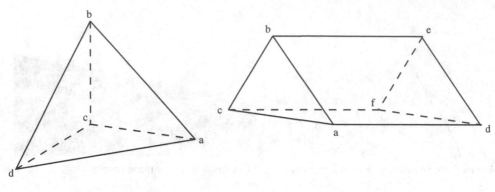

TETRAHEDRAL ELEMENT   PENTAHEDRAL ELEMENT

Figure 6.30. Some 3D polyhedral cells.

These techniques for sparse matrices become productive when the number of elements is large.

4. It may surprise the reader to note that the unstructured grid procedure is the most general. Since the procedure can handle any polygonal cells (in two dimensions), the Cartesian and curvilinear grids are already included. In the latter cases, however, the advantages of an (I, J) structure must be sacrificed.

5. The procedure for unstructured grids developed in this chapter can be straight-forwardly extended to 3D polyhedral cells (see Figure 6.30). The only difference in three dimensions is that all evaluations with $i = 1, 2$ must now be carried out over $i = 1, 2$, and 3. By way of illustration, consider the line structure at the triangular cell face of a tetrahedral cell shown in Figure 6.31(a). Two lines

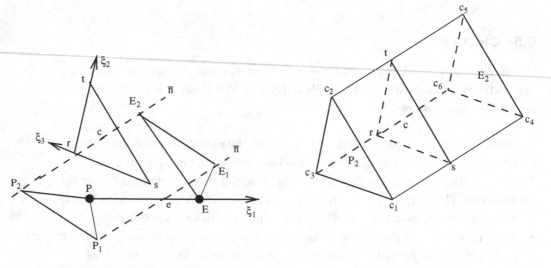

a) LINE STRUCTURE NEAR A CELL FACE   b) CELL FACE CONTROL VOLUME

Figure 6.31. Construction at the polygonal cell face.

normal to the cell face are drawn through c and e. Now, imagine a plane through P parallel to the cell face. This plane will orthogonally intersect the two normals at P_1 and P_2. A similar face-parallel plane through E will intersect the two normals at E_1 and E_2. Necessary evaluations of face-normal transport can now be carried out along the line P_2-c-E_2. Similarly, the construction of a *control volume* at the cell face is shown in Figure 6.31(b) when the cell face is triangular. To evaluate β_1^i, while direction ξ_1 is along PE, directions ξ_2 and ξ_3 may be chosen along any two sides of the triangle rst with origin at r, s, or, t. The actual directions are determined by requiring that Jacobian J be positive. Similarly, to affect vector boundary conditions, two tangent vectors $\vec{t}_1$ and $\vec{t}_2$ must be defined at the boundary cell face. Out of these, $\vec{t}_1$ (say) may be chosen along PP_2 and direction of $\vec{t}_2$ can be determined using the direction of the normal to the boundary cell face so as to form an orthogonal frame $\vec{t}_1, \vec{t}_2, \vec{n}$. The reader may find these figures useful for developing a 3D unstructured grid procedure [18].

6. Because of its generality, commercial codes are increasingly adopting unstructured grids. Although generality is welcome, the codes must rely heavily on polyhedral mesh generators as well as on creation of special routines for processing of computed results. Such postprocessors typically create contour, vector, and/or surface plots. For comparison of computed results with experimental data, however, one often needs to resort to *interpolations*. The reader will appreciate this difficulty because whereas most detailed measurements in a flow are carried out along a single straight line at a time, the grid nodes generated by packages such as ANSYS may not fall on a single line (in two dimensions) or even in a single plane (in three dimensions).

7. Despite the above-mentioned difficulty, unstructured grid codes are most versatile and, therefore, suitable for complex domains encountered in industrial and environmental applications.

EXERCISES

1. Derive expressions for β_j^i ($i = 1, 2, 3$ and $j = 1, 2, 3$) for a 3D curvilinear grid.

2. Using Equations 6.24 and 6.27, express dA_i and dV for a 3D curvilinear grid.

3. Starting with the p' equation in Cartesian coordinates (see Chapter 5), derive Equation 6.39. Identify the neglected terms in Equation 6.39 and explain how the effect of these terms can be recovered in a predictor–corrector fashion.

4. Analogous to Equation 6.42, derive an expression for $\overline{p}_{x_2,P}$.

5. Derive Equations 6.91, 6.92, and 6.93.

6. Derive Equation 6.113.

7. Using Equations 6.121 and 6.122, derive explicit symmetry boundary conditions for $u_{1,B}$ and $u_{2,B}$.

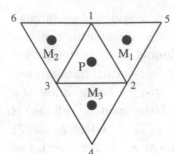

Figure 6.32. Neighbouring cells of an unstructured mesh.

8. Using Equations 6.125 and 6.126, derive explicit exit boundary conditions for $u_{1,B}$ and $u_{2,B}$.

9. A boundary receives radiant influx $F_B = \epsilon\sigma(T_\infty^4 - T_B^4)$. Derive expressions for Su and Sp for the node adjacent to this boundary and evaluate T_B.

10. Derive an exact expression for AP_{ck}^u by control-volume discretisation over cell-face control volume c_1–c_2–c_3–c_4 shown in Figure 6.13.

11. Show that $x_{2,E_2} - x_{2,P_2} = |\sum_{i=1}^{2}\beta_1^i(x_{i,E} - x_{i,P})|\beta_1^2/A_c^2$.

12. Verify Equations 6.139 and 6.140 in the evaluation of $\overline{p}_{x_2,P}$.

13. Starting with Equation 6.62, derive an expression for total convective–diffusive transport at the cell face of a tetrahedral element.

14. In Exercise 13, if the cell face were a boundary face, how would you determine the tangent vector $\vec{t}_2$ if $\vec{t}_1$ is along PP_2?

15. Carry out discretisation of convection terms using a TVD scheme on an unstructured mesh.

16. Consider node P surrounded by nodes M_1, M_2, and M_3 of an unstructured mesh shown in Figure 6.32. Each element is a perfect equilateral triangle (each side 1 cm). Table 6.2 gives coordinates of vertices surrounding these nodes. In a particular problem, the fluid properties ($\rho = 1.2$ kg/m^3 and viscosity $\mu = 15 \times 10^{-6}$ N-s/m^2) are assumed constant so that the equations for flow and energy transfer are decoupled. Steady state prevails. The converged velocity distributions (u and v) are shown in Table 6.3.

    Now, the energy equation is being solved and the prevailing temperatures at nodes neighbouring P are as shown in Table 6.3. Take $\Gamma^T = \mu/Pr$ with $Pr = 0.7$. The source term in the energy equation is zero. The convection

Table 6.2: Coordinates of vertices.

	1	2	3	4	5	6
x (cm)	0.5	1.0	0.0	0.5	1.5	−0.5
y (cm)	0.866	0.0	0.0	−0.866	0.866	0.866

Table 6.3: Current distribution of $u, v,$ and T.

Φ	P	M_1	M_2	M_3
u (m / s)	1.1	2.1	−0.3	−0.8
v (m / s)	−0.8	−1.0	−1.5	−0.8
T (°C)	?	65	80	72

terms are discretised using UDS. The equation is being solved with $\alpha_T = 1$. The objective of this problem is to determine T_P.

Tabulate intermediate calculations (in consistent units) to your answer in the form of Table 6.4 and, hence, determine T_P. Does T_P weigh heavily in favour of T_{M_2}? If yes, explain why.

17. An analyst computes flow over a cylinder placed between two parallel plates as shown in Figure 5.28 using an unstructured mesh. The objective is to predict the drag coefficient (C_D) of the cylinder as a function of Reynolds number. The definition of C_D is

$$C_D = \frac{F_{\text{pres}} + F_{\text{fric}}}{0.5 \rho U_o^2 A},$$

where F_{pres} and F_{fric} are net pressure and frictional forces, respectively, acting on the cylinder in the negative x_1 direction, U_o is the uniform axial velocity at the channel entrance, and the cylinder projected area $A = D \times 1$.

After solving for the flow, the analyst evaluates the forces as

$$F_{\text{pres}} = 2 \times \sum_{K_B} (p_B - \overline{p}_{\text{in}}) \beta_1^1,$$

$$F_{\text{fric}} = -2 \times \sum_{K_B} \mu \left[\frac{(u_1 + \Delta u_1)_P \, l_{x_1} + (u_2 + \Delta u_2)_P \, l_{x_2}}{l_{P_2 B} \sqrt{l_{x_1}^2 + l_{x_2}^2}} \right] \beta_1^1,$$

where $\overline{p}_{\text{in}}$ is the average pressure at the channel entrance and K_B are total number of cells near the cylinder boundary (see Figure 6.33). Examine whether the analyst's evaluations are correct.

18. In Exercise 17, heat transfer from the cylinder is considered with a constant wall temperature boundary condition. How will you evaluate *local* and

Table 6.4: Intermediate tabulation – energy equation.

Face k	β_1^1	β_1^2	f_m	A_{fk}	$l_{P_2 E_2}$	C_{ck}	f_{ck}	d_{ck}	AE_k
1									
2									
3									

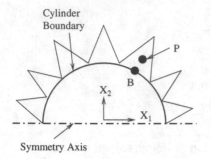

Figure 6.33. Cells near the cylinder boundary.

averaged heat transfer coefficients at the cylinder surface after a converged temperature solution is available? The temperature of the fluid entering the channel is T_{in} whereas the channel walls are maintained at T_{wc}. Write the expressions in discretised form. The heat transfer coefficient is defined as $h = q_w/(T_w - T_{ref})$. What should be the relevant reference temperature T_{ref} for this problem?

19. In the study of boundary layer development in the presence of favourable pressure gradients, an apparatus shown in Figure 6.34 is constructed. It is then assumed that in the presence of a sloping wall, the local free-stream velocity varies as $U_\infty(x) = U_0(1 + x/L)$. An analyst desires to verify this assumption by carrying out computation of the flow from entry to exit as an elliptic flow and allowing for the presence of the plate of thickness t. The following information is given: $U_0 = 1.8$ m/s, $L = 1$ m, $H = 0.7$ m, and air is at $30°C$ and 1 atm.

   (a) Write the equations and the boundary conditions governing the flow. Hence, identify the relevant Φs assuming turbulent air flow.

   (b) Which turbulence model will you use? HRE or LRE?

   (c) Which type of grid will you prefer? Curvilinear or unstructured?

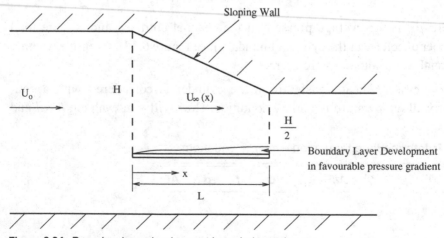

Figure 6.34. Boundary layer development in a wind tunnel.

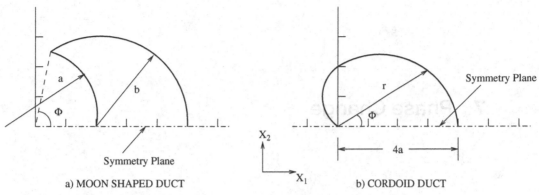

a) MOON SHAPED DUCT   b) CORDOID DUCT

Figure 6.35. Complex ducts.

20. Consider fully developed laminar flow through the two complex ducts shown in Figure 6.35. The flow is in the x_3 direction. The figure shows half cross sections in both cases with symmetry planes parallel to the x_1 axis. It is desired to predict $f \times Re$ for the ducts. The geometric details are as follows:

$$\text{Moon-shaped duct: } a = b = 3 \text{ units and } \Phi = 60°,$$

$$\text{Cordoid duct: } r = 2a\,(1 + \cos \Phi), \, a = 2 \text{ units}, \, 0 < \Phi < \pi.$$

What type of grid will you prefer for computation? Curvilinear or unstructured? Draw a hand sketch to explain the reasons for your choice.

7   Phase Change

7.1   Introduction

There is hardly a product that, during its manufacture, does not undergo a process of melting and solidification. Engineering processes such as casting, welding, surface hardening or alloying, and crystallisation involve *phase change*. The processes of freezing and thawing are of interest in processing of foods. Phase-change materials (PCMs) are used in energy storage devices that enable storage and retrieval of energy at nearly constant temperature.

The phenomenon of melting or solidification is brought about by a process of *latent heat* (λ) transfer at the *interface* between solid and liquid phases. For a *pure substance*, throughout this process, the temperature T_m (melting point) of the interface remains constant whereas in the liquid and solid phases, the temperatures vary with time. Both λ and T_m are properties of a pure substance. Within each of the single phases, heat transfer is essentially governed by a process of unsteady heat conduction, although, under certain circumstances, convection may also be present in the liquid phase under the action of body (buoyancy, for example) or surface (surface tension) forces.

There are two approaches to solving phase-change problems:

1. the variable domain formulation and
2. the fixed domain (or fixed-grid) formulation.

In the first approach, which has several variants, two energy equations are solved in the solid and the liquid phases with temperatures T_s and T_l, respectively, as dependent variables. In addition to the initial (i.e., at $t = 0$) and the domain boundary conditions, the following *interface* conditions are also invoked to match the temperatures of the two phases:

$$T_s = T_l = T_m, \tag{7.1}$$

$$k_s \frac{\partial T_s}{\partial n}\bigg|_i - k_l \frac{\partial T_l}{\partial n}\bigg|_i = \rho \lambda V_i, \tag{7.2}$$

where n is normal to the interface and V_i is the instantaneous velocity of the interface in the direction of the normal. In a finite domain, the solid and liquid regions thus enlarge or contract as time progresses. Hence, we use the designation *variable domain formulation*. The interface, of course, moves through the domain and, at a given instant, may assume arbitrary shape. The arbitrariness may arise from the boundary shape, boundary conditions, or the presence of convection in the liquid phase. The variable domain formulation thus requires *tracking* of the interface location at every instant of time to effect condition (7.2). In complex three-dimensional domains, such tracking can turn out to be very cumbersome.

In this chapter, only the fixed domain formulation will be considered. This formulation treats *enthalpy h* (sensible + latent heat) rather than temperature T as the main dependent variable in the energy equation. In the absence of internal heat generation, this equation can be written as

$$\frac{\partial(\rho h)}{\partial t} + \frac{\partial}{\partial x_j}(\rho u_j h) = \frac{\partial}{\partial x_j}\left(K\frac{\partial T}{\partial x_j}\right), \tag{7.3}$$

where the velocity u_j may be finite only in the liquid phase and zero in the solid phase. The equation is applicable to both solid and liquid phases and, therefore, to the entire domain including the interface. Thus, the interface condition (7.2) is already satisfied. Equation 7.3, however, contains two dependent variables (h and T) and a set of relations (known as the equations of state) between them must be specified. With this specification, the equation can be readily adapted to computations on a *fixed* grid through which the interface moves with time. Thus, the phase-change problems too can be computed with a generalised computer code. This fixed-grid formulation is also referred to as the *enthalpy* formulation in the literature.

There are a variety of phase-change problems. For example, in casting, only the total solidification time may be of interest; the domain is finite. In such problems, the interface need not be explicitly tracked. In contrast, in problems such as welding and surface hardening, it is important to identify the heat-affected zone and interface tracking is essential. In impure materials and alloys, latent heat transfer takes place over a range of temperatures ($T_m - \epsilon < T < T_m + \epsilon$) that demarcate what is known as the *mushy* zone. The properties of the mushy zone, however, must be known or modelled. There are other problems in which the thermo-physical properties of the two phases not only are different (ice water, for example) but are nonlinear functions of temperature, concentration, velocity gradients (in liquid phase), and/or local porosities. Equation 7.3 can readily capture such a variety.

The problem of solving Equation 7.3 through discretised equations is not straightforward; therefore, in the next two sections, only 1D problems will be considered to explain the main ideas. This will provide sufficient grounding to the reader to understand extensions to multidimensions through indicated references.

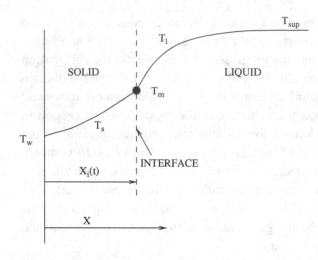

Figure 7.1. 1D phase-change problem.

7.2 1D Problems for Pure Substances

7.2.1 Exact Solution

It is important to note that there are very few exact solutions to phase-change problems even in one dimension. To appreciate the nature of the solution, consider the problem shown in Figure 7.1. An initially $(t = 0)$ superheated liquid $(T_{sup} > T_m)$ in a semi-infinite domain is subjected to temperature T_w $(< T_m)$ at $x = 0$ and this temperature is maintained for all times $t > 0$. Solidification commences instantly and the interface moves to the right. The instantaneous location of the interface $X_i(t)$ is shown in the figure. The task is to predict velocity $dX_i(t)/dt$ as a function of time and the temperature distributions in each phase as a function of x and t.

The governing equation for this problem will be

$$\frac{\partial(\rho h)}{\partial t} = \frac{\partial}{\partial x}\left(K\frac{\partial T}{\partial x}\right), \tag{7.4}$$

with $T(x, 0) = T_{sup}$, $T(0, t) = T_w$, and $T(\infty, t) = T_{sup}$. The liquid is of course stagnant. The exact solution for this problem was developed by von Neumann [23]. The solutions for the solid and liquid phases read as

$$\frac{T_s - T_m}{T_w - T_m} = 1 - \frac{\mathrm{erf}(x/\sqrt{4\,\bar{\alpha}_s\,t})}{\mathrm{erf}(X_i/\sqrt{4\,\bar{\alpha}_s\,t})}, \tag{7.5}$$

$$\frac{T_l - T_m}{T_{sup} - T_m} = 1 - \frac{\mathrm{erfc}(x/\sqrt{4\,\bar{\alpha}_l\,t})}{\mathrm{erfc}(X_i/\sqrt{4\,\bar{\alpha}_l\,t})}, \tag{7.6}$$

where α is the thermal diffusivity and suffixes s and l refer to solid and liquid phases, respectively. Now, since these solutions hold for all values of X_i, by inspection, we must have

$$X_i \propto \sqrt{t} \quad \text{or} \quad X_i = C\sqrt{t}, \tag{7.7}$$

where C can be determined from the interface condition (7.2). The transcendental equation for determination of C thus becomes

$$\frac{\rho \lambda C}{2} = \frac{T_m - T_w}{\text{erf}(C/\sqrt{4\alpha_s})} \frac{K_s}{\sqrt{\pi \alpha_s}} \exp(-C^2/4\alpha_s)$$

$$+ \frac{T_m - T_{sup}}{\text{erfc}(C/\sqrt{4\alpha_l})} \frac{K_l}{\sqrt{\pi \alpha_l}} \exp(-C^2/4\alpha_l). \tag{7.8}$$

This transcendental equation shows that $C = C(T_m - T_w, T_m - T_{sup}, K_s, K_l, \alpha_s, \alpha_l)$. Thus, C will be different for each initial and boundary condition and for each specification of physical properties. The value of C must be iteratively determined to calculate $d X_i (t)/d t$ from Equation 7.7 and hence to calculate the temperature as a function of x and t from Equations 7.5 and 7.6. It can be shown that the system is governed by a dimensionless number, called the Stefan number, which is defined as

$$St = \frac{C_{ps}(T_m - T_w)}{\lambda}. \tag{7.9}$$

The larger the value of St, the faster is the interface movement. A further point to note is that, although the temperature profiles show discontinuity at the interface, they are smooth within each phase and the variation of T with t at any x is also continuous and smooth.

7.2.2 Simple Numerical Solution

It might appear that it is a straightforward matter to discretise Equation 7.4 to obtain a numerical solution. However, there is a difficulty associated with predicting continuous temperature histories when a numerical solution is obtained. To appreciate the difficulty, we assume uniform and equal properties for both phases (i.e., $\rho_s = \rho_l = \rho$, $C_{ps} = C_{pl} = C_p$, and $K_s = K_l = K$). Thus, Equation 7.4 can be written as

$$\frac{\partial \Phi}{\partial \tau} = \frac{\partial^2 \theta}{\partial X^2}, \tag{7.10}$$

where

$$\Phi = \frac{h - h_s}{\lambda} \quad \text{(dimensionless enthalpy)}, \tag{7.11}$$

$$\theta = \frac{C_p(T - T_m)}{\lambda} \quad \text{(dimensionless temperature)}, \tag{7.12}$$

$$\tau = \frac{\alpha t}{L^2} \quad \text{(dimensionless time)}, \tag{7.13}$$

$$X = \frac{x}{L} \quad \text{(dimensionless length)}. \tag{7.14}$$

Table 7.1: Equations of state.

State	$T = f(h)$	$h = f(T)$
Solid	$T = h/C_p$ for $h < h_s$	$h = C_p T$ for $T < T_m$
Liquid	$T = (h - \lambda)/C_p$ for $h > h_l$	$h = C_p T + \lambda$ for $T > T_m$
Interface	$T = T_m$ for $h_s < h < h_l$	$h = C_p T_m + h_{ps}(t)$ $\int_t^{t+\Delta t} (d h_{ps}/d t)\, d t = \lambda$

In these equations, L is the domain length where the boundary condition corresponding to $x = \infty$ is specified and $h_s = C_p T_m$ is the *solidus* enthalpy. There are two ways to connect h to T (or Φ to θ) via the equations of state, as shown in Table 7.1 and Figure 7.2. In Table 7.1, $h_l = C_p T_m + \lambda$ is the liquidus enthalpy and $h_{ps}(t)$ is the *psuedo-enthalpy* in whose definition Δt is not a priori known.

When $h = f(T)$ relationships are used, clearly one would require a procedure for determining the integral constraint at the interface. Such a procedure is developed in [85]. We shall, however, consider $T = f(h)$ relationships so that

$$\theta = \Phi \quad \text{for} \quad \Phi \le 0 \quad \text{(solid)}, \tag{7.15}$$

$$\theta = 0 \quad \text{for} \quad 0 \le \Phi \le 1 \quad \text{(interface)}, \tag{7.16}$$

$$\theta = \Phi - 1 \quad \text{for} \quad \Phi \ge 1 \quad \text{(liquid)}. \tag{7.17}$$

Now, assuming the IOCV method and using a uniform grid, it is a simple matter to show that

$$\Phi_j^{l+1} = \frac{\Delta \tau}{\Delta X^2} \left(\theta_{j+1}^{l+1} - 2\theta_j^{l+1} + \theta_{j-1}^{l+1} \right) + \Phi_j^o, \tag{7.18}$$

where superscript n is dropped for convenience, but superscript $l + 1$ is retained to indicate that Equation 7.18 must be solved iteratively to satisfy the equations of

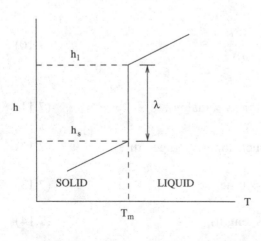

Figure 7.2. Equation of state for a pure substance.

state. The overall calculation procedure will be as follows:

1. At $\tau = 0$, specify initial condition θ_j^o for $j = 1$ to N. Hence, evaluate Φ_j^o. Set $\theta_j = \theta_j^o$.
2. Choose $\Delta\tau$ to begin a new step.
3. Solve Equation 7.18 once using the GS method to obtain the Φ_j^{l+1} distribution.
4. Determine θ_j^{l+1} using equations of state (7.15) to (7.17) and return to step 3 to carry out the next iteration.
5. After a few iterations, the change in Φ_j^{l+1} between successive iterations will be small and convergence is obtained.
6. Set $\Phi_j^o = \Phi_j$ and return to step 2 to execute the next time step.

Problem 1

To appreciate the nature of the numerical solution, consider a problem with the following specifications:

$$\rho = 1\,\text{kg/m}^3,\ C_p = 2.5\,\text{MJ/kg-K},\ K = 2W/m\text{-}K,$$

$$\lambda = 100\,\text{MJ/kg},\ T_m = 0°C,\ L - 1m,$$

$$T(x, 0) = T_{\text{sup}} = 2°C,\ \ \text{and}\ \ T_w = T(0, t) = -10°C$$

For this problem $St = 0.25$ and, as evaluated from Equation 7.8, $C = 5.767 \times 10^{-4}$. A numerical solution is executed with initial conditions $\theta(\tau = 0) = 0.05$ and $\Phi(\tau = 0) = 1.05$. The boundary condition is $\theta(X = 0) = -0.25$. The time step is determined from $\Delta\tau/\Delta X^2 = 0.2$ and the computations are carried out till $\tau = 1.6$ (or nearly 23 days). Two grid spacings are considered: $\Delta X = 0.2$ ($N = 7$) and $\Delta X = 0.0769$ ($N = 15$). At each time step, a converged solution is obtained in 5–11 iterations. The exact and the numerical solutions for temperature at $x = 0.5$ m are plotted in Figure 7.3 as a function of time. The figure shows a *wavy* temperature history. The waviness, however, decreases with refinement of the grid size. When ΔX is reduced still further so that $N = 51$ (say), the results (not shown) indicate that the exact and the numerical solutions nearly coincide. That is, the essentially wavy solution now appears smooth, albeit at the expense of significantly increased computer time. A few comments are therefore in order:

1. The numerical procedure is very simple and can be easily extended to multidimensional problems. However, to obtain non-wavy solutions, an extremely fine mesh size is required. This can be very uneconomical.
2. Why does waviness occur? This can be appreciated from Figure 7.4, where a phase-change node j is considered. When the interface resides within the control volume surrounding node j (so that $0 < \Phi_j < 1$), $\theta_j = 0$ (see Equation 7.16).

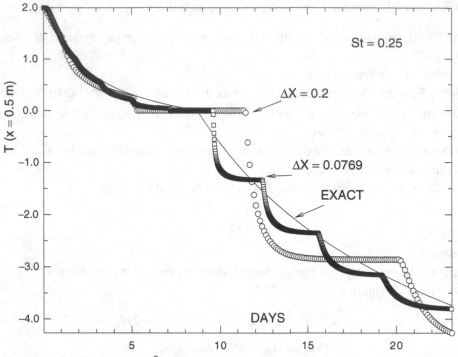

Figure 7.3. Solution for $\Delta\tau/\Delta X^2 = 0.2$.

Thus, throughout the period of interface transit through the control volume, the nodal temperature at the phase-change node remains *stationary* at $\theta_j = 0$. As a result, the temperature history demonstrates a wavy pattern. However, when $\Delta x \to 0$ (or grid spacing is reduced) the transit time itself is reduced and hence the predicted history appears smooth.

3. The calculation procedure, of course, *necessitates* a point-by-point GS iteration method for solution of Equation 7.18. This is because *bookkeeping* is required in step 4 of the procedure for each node to identify whether the node is in solid ($\Phi_j < 0$), in liquid ($\Phi_j > 1$), or undergoing phase change ($0 < \Phi_j < 1$). This bookkeeping can again be expensive in terms of computer time. It also prevents use of a line-by-line procedure such as the TDMA.

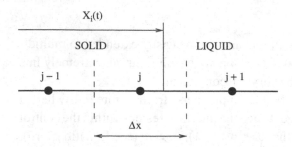

Figure 7.4. Typical phase-change node.

4. The interface location can be identified from the location of $\theta = 0$, but, as already explained, this will again predict a wavy interface history. Instead, one may use variable Φ to predict the interface history. This is because Φ_j is nothing but the *liquid fraction* of the control volume surrounding phase-change node j. Thus, at any time instant, one may simply add ΔX for all nodes for which $\Phi_j < 0$ (i.e., solid nodes) and further add $(1 - \Phi_j) \Delta X$ for the node for which $0 < \Phi_j < 1$ and ignore all nodes for which $\Phi_j > 1$. The sum will readily predict the instantaneous value of X_i and this prediction will appear smooth but not accurate on a coarse grid. This alternative procedure will again require bookkeeping.

These comments indicate that the simple procedure needs refinement in terms of both economy and convenience.

7.2.3 Numerical Solution Using TDMA

To eliminate the bookkeeping requirements, the $\theta \sim \Phi$ relations (7.15) to (7.17) must be generalized [11] by writing

$$\theta = \Phi + \Psi', \tag{7.19}$$

where

$$\Phi' = \frac{1}{2} \left[|1 - \Phi| - |\Phi| - 1 \right]. \tag{7.20}$$

Equation 7.20 ensures that $\Phi' = 0$ in solid ($\Phi < 0$), $\Phi' = -\Phi$ during phase change ($0 < \Phi < 1$), and $\Phi' = -1$ in liquid ($\Phi > 1$). Using Equation 7.19, we can reexpress Equation 7.10 as

$$\frac{\partial \Phi}{\partial \tau} = \frac{\partial^2 \Phi}{\partial X^2} + \frac{\partial^2 \Phi'}{\partial X^2} \tag{7.21}$$

and the discretised version will read as[1]

$$\left(1 + 2 \frac{\Delta \tau}{\Delta X^2} \right) \Phi_j^{l+1} = \frac{\Delta \tau}{\Delta X^2} \left(\Phi_{j+1}^{l+1} + \Phi_{j-1}^{l+1} \right)$$

$$+ \frac{\Delta \tau}{\Delta X^2} \left(\Phi_{j+1}' - 2\Phi_j' + \Phi_{j-1}' \right) + \Phi_j^o, \tag{7.22}$$

where Φ' values lag behind Φ values by one iteration. Thus, in step 4 of the simple numerical procedure described in the previous subsection, Φ_j' (rather than θ_j) are evaluated using Equation 7.20 and the bookkeeping requirement is eliminated. The

---

[1] It is assumed that the reader will be able to make necessary changes to the discretised equation for $j = 2$ and $j = N - 1$ nodes to account for any type of boundary condition.

introduction of the variable Φ' yields two further advantages:

1. The terms containing Φ' and Φ° can be treated as sources. Thus, at the current iteration level, Equation 7.22 can be solved by TDMA. This can achieve considerable economy in computer time. For example, for the problem considered in the previous subsection, with $N = 51$, the TDMA solution turns out to be nearly 2.5 times faster than the GS solution.
2. It is easy to recognize that at each time step, when a converged solution is obtained, $X_i(\tau)$ can be estimated from the simple formula

$$X_i = \sum_{j=2}^{N-1} (1 + \Phi'_j) \Delta X. \qquad (7.23)$$

This is because $(1 + \Phi'_j)$ represents the solid fraction for each node j. Again, the bookkeeping requirement is eliminated.

Although useful for obtaining faster solutions on fine grids, the introduction of the Φ' variable does not eliminate the problem of wavy temperature histories on coarse grids. This is because the replacement indicated in Equation 7.19 still renders $\theta = 0$ at the phase-change node $(0 < \Phi_j < 1)$. In the next subsection, it will be shown that accurate solutions can be obtained even on coarser grids while still employing the TDMA procedure. Thus, we seek an economic solution that combines the beneficial effects of computations at fewer nodes with the speed of the line-by-line procedure.

7.2.4 Accurate Solutions on a Coarse Grid

To prevent θ from remaining stationary at zero at the phase-change node, Equation 7.19 is rewritten as

$$\theta = \Phi + \Phi'', \qquad (7.24)$$

where

$$\Phi'' = \Phi' + \theta_{pc}, \qquad (7.25)$$

with, θ_{pc} denoting the nodal value of θ at the phase-change node $0 < \Phi_j < 1$. Note that $\theta_{pc} = 0$ at all single phase nodes. Making these substitutions in Equation 7.18 leads to

$$\left(1 + 2\,\frac{\Delta\tau}{\Delta X^2}\right)\Phi_j^{l+1} = \frac{\Delta\tau}{\Delta X^2}\left(\Phi_{j+1}^{l+1} + \Phi_{j-1}^{l+1}\right)$$

$$+ \frac{\Delta\tau}{\Delta X^2}\left(\Phi''_{j+1} - 2\,\Phi''_j + \Phi''_{j-1}\right) + \Phi_j^\circ. \qquad (7.26)$$

This equation is the same as Equation 7.22 except that Φ' is replaced by Φ'' and the latter will again lag behind Φ by one iteration. Equation 7.24 is therefore amenable to solution by TDMA.

To make further progress, a procedure for evaluating Φ'' or, in effect, θ_{pc} must be set out since Φ' can be evaluated from its definition (7.20). Thus, consider Figure 7.4 again and define

$$\Delta X_i = X_i - X_j, \tag{7.27}$$

where X_i is the location of the interface where θ is truly zero and X_j is the coordinate of node j. At the time instant considered in the figure, therefore, ΔX_i is positive and we may evaluate $\theta_{pc,j}$ by *linear* interpolation as

$$\theta_{pc,j} = \left[\frac{\Delta X_i}{\Delta X_i + \Delta X} \right] \theta_{j-1}. \tag{7.28}$$

At another earlier time instant, ΔX_i may be negative ($X_i < X_j$) and we may write

$$\theta_{pc,j} = \left[\frac{|\Delta X_i|}{|\Delta X_i| + \Delta X} \right] \theta_{j+1}. \tag{7.29}$$

Note, however, that for both positive or negative values of ΔX_i

$$\Delta X_i = X_i - X_j = (0.5 - \Phi_{pc,j}) \Delta X = (0.5 + \Phi'_{pc,j}) \Delta X \tag{7.30}$$

since $\Phi_{pc,j} = -\Phi'_{pc,j}$ at the phase change node. Equations 7.28 and 7.29 therefore can be generalised to read as

$$\theta_{pc,j} = \frac{F}{2} \left[(A + |A|) \theta_{j-1} - (A - |A|) \theta_{j+1} \right], \tag{7.31}$$

where

$$A = \frac{0.5 + \Phi'_{pc,j}}{|0.5 + \Phi'_{pc,j}| + 1} \tag{7.32}$$

and

$$F = -\frac{(1 + \Phi'_j) \Phi'_j}{(1 - \Phi_j) \Phi_j}. \tag{7.33}$$

In these equations, $F = 0$ at the single phase nodes (rendering $\theta_{pc} = 0$) but $F = 1$ at the phase-change node as desired. Thus, the phase-change node temperature can be evaluated without bookkeeping. Therefore, in step 4 of our calculation procedure, Φ''_j is also evaluated without bookkeeping.

Problem 1 of Section 7.2.2 is now solved again for the *coarse* grid with $N = 7$ (or $\Delta X = 0.2$) for $St = 0.25$, and the predicted temperature history is shown in Figure 7.5. Now, even the coarse grid solution is nearly accurate. In the same figure, computations for $St = 1$ ($C = 1.075 \times 10^{-3}$) and $St = 3$ ($C = 1.6 \times 10^{-3}$) are also shown and the grids used are indicated in the figure. Again, smooth histories

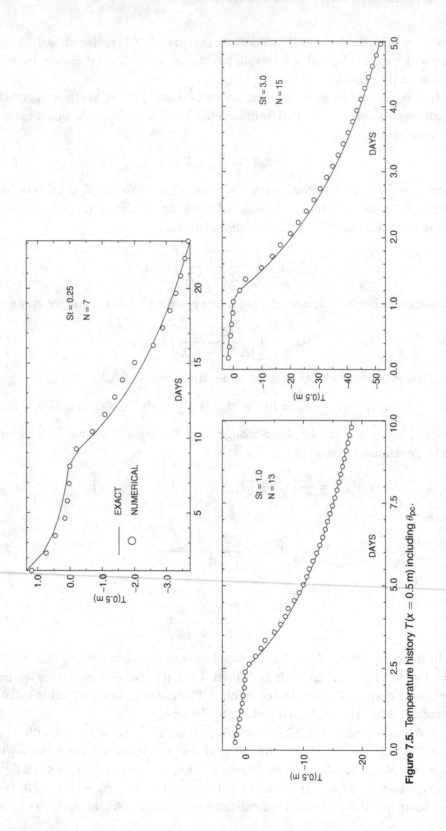

Figure 7.5. Temperature history $T(x = 0.5\,\text{m})$ including θ_{pc}.

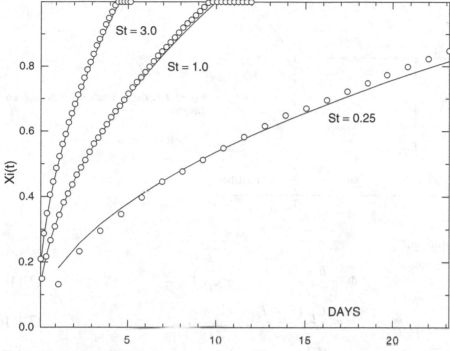

Figure 7.6. Solutions for $X_i(t)$.

are predicted that agree with the exact solution well. In each case, solutions are obtained with $\Delta\tau/\Delta X^2 = 2$, which is 10 times larger than that used in Figure 7.3. Thus, inclusion of $\theta_{pc,j}$ permits the use of coarse grids and allows large time steps and yet yields accurate solutions. This finding is particularly important for multi-dimensional problems. Figure 7.6 shows the variation of X_i (as calculated using Equation 7.23) with time. It is seen that as the Stefan number increases, the interface moves faster. Notice that for $St = 1$ and $St = 3$, the computations are carried on even after the complete domain is solidified; hence, the interface location appears to remain stationary at 1 m.

7.3 1D Problems for Impure Substances

In impure materials or alloys, phase change takes place over a range of temperatures $T_s < T < T_l$ where T_s and T_l may be termed as solidus and liquidus temperature, respectively. Here, we shall permit different properties of solid and liquid phases. The $h \sim T$ relation, therefore, may appear as shown in Figure 7.7. In this figure, the region (also called the *mushy* region) between T_s and the fusion temperature T_m is shown blank because the $h \sim T$ relation may take a variety of forms in different materials.

The energy equation (7.4) will again be applicable. To account for different properties of the two phases, however, the following dimensionless variables

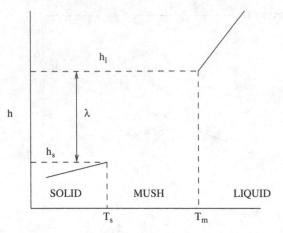

Figure 7.7. $h \sim T$ relation for an impure material.

are employed:

$$\Phi = \frac{h - h_s}{\lambda}, \qquad h_s = C_{ps} T_s, \tag{7.34}$$

$$\theta = \frac{C_{ps}(T - T_s)}{\lambda}, \tag{7.35}$$

$$\tau = \frac{\alpha_s t}{L^2}, \tag{7.36}$$

$$X = \frac{x}{L}, \tag{7.37}$$

$$\rho^* = \frac{\rho}{\rho_s}, \qquad k^* = \frac{K}{K_s}, \qquad C_p^* = \frac{C_p}{C_{ps}}. \tag{7.38}$$

Therefore, Equation 7.4 can be written as

$$\frac{\partial(\rho^* \Phi)}{\partial \tau} = \frac{\partial}{\partial X}\left(k^* \frac{\partial \theta}{\partial X}\right) \tag{7.39}$$

and the equations of state will take the form

$$\theta = \Phi \qquad \text{for} \qquad \Phi \leq 0, \tag{7.40}$$

$$\theta = f(\Phi) \qquad \text{for} \qquad 0 < \Phi < 1, \tag{7.41}$$

$$\theta = \theta_m + \frac{C_{ps}}{C_{pl}}(\Phi - 1) \qquad \text{for} \qquad \Phi > 1. \tag{7.42}$$

For alloys, function $f(\Phi)$ may take a variety of forms. For Al–4.5% Cu alloy, for example, Voller and Swaminathan [85] have used the following general relationship:

$$\Phi = \left[\frac{\theta - \theta_s}{\theta_l - \theta_s}\right]^n \qquad \text{for} \qquad \theta_s < \theta < \theta_l \tag{7.43}$$

and

$$\Phi = 1 \qquad \text{for} \qquad \theta_1 < \theta < \theta_m, \tag{7.44}$$

where $\theta_s < \theta_1 < \theta_m$, and the values of these temperatures and n (0.2 to 0.5) are known. In another form, known as Schiel's equation, the relationship is given by

$$\theta = \theta_s \qquad \text{for} \qquad 0 < \Phi < \Phi_s, \tag{7.45}$$

$$\Phi = \left[\frac{\theta - \theta_m}{\theta_1 - \theta_m} \right]^{-\beta} \qquad \text{for} \qquad \Phi_s < \Phi < 1, \tag{7.46}$$

$$\Phi = 1 \qquad \text{for} \qquad \theta_1 < \theta < \theta_m, \tag{7.47}$$

where $\beta = (1 - \gamma)^{-1}$ and γ is the partition coefficient. The values of γ, ϕ_s, and θ_1 are known.

The discretised version of Equation 7.39 will read as

$$\left[\frac{\rho_P^* \Delta X}{\Delta \tau} + \frac{k_e^* + k_w^*}{\Delta X} \right] \Phi_P = \left(\frac{k_e^*}{\Delta X} \right) \Phi_E + \left(\frac{k_w^*}{\Delta X} \right) \Phi_W$$

$$+ \left(\frac{k_e^*}{\Delta X} \right) (\Phi_E'' - \Phi_P'') + \left(\frac{k_w^*}{\Delta X} \right) (\Phi_W'' - \Phi_P'')$$

$$+ \frac{\rho_P^* \Delta X}{\Delta \tau} \Phi_P^o. \tag{7.48}$$

The trick now is to correctly interpret function $f(\Phi)$ so as to calculate θ_{pc} since Φ' (see Equation 7.20) can be easily calculated from Φ. This will enable calculation of Φ'' (see Equation 7.25).

Problem 2
To illustrate the procedure, consider a specific case of Al–4.5% Cu alloy for which the data are as follows and Schiel's equation is used:

$$K_s = 200 \text{ W/m-K}, \; K_1 = 90 \text{ W/m-K},$$

$$C_{ps} = 900 \text{ J/kg-K}, \; C_{pl} = 1,100 \text{ J/kg-K},$$

$$\rho_s = \rho_1 = 2,800 \text{ kg/m}^3,$$

$$\lambda = 3.9 \times 10^5 \text{ J/kg}, \; L = 0.5 \text{ m},$$

and

$$T_s = 821 \text{ K}, \; T_1 = 919 \text{ K}, \; T_m = 933 \text{ K}.$$

The initial state is superheated $T_{in} = 969$ K and $T_w (x = 0) = 573$ K, with $\gamma = 0.14$ (or $\beta = 1.163$).

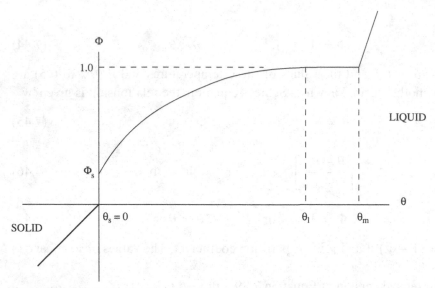

Figure 7.8. Schiel's function.

Thus, solidification commences instantly and calculations can be executed with

$$\theta_{in} = 0.341538, \; \Phi_{in} = 1.10154, \; \Phi'_{in} = -1,$$

$$\theta_w = \Phi_w = -0.5723,$$

$$\theta_s = 0, \; \Phi_s = 0.089,$$

$$\theta_l = 0.226154, \; \Phi_l = 1.0,$$

$$\theta_m = 0.258462, \; \Phi_m = 1.0.$$

Figure 7.8 shows the Schiel's function. We now specify θ_{pc} for the range $0 < \Phi < 1$ for which $\Phi' = -\Phi$.

$0 < \Phi < \Phi_s$: In this range, $\theta_s = 0$ remains stationary. Therefore, we may employ Equation 7.31.

$\Phi_s < \Phi < \Phi_l$: In this range, from Equation 7.46,

$$\theta_{pc,j} = \theta_m + (\theta_l - \theta_m)\,\Phi^{-1/\beta}. \tag{7.49}$$

$\theta_l < \theta < \theta_m$: In this range, Φ remains constant at 1. Therefore, the right-hand side of Equation 7.39 can be equated to zero. Therefore, the solution in discretised form is

$$\theta_{pc,j} = \frac{k_e^* \theta_E + k_w^* \theta_W}{k_e^* + k_w^*}. \tag{7.50}$$

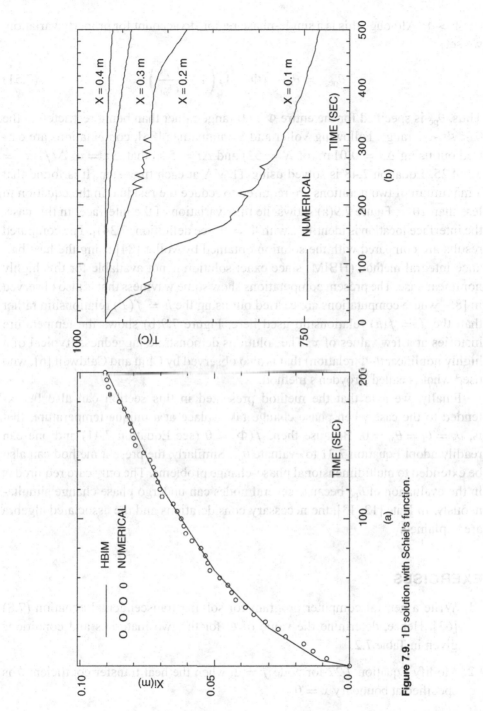

Figure 7.9. 1D solution with Schiel's function.

$\Phi > 1$: Although this is a single-phase region, to account for property variation, we set

$$\theta_{pc,j} = \theta_m - (\Phi - 1)\left(1 - \frac{C_{ps}}{C_{pl}}\right). \qquad (7.51)$$

Thus, θ_{pc} is specified for the entire $\Phi > 0$ range rather than being restricted to the $0 < \Phi < 1$ range. Following Voller and Swaminathan [85], computations are carried out using $\Delta x = 0.01$ m (or $N = 52$) and $\Delta t = 5$ so that $\Delta \tau = \alpha_s \Delta t / \Delta x^2 = 3.96825$. Equation 7.48 is solved using TDMA at each time step. It is found that a maximum of two iterations are required to reduce the residual in the equation to less than 10^{-5}. Figure 7.9(a) shows the time variation of the interface. In this case, the interface location is identified with $\Phi = 0$ [see definition (7.34)]. The computed results are compared with the solution obtained by Voller [84] using the heat balance integral method (HBIM) since exact solution is not available for this highly nonlinear case. The present computations show some waviness that is also observed in [85] where computations are carried out using the $h = f(T)$ relationship rather than the $T = f(h)$ relationship used here. Figure 7.9(b) shows the temperature histories at a few values of x. The solutions demonstrate jaggedness (typical of a highly nonlinear θ–Φ relation) that is also observed by Chiu and Caldwell [6], who used what is called Broyden's method.

Finally, we note that the method presented in this section can also be extended to the case when phase change takes place at a unique temperature, that is, $\theta_s = \theta_l = \theta_m = 0$. Because then, $f(\Phi) = 0$ (see Equation 7.41) and one can readily adopt Equation 7.31 to evaluate θ_{pc}. Similarly, the present method can also be extended to multidimensional phase-change problems. The only care required is in the evaluation of θ_{pc} because several nodes can undergo phase change simultaneously. In Date [12, 13], the necessary considerations and the associated algebra are explained.

EXERCISES

1. Write a general computer program for solving transcendental equation (7.8) [63]. Hence, determine the value of C for the two materials and conditions given in Table 7.2.

2. Modify Equation 7.22 for node $j = 2$, when the heat transfer coefficient h is specified at boundary $x = 0$.

3. Show the validity of Equation 7.23 in a solidification problem.

4. With respect to Figure 7.4, demonstrate the correctness of Equation 7.30 and hence of Equation 7.31.

5. Show the correctness of Equation 7.51 for $\Phi > 1$.

Table 7.2: Properties for Exercise 1.

ρ	C_{ps}	C_{pl}	k_s	k_l	λ	T_w	T_m	T_l
2,180	1,549	1,549	0.49	0.49	1.37×10^5	200	220	230
2,800	900	1,100	200	90	3.9×10^5	573	933	933

6. In an energy storage device, a PCM is sandwiched between two streams of heat transfer fluid (HTF) as shown in Figure 7.10. The HTF flows at 200°C with heat transfer coefficient 300 W/m²-K. The PCM is initially in a saturated state ($T_m = 220$°C) and its thickness is 8 cm. Estimate the time for heat (sensible + latent) recovery and the quantity recovered. The PCM properties are as follows: $\rho = 2,180$ kg/m³, $C_p = 1,549$ J/kg-K, $K = 0.49$ W/m-K, and $\lambda = 1.37 \times 10^5$ J/kg.

7. Consider solidification of a PCM contained in a spherical vessel of radius R. Initially, the PCM is at temperature $T_{in} = T_m$. The vessel wall temperature is $T_w < T_m$ and held constant with respect to time. Assuming only radial heat transfer, the applicable energy equation is

$$A \frac{\partial(\rho h)}{\partial t} = \frac{\partial}{\partial r}\left(K A \frac{\partial T}{\partial r}\right),$$

where $A = 4\pi r^2$.

(a) Nondimensionalise this equation assuming constant properties.

(b) Discretise the equation and write a computer program to solve the discretised equations. Use of a nonuniform grid with closer spacings near $r = R$ and $r = 0$ is desirable. Take $\rho = C_p = k = \lambda = 1$, $R = 1$, and $T_m = 0$ and compute for $T_w = -0.1, -1.0$, and, 10.0.

(c) Plot the variation of interface location R_i/R as a function of dimensionless time in each case and estimate total solidification time. Compare your results with those of [7].

8. Repeat Exercise 7 for a superheated PCM so that $T_{in} > T_m$. Take $T_w = -1.0$ and use three values of T_{in}: 0.1, 1, and 2.

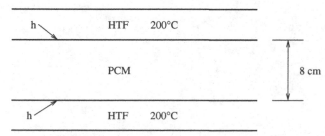

Figure 7.10. Phase-Change Energy Storage Device – Exercise 6.

9. Repeat Exercise 7 assuming that the convective heat transfer coefficient h and associated ambient temperature $T_\infty < T_m$ are specified at $r = R$ and T_w is unknown. Show that, in this problem, the interface movement is governed by two parameters: the Stefan number $St = C_p (T_m - T_\infty)/\lambda$ and the Biot number $Bi = h R/K_{PCM}$. Assume $T_\infty = -1$ and $T_m = 0$ and compute for $Bi = 1, 5,$ and 10. Plot the variation of T_w and R_i with time in each case.

8 Numerical Grid Generation

8.1 Introduction

As mentioned in Chapter 6, curvilinear grid generation for 2D domains involves specification of functions

$$x_1 = x_1(\xi_1, \xi_2), \qquad x_2 = x_2(\xi_1, \xi_2), \qquad (8.1)$$

where ξ_1, ξ_2 are curvilinear coordinates and x_1, x_2 are Cartesian coordinates. These two functions can be generated in two ways: (1) by algebraic specification or (2) by differential specification.

Algebraic specification is typically employed in 1D problems but can also be employed in 2D problems when the domain is simple (Section 8.2). For complex domains, however, differential grid generation is preferred. In this type, functions (8.1) are generated by solving differential equations with dependent variables x_1 and x_2. The differential equations can be of parabolic, hyperbolic, or elliptic type [81]. However, we shall consider the most commonly used elliptic grid generation technique (Sections 8.3 and 8.4)

The unstructured meshes again can be generated in a variety of ways. Two types will be considered: (1) generation by exploiting structuredness and (2) automatic mesh generation (Section 8.5).

8.2 Algebraic Grid Generation

8.2.1 1D Domains

The objective of grid generation is to locate nodes such that they are closely spaced in regions where the dependent variable Φ in the transport equations is expected to have steep gradients and sparsely spaced in regions where the gradients are small. This ensures that accurate solutions are economically obtained.

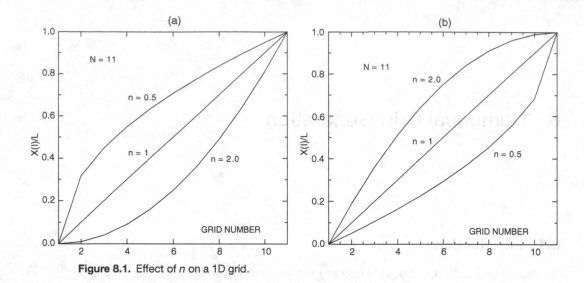

Figure 8.1. Effect of *n* on a 1D grid.

Consider a 1D domain of length L with N nodes so that there are $N - 2$ control volumes. One may now specify either the node coordinates $x(i)$ or the cell-face coordinates $x_c(i)$, where the latter occupies location of cell face w to the west of node P. Two useful algebraic formulas for node-coordinate determination are

$$\frac{x(i)}{L} = \left[\frac{i-1}{N-1}\right]^n, \tag{8.2}$$

and

$$\frac{x(i)}{L} = 1 - \left[1 - \frac{i-1}{N-1}\right]^n, \tag{8.3}$$

where n takes arbitrary positive value. For $n = 1$, these relationships are linear, implying uniform node spacing. According to relation (8.2), when $n > 1$, the grid is fine near $x = 0$ and becomes progressively coarser towards $x = L$ [see Figure 8.1(a)]. When $n < 1$, however, the grid is coarse near $x = 0$ and becomes uniformly fine near $x = L$. Relation (8.3) is employed when these trends are to be reversed [see Figure 8.1(b)]. In either case, once the x coordinates are known, the cell-face coordinates x_c can be determined by requiring that the cell face be *midway* between the adjacent nodes, as has been our preferred practice. Conversely, one can specify node coordinates $x_c(i)$ via formulas of this type and then determine the x coordinates.

8.2.2 2D Domains

In 2D domains, often the shape of the domain boundaries as well as the coordinates can be specified by algebraic equations. One such example is that of an eccentric

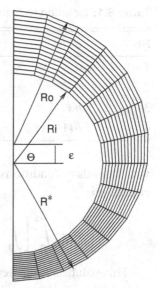

Figure 8.2. Eccentric annulus.

annulus shown in Figure 8.2. In this case, the grid coordinates can be generated from

$$x_1 = R \cos\theta, \qquad x_2 = R \sin\theta, \tag{8.4}$$

$$R^* = -\epsilon \sin\theta + \sqrt{R_0^2 - (\epsilon \cos\theta)^2}, \tag{8.5}$$

where $-\pi/2 \le \theta \le \pi/2$, $R_i \le R \le R^*$, and ϵ is eccentricity. When $\epsilon = 0$, a concentric annulus is generated. Shah and London [66] have given results for fully developed laminar flow and heat transfer in several ducts of noncircular cross section. The domains of such ducts (sine, ellipsoid, cordoid, etc.) can be mapped by relationships of the type given here.

8.3 Differential Grid Generation

8.3.1 1D Domains

In algebraic specification, the fineness of grid spacings could be controlled using formulas (8.2) and (8.3). This can also be done by solving a differential equation. To understand the main ideas, consider the differential conduction equation

$$\frac{d^2 T}{d x^2} + \frac{q'''}{k} = 0, \tag{8.6}$$

Table 8.1: Solution to Equation 8.7.

No.	$q'''(x)$	T
1	0	x
2	a	$x\left[1 - \frac{a}{2k}(1-x)\right]$
3	bx	$x\left[1 - \frac{b}{6k}(1-x^2)\right]$
4	$b(1-x)$	$x\left[1 - \frac{b}{3k}\left\{1 - \frac{x}{2}(3-x)\right\}\right]$

with boundary conditions $T = 0$ at $x = 0$ and $T = 1$ at $x = 1$. The solution to the equation is

$$T = -\int_0^x \left[\int_0^x \frac{q'''}{k}\,dx\right]dx + \left[1 + \int_0^1 \left(\int_0^x \frac{q'''}{k}\,dx\right)dx\right]x. \quad (8.7)$$

This solution is now evaluated for different assumptions for the variation of q''' with x. The solutions are shown in Table 8.1 and Figure 8.3 with $a = 2$, $b = 3$, and conductivity $k = 1$ in all cases. Clearly, the variation of T is controlled by variation of q''' with x.

Now, to make Equation 8.6 a *determinant* of grid node locations, we simply *interchange* the roles of x and T. Thus, the solution for $q''' = bx$, for example, is

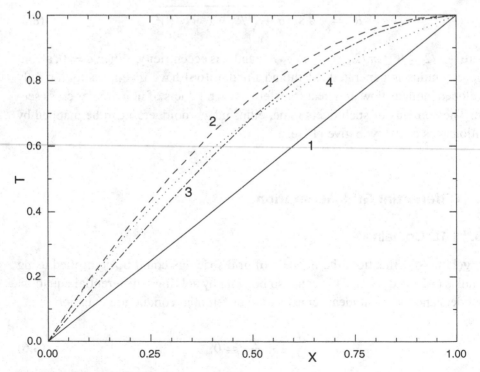

Figure 8.3. Effect of $q'''(x)$ function.

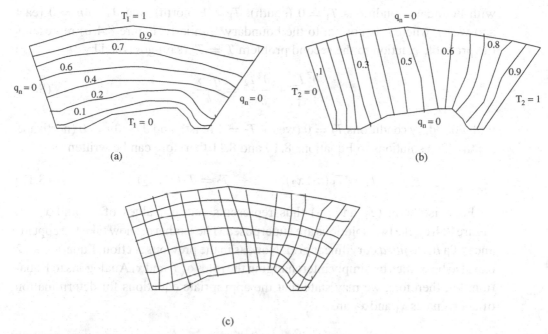

Figure 8.4. Differential construction of a 2D curvilinear grid.

taken as

$$x = T \left[1 - \frac{b}{6k} (1 - T^2) \right], \qquad 0 \le T \le 1. \tag{8.8}$$

By assigning different values to T in the specified range, we can get as many values of $x(i)$ as desired. To generalise this idea, we may state that the appropriate equation for determination of node coordinates is

$$\frac{d^2 \xi}{d x^2} = c(\xi), \tag{8.9}$$

where $c(\xi)$ is a *stretching* function to be specified by the analyst. To generate a solution of the form shown in Equation 8.8, Equation 8.9 must be *inverted*. This matter will be discussed in Section 8.3.3.

8.3.2 2D Domains

To understand the extension of the aforementioned notion to 2D domains, consider the domain shown in Figure 8.4. We now consider two problems with different boundary conditions. Figure 8.4(a) shows the *probable* solution to the first problem $T = T_1$ (say) governed by

$$\frac{\partial^2 T_1}{\partial x_1^2} + \frac{\partial^2 T_1}{\partial x_2^2} = \frac{q'''_1}{k}, \tag{8.10}$$

with boundary conditions $T_1 = 0$ (south), $T_1 = 1$ (north), and $\partial T_1/\partial n = 0$ (east and west), where n is normal to the boundary. Similarly, Figure 8.4(b) represents the probable solution to the second problem $T = T_2$ (say) governed by

$$\frac{\partial^2 T_2}{\partial x_1^2} + \frac{\partial^2 T_2}{\partial x_2^2} = \frac{q'''_2}{k}, \tag{8.11}$$

with boundary conditions $T_2 = 0$ (west), $T_2 = 1$ (east), and $\partial T_2/\partial n = 0$ (north and south). The solutions to Equations 8.10 and 8.11 therefore can be written as

$$T_1 = T_1(x_1, x_2), \qquad\qquad T_2 = T_2(x_1, x_2). \tag{8.12}$$

Each isotherm (T_1 and T_2) thus represents sets of values of x_1 and x_2. In Figure 8.4(c), the two solutions are superposed. The isotherms now take the appearance of a *body-fitted* curvilinear grid. Now, as in the previous section, Equation 8.12 can also be written by simply interchanging the roles of T and x. Analogous to Equation 8.9, therefore, we may state that the appropriate equations for determination of coordinates x_1 and x_2 are

$$\nabla^2 \xi_1 = \frac{\partial^2 \xi_1}{\partial x_1^2} + \frac{\partial^2 \xi_1}{\partial x_2^2} = P(\xi_1, \xi_2), \tag{8.13}$$

$$\nabla^2 \xi_2 = \frac{\partial^2 \xi_2}{\partial x_1^2} + \frac{\partial^2 \xi_2}{\partial x_2^2} = Q(\xi_1, \xi_2), \tag{8.14}$$

where ξ_1 and ξ_2 are curvilinear coordinates and P and Q are stretching functions.

8.3.3 Inversion of Determinant Equations

To make Equations 8.9 (in the 1D domain) and 8.13 and 8.14 (in 2D domains) determinants of Cartesian coordinates, they must be inverted. Thus, for the *1D domain*, we have

$$\frac{\partial}{\partial x} = \frac{\partial \xi}{\partial x} \frac{\partial}{\partial \xi}. \tag{8.15}$$

Now, if directions x and ξ coincide ($\partial \xi/\partial x = 1$) then Equation 8.9 can be written as

$$\frac{\partial^2 x}{\partial \xi^2} = C, \tag{8.16}$$

with $x = 0$ at $\xi = 0$ and $x = L$ at $\xi = 1$. Grid coordinates $x(i)$ can now be determined for various choices of C.

For *2D domains*, however, the matter is not so simple and requires vector analysis. Thus, we recall that a *covariant* base vector (tangent to coordinate direction ξ_i) is defined as

$$\vec{a}_i = \frac{d\vec{r}}{d\xi_i} = \vec{i}\,\frac{\partial x_1}{\partial \xi_i} + \vec{j}\,\frac{\partial x_2}{\partial \xi_i} + \vec{k}\,\frac{\partial x_3}{\partial \xi_i}. \qquad (8.17)$$

Similarly, the *contravariant* base vector (normal to coordinate surface ξ_i = constant) is defined as

$$\vec{a}^i = \nabla \xi_i = \vec{i}\,\frac{\partial \xi_i}{\partial x_1} + \vec{j}\,\frac{\partial \xi_i}{\partial x_2} + \vec{k}\,\frac{\partial \xi_i}{\partial x_3} = \vec{a}_j \times \vec{a}_k / J, \qquad (8.18)$$

where J is the Jacobian. Now, from Green's theorem [70], for any quantity (vector or scalar) Φ,

$$\nabla \Phi = \frac{1}{J} \sum_{i=1}^{3} \frac{\partial}{\partial \xi_i} \left[\vec{a}_j \times \vec{a}_k \right] \cdot \Phi = \sum_{i=1}^{3} \frac{\partial}{\partial \xi_i} \left[\vec{a}^i \, \Phi \right] = \sum_{i=1}^{3} \vec{a}^i \, \frac{\partial \Phi}{\partial \xi_i} \qquad (8.19)$$

since $\partial \vec{a}^i / \partial \xi_i = 0$. Therefore,

$$\nabla^2 \Phi = \nabla \cdot \nabla \Phi = \left[\sum_{i=1}^{3} \vec{a}^i \, \frac{\partial}{\partial \xi_i} \right] \cdot \left[\sum_{l=1}^{3} \vec{a}^l \, \frac{\partial \Phi}{\partial \xi_l} \right]$$

$$= \sum_{i=1}^{3} \sum_{l=1}^{3} \vec{a}^i \cdot \vec{a}^l \, \frac{\partial}{\partial \xi_i} \left(\frac{\partial \Phi}{\partial \xi_l} \right) + \sum_{i=1}^{3} \sum_{l=1}^{3} \vec{a}^i \, \frac{\partial \vec{a}^l}{\partial \xi_i} \, \frac{\partial \Phi}{\partial \xi_l}. \qquad (8.20)$$

If we now set $\Phi = \xi_l$ (a scalar), then

$$\nabla^2 \xi_l = \sum_{i=1}^{3} \vec{a}^i \, \frac{\partial \vec{a}^l}{\partial \xi_i}. \qquad (8.21)$$

Substituting Equation 8.21 in Equation 8.20 gives,

$$\nabla^2 \Phi = \sum_{i=1}^{3} \sum_{l=1}^{3} \vec{a}^i \cdot \vec{a}^l \, \frac{\partial}{\partial \xi_i} \left(\frac{\partial \Phi}{\partial \xi_l} \right) + \sum_{l=1}^{3} \nabla^2 \xi_l \, \frac{\partial \Phi}{\partial \xi_l}. \qquad (8.22)$$

In two dimensions ($\partial / \partial x_3 = \partial / \partial \xi_3 = 0$), Equation 8.22 will read as

$$\nabla^2 \Phi = \vec{a}^1 \cdot \vec{a}^1 \, \frac{\partial^2 \Phi}{\partial \xi_1^2} + 2\vec{a}^1 \cdot \vec{a}^2 \, \frac{\partial^2 \Phi}{\partial \xi_1 \partial \xi_2} + \vec{a}^2 \cdot \vec{a}^2 \, \frac{\partial^2 \Phi}{\partial \xi_2^2}$$

$$+ \nabla^2 \xi_1 \, \frac{\partial \Phi}{\partial \xi_1} + \nabla^2 \xi_2 \, \frac{\partial \Phi}{\partial \xi_2}. \qquad (8.23)$$

The dot products are now easily evaluated from Equations 8.17 and 8.18. Thus

$$\vec{a}^1 \cdot \vec{a}^1 = \frac{1}{J^2} \left[\left(\frac{\partial x_1}{\partial \xi_2} \right)^2 + \left(\frac{\partial x_2}{\partial \xi_2} \right)^2 \right] = \alpha / J^2,$$

$$\vec{a}^1 \cdot \vec{a}^2 = -\frac{1}{J^2} \left[\frac{\partial x_1}{\partial \xi_1} \frac{\partial x_1}{\partial \xi_2} + \frac{\partial x_2}{\partial \xi_1} \frac{\partial x_2}{\partial \xi_2} \right] = -\beta / J^2,$$

$$\vec{a}^2 \cdot \vec{a}^2 = \frac{1}{J^2} \left[\left(\frac{\partial x_1}{\partial \xi_1} \right)^2 + \left(\frac{\partial x_2}{\partial \xi_1} \right)^2 \right] = \gamma / J^2. \tag{8.24}$$

Employing these relations and using Equations 8.13 and 8.14, we can show that

$$\nabla^2 \Phi = \frac{1}{J^2} \left[\alpha \frac{\partial^2 \Phi}{\partial \xi_1^2} - 2\beta \frac{\partial^2 \Phi}{\partial \xi_1 \partial \xi_2} + \gamma \frac{\partial^2 \Phi}{\partial \xi_2^2} \right] + P \frac{\partial \Phi}{\partial \xi_1} + Q \frac{\partial \Phi}{\partial \xi_2}. \tag{8.25}$$

We now replace Φ by x_1 and x_2 and note that $\nabla^2 x_1 = \nabla^2 x_2 = 0$. Then, the equations for x_1 and x_2 will read as

$$\alpha \frac{\partial^2 x_1}{\partial \xi_1^2} - 2\beta \frac{\partial^2 x_1}{\partial \xi_1 \partial \xi_2} + \gamma \frac{\partial^2 x_1}{\partial \xi_2^2} = -J^2 \left[P \frac{\partial x_1}{\partial \xi_1} + Q \frac{\partial x_1}{\partial \xi_2} \right], \tag{8.26}$$

$$\alpha \frac{\partial^2 x_2}{\partial \xi_1^2} - 2\beta \frac{\partial^2 x_2}{\partial \xi_1 \partial \xi_2} + \gamma \frac{\partial^2 x_2}{\partial \xi_2^2} = -J^2 \left[P \frac{\partial x_2}{\partial \xi_1} + Q \frac{\partial x_2}{\partial \xi_2} \right], \tag{8.27}$$

where

$$J = \frac{\partial x_1}{\partial \xi_1} \frac{\partial x_2}{\partial \xi_2} - \frac{\partial x_2}{\partial \xi_1} \frac{\partial x_1}{\partial \xi_2}. \tag{8.28}$$

To determine functions (8.1), therefore, Equations 8.26 and 8.27 must be solved simultaneously with the boundary conditions specified at $\xi_1 = 0$, $\xi_1 = \xi_{1\max}$, $\xi_2 = 0$, and $\xi_2 = \xi_{2\max}$. Note that Equations 8.26 and 8.27 are coupled and nonlinear because α, β, and γ are themselves functions of dependent variables x_1 and x_2. Further, we note that the equations contain both the first and second derivatives and, if $-J^2 P$ and $-J^2 Q$ are regarded as *velocities*, the equations have the structure of a general transport equation.

It might appear that Equations 8.26 and 8.27 can be easily discretised and solved. However, there is a difficulty associated with the application of boundary conditions. The difficulty can be understood as follows. In fluid flow problems, we would often desire that the grid lines intersect orthogonally with the boundary. Thus, at the north and south boundaries, for example, we would desire that $\partial x_1/\partial \xi_2 = 0$. However, once this specification is made, we cannot specify x_1 on these boundaries. This is because if Dirichlet and Neumann boundary conditions are specified at the *same*

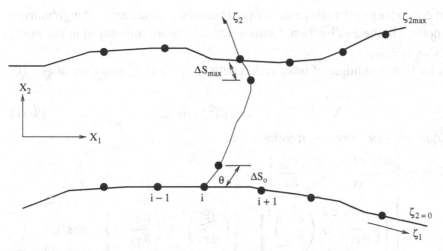

Figure 8.5. Grid line construction – Sorenson's method.

boundary, than the problem becomes *overspecified* or ill-posed. Therefore, we can specify either the value of $\partial x_1/\partial \xi_2$ or of x_1. However, if only one of these two boundary conditions is specified then the converged solutions to Equations 8.26 and 8.27 often demonstrate grid-node *clustering* in some portions of the domain and highly sparse node distributions in other regions.

Ideally, one would like to have complete freedom to choose x_1 and x_2 locations on the boundaries and yet achieve orthogonal intersection (or at any other desired angle) of the grid lines with the boundaries. The method of Sorenson [71] allows precisely this freedom. The method is described in the next section.

8.4 Sorenson's Method

8.4.1 Main Specifications

Sorenson's method permits coordinate and coordinate-gradient specification for the same variable x_1 or x_2 at *two* of the four boundaries of the domain. Thus, let $\xi_2 = 0$ (south) and $\xi_2 = \xi_{2\text{max}}$ (north) be these two boundaries as shown in Figure 8.5. We now define

$$\Delta s_0 = \left[\Delta x_1^2 + \Delta x_2^2 \right]^{0.5}_{\xi_2=0} \tag{8.29}$$

or, in the limit,

$$\frac{d s_0}{d \xi_2} = \left[\left(\frac{\partial x_1}{\partial \xi_2} \right)^2 + \left(\frac{\partial x_2}{\partial \xi_2} \right)^2 \right]^{0.5}_{\xi_2=0}. \tag{8.30}$$

Note that Δs_0 is the physical distance between boundary node 1 and its neighbouring interior node 2 in the ξ_2 direction. Similar definitions are introduced at the north boundary $\xi_2 = \xi_{2\max}$.

Now, let θ_0 be the angle of intersection between ξ_1 and ξ_2 grid lines at $\xi_2 = 0$. Then,

$$\nabla \xi_1 \cdot \nabla \xi_2 = |\vec{a}^1||\vec{a}^2| \cos\theta_0. \tag{8.31}$$

Using Equation 8.19, however, it follows that

$$\nabla \xi_1 \cdot \nabla \xi_2 = \left[\frac{\partial \xi_1}{\partial x_1}\frac{\partial \xi_2}{\partial x_1} + \frac{\partial \xi_1}{\partial x_2}\frac{\partial \xi_2}{\partial x_2}\right]_{\xi_2=0}$$

$$= \left[\left\{\left(\frac{\partial \xi_1}{\partial x_1}\right)^2 + \left(\frac{\partial \xi_1}{\partial x_2}\right)^2\right\}^{0.5}\left\{\left(\frac{\partial \xi_2}{\partial x_1}\right)^2 + \left(\frac{\partial \xi_2}{\partial x_2}\right)^2\right\}^{0.5}\cos\theta_0\right]_{\xi_2=0}, \tag{8.32}$$

but, from the definitions of β_i^j introduced in Chapter 6,

$$\frac{\partial \xi_1}{\partial x_1} = \frac{1}{J}\frac{\partial x_2}{\partial \xi_2}, \qquad \frac{\partial \xi_2}{\partial x_1} = -\frac{1}{J}\frac{\partial x_2}{\partial \xi_1},$$

$$\frac{\partial \xi_1}{\partial x_2} = -\frac{1}{J}\frac{\partial x_1}{\partial \xi_2}, \qquad \frac{\partial \xi_2}{\partial x_2} = \frac{1}{J}\frac{\partial x_1}{\partial \xi_1}. \tag{8.33}$$

Substituting these definitions and using Equation 8.30, we can write Equation 8.32 as

$$-\left[\frac{\partial x_2}{\partial \xi_1}\frac{\partial x_2}{\partial \xi_2} + \frac{\partial x_1}{\partial \xi_1}\frac{\partial x_1}{\partial \xi_2}\right]_{\xi_2=0} = \left[\frac{d s_0}{d\xi_2}\left\{\left(\frac{\partial x_2}{\partial \xi_1}\right)^2 + \left(\frac{\partial x_1}{\partial \xi_1}\right)^2\right\}^{0.5}\cos\theta_0\right]_{\xi_2=0}. \tag{8.34}$$

Evaluation of $\partial x_1/\partial \xi_2$ and $\partial x_2/\partial \xi_2$

To make further progress, we must evaluate $\partial x_1/\partial \xi_2$ and $\partial x_2/\partial \xi_2$ at $\xi_2 = 0$. This can be done using Equation 8.34. Thus,

$$\cos\theta_0 = -\left[\frac{\partial x_2}{\partial \xi_1}\frac{\partial x_2}{\partial \xi_2} + \frac{\partial x_1}{\partial \xi_1}\frac{\partial x_1}{\partial \xi_2}\right]\left[\frac{d s_0}{d\xi_2}\left\{\left(\frac{\partial x_2}{\partial \xi_1}\right)^2 + \left(\frac{\partial x_1}{\partial \xi_1}\right)^2\right\}^{0.5}\right]^{-1}_{\xi_2=0}. \tag{8.35}$$

Therefore, since $\sin\theta_0 = \sqrt{1 - \cos^2\theta_0}$,

$$\sin\theta_0 = \left[\frac{\partial x_2}{\partial \xi_2}\frac{\partial x_1}{\partial \xi_1} - \frac{\partial x_1}{\partial \xi_2}\frac{\partial x_2}{\partial \xi_1}\right]\left[\frac{d s_0}{d\xi_2}\left\{\left(\frac{\partial x_2}{\partial \xi_1}\right)^2 + \left(\frac{\partial x_1}{\partial \xi_1}\right)^2\right\}^{0.5}\right]^{-1}_{\xi_2=0}. \tag{8.36}$$

Now, solving Equations 8.35 and 8.36 simultaneously, we can show that

$$\frac{\partial x_1}{\partial \xi_2}\Big|_{\xi_2=0} = -\frac{d\, s_0}{d\, \xi_2} \left\{ \frac{\partial x_1}{\partial \xi_1} \cos \theta_0 + \frac{\partial x_2}{\partial \xi_1} \sin \theta_0 \right\} \left[\left(\frac{\partial x_2}{\partial \xi_1} \right)^2 + \left(\frac{\partial x_1}{\partial \xi_1} \right)^2 \right]^{-0.5}_{\xi_2=0},$$

(8.37)

$$\frac{\partial x_2}{\partial \xi_2}\Big|_{\xi_2=0} = \frac{d\, s_0}{d\, \xi_2} \left\{ \frac{\partial x_1}{\partial \xi_1} \sin \theta_0 - \frac{\partial x_2}{\partial \xi_1} \cos \theta_0 \right\} \left[\left(\frac{\partial x_2}{\partial \xi_1} \right)^2 + \left(\frac{\partial x_1}{\partial \xi_1} \right)^2 \right]^{-0.5}_{\xi_2=0}.$$

(8.38)

Identical expressions can be developed for $\partial x_1/\partial \xi_2$ and $\partial x_2/\partial \xi_2$ at $\xi_2 = \xi_{2max}$.

8.4.2 Stretching Functions

Sorenson [71] defines P and Q functions as

$$P(\xi_1, \xi_2) = P(\xi_1, 0) \exp(-a\,\xi_2) + P(\xi_1, \xi_{2max}) \exp\{-c\,(\xi_{2max} - \xi_2)\}, \quad (8.39)$$

$$Q(\xi_1, \xi_2) = Q(\xi_1, 0) \exp(-b\,\xi_2) + Q(\xi_1, \xi_{2max}) \exp\{-d\,(\xi_{2max} - \xi_2)\}, \quad (8.40)$$

where a, b, c, and d are positive constants to be chosen by the analyst. Now, for convenience, we introduce the following symbols:

$$L_1 = -\text{(LHS of Equation 8.26)}/J^2$$

and

$$L_2 = -\text{(LHS of Equation 8.27)}/J^2.$$

Thus,

$$L_1(\xi_2 = 0) = P(\xi_1, 0) \frac{\partial x_1}{\partial \xi_1}\Big|_{\xi_2=0} + Q(\xi_1, 0) \frac{\partial x_1}{\partial \xi_2}\Big|_{\xi_2=0}, \quad (8.41)$$

$$L_2(\xi_2 = 0) = P(\xi_1, 0) \frac{\partial x_2}{\partial \xi_1}\Big|_{\xi_2=0} + Q(\xi_1, 0) \frac{\partial x_2}{\partial \xi_2}\Big|_{\xi_2=0}. \quad (8.42)$$

Therefore, using the definition of J (see Equation 8.28), we get

$$P(\xi_1, 0) = \frac{1}{J} \left[L_1 \frac{\partial x_2}{\partial \xi_2} - L_2 \frac{\partial x_1}{\partial \xi_2} \right]_{\xi_2=0}, \quad (8.43)$$

$$Q(\xi_1, 0) = \frac{1}{J} \left[L_2 \frac{\partial x_1}{\partial \xi_1} - L_2 \frac{\partial x_2}{\partial \xi_1} \right]_{\xi_2=0}. \quad (8.44)$$

Identical expressions again emerge for $P(\xi_1, \xi_{2max})$ and $Q(\xi_1, \xi_{2max})$. One can thus prescribe P and Q functions over the whole domain using Equations 8.39 and 8.40.

8.4.3 Discretisation

Equations 8.26 and 8.27 can be written in the following general form:

$$P^* \frac{\partial \Phi}{\partial \xi_1} - \alpha \frac{\partial^2 \Phi}{\partial \xi_1^2} + Q^* \frac{\partial \Phi}{\partial \xi_2} - \gamma \frac{\partial^2 \Phi}{\partial \xi_2^2} = -2\beta \frac{\partial}{\partial \xi_1} \left(\frac{\partial \Phi}{\partial \xi_2} \right), \qquad (8.45)$$

where $\Phi = x_1, x_2$, $P^* = -PJ^2$, and $Q^* = -QJ^2$. Equation 8.45, being of the conduction–convection type, can be discretised using the UDS to yield

$$AP\, \Phi_{\mathrm{P}} = AE\, \Phi_{\mathrm{E}} + AW\, \Phi_{\mathrm{W}} + AN\, \Phi_{\mathrm{N}} + AS\, \Phi_{\mathrm{S}} + S, \qquad (8.46)$$

where

$$AE = \alpha_{\mathrm{P}} + \frac{1}{2}(|P_{\mathrm{P}}^*| - P_{\mathrm{P}}^*),$$

$$AW = \alpha_{\mathrm{P}} + \frac{1}{2}(|P_{\mathrm{P}}^*| + P_{\mathrm{P}}^*),$$

$$AN = \gamma_{\mathrm{P}} + \frac{1}{2}(|Q_{\mathrm{P}}^*| - Q_{\mathrm{P}}^*),$$

$$AS = \gamma_{\mathrm{P}} + \frac{1}{2}(|Q_{\mathrm{P}}^*| + Q_{\mathrm{P}}^*),$$

$$AP = AE + AW + AN + AS,$$

$$S = -2\beta_{\mathrm{P}}(\Phi_{\mathrm{ne}} - \Phi_{\mathrm{nw}} - \Phi_{\mathrm{se}} + \Phi_{\mathrm{sw}}). \qquad (8.47)$$

Equation 8.46 can be solved using the ADI method.

8.4.4 Solution Procedure

Sorenson's method can be implemented through the following steps.

Initialisation
1. Choose coordinates $x_1(\xi_1, 0)$, $x_2(\xi_1, 0)$, $x_1(\xi_1, \xi_{2max})$, and $x_2(\xi_1, \xi_{2max})$ on the south and north boundaries, respectively. Also specify $x_1(0, \xi_2)$ (west) and $x_1(\xi_{1max}, \xi_2)$ (east).
2. Specify[1] Δs_0 and Δs_{max} and θ_0 and θ_{max}. For orthogonal intersection, $\theta = \pi/2$.
3. Let $P(\xi_1, \xi_2) = Q(\xi_1, \xi_2) = 0$.

---

[1] It will be appreciated that this liberty to specify Δs_0 and Δs_{max} can be very useful when south and north boundaries are *walls* and the HRE e–ϵ turbulence model is employed. One can therefore place the first node away from the wall in the range $30 < y^+ < 100$.

4. From the known coordinates on the south and the north boundaries, interpolate $x_1(\xi_1, \xi_2)$ and $x_2(\xi_1, \xi_2)$ to serve as the initial guess. Usually, linear interpolation between corresponding points on the south and north boundary for each ξ_1 suffices.

5. Now evaluate $\partial x_1/\partial\xi_1$, $\partial x_2/\partial\xi_1$, $\partial x_1/\partial\xi_2$, and $\partial x_2/\partial\xi_2$ at $\xi_2 = 0$ and $\xi_2 = \xi_{2max}$. These remain *fixed* for all subsequent operations.

Iterations Begin

6. Evaluate L_1 and L_2 at $\xi_2 = 0$ and $\xi_2 = \xi_{2max}$. In these evaluations, the second-order derivatives at $\xi_2 = 0$, for example, are represented as follows:

$$\frac{\partial^2 \Phi}{\partial \xi_2^2} = \frac{1}{2}(-7\,\Phi_{i,1} + 8\,\Phi_{i,2} - \Phi_{i,3}) - 3\,(\Phi_{i,2} - \Phi_{i,1}), \qquad (8.48)$$

$$\frac{\partial^2 \Phi}{\partial \xi_1^2} = \Phi_{i+1,1} - 2\,\Phi_{i,1} - \Phi_{i-1,1}, \qquad (8.49)$$

$$\frac{\partial}{\partial \xi_1}\left(\frac{\partial \Phi}{\partial \xi_2}\right) = \frac{1}{2}\left(\left.\frac{\partial \Phi}{\partial \xi_2}\right|_{i+1} - \left.\frac{\partial \Phi}{\partial \xi_2}\right|_{i-1}\right). \qquad (8.50)$$

7. Use equations such as 8.43 and 8.44 to evaluate $P(\xi_1, 0)$, $Q(\xi_1, 0)$, $P(\xi_1, \xi_{2max})$, and $Q(\xi_1, \xi_{2max})$.

8. Using the preceding information and already chosen[2] constants a, b, c, and d, evaluate $P(\xi_1, \xi_2)$ and $Q(\xi_1, \xi_2)$ at all nodes in the domain. Between iterations, underrelaxation in evaluation of P and Q is advised.

9. Specify boundary conditions for x_2 at the west and east boundaries. Here, care must be taken to take account of the type of grid being generated. If an H- or C-type grid is being generated, one must specify the x_2 from known equations of the west and east boundaries since x_1 values are already known (see step 1). Alternatively, one may specify the $\partial x_2/\partial\xi_1$ condition to let the $\xi_2 = $ constant line intersect the boundary at a desired angle. If an O-type grid is being generated then one specifies *periodic* condition $\Phi(0, \xi_2) = \Phi(\xi_{1max}, \xi_2)$.

10. Solve Equation 8.45 for $\Phi = x_1, x_2$ and check convergence.

11. If the convergence criterion is not met, go to step 6.

8.4.5 Applications

H Grid

Figure 8.6 shows the grid for a flow between parallel plates with a constriction. South ($x_2 = 0$) is the axis of symmetry, north is a wall, west ($x_1 = -8$) is the

---

[2] Typically, $a = b = c = d = 0.7$. If too small a value is used (0.2, say), the effect of the constants decays slowly away from the south/north boundaries. If too large a value is chosen, the effect decays very rapidly.

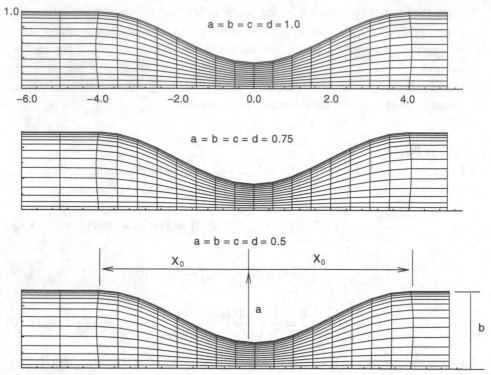

Figure 8.6. Example of H – grid.

inflow boundary, and east ($x_1 = 20$) is the exit boundary. The channel half-width is $b = 1$ and the constriction height is δ. The constriction profile for the range $-x_0 < x_1 < x_0$ is given by

$$\frac{x_2}{b} = 1 - \frac{\delta}{2b}\left(1 + \cos\frac{\pi x_1}{x_0}\right).$$

The figure shows the grid generated with $\Delta s_0 = \Delta s_{\max} = 0.035$, $\delta/b = 2/3$, and $x_0/b = 4$. The grids, 32 in the ξ_1 direction and 15 in the ξ_2 direction, are generated using the following boundary conditions.

South: $x_2 = 0$, for $-8 < x_1 < 20$.

North: $x_2 = 1$ for $-8 < x_1 < -x_0$, $x_2 = f(x_1)$ for $-x_0 < x_1 < x_0$ and, $x_2 = 1$ for $x_0 < x_1 < 20$, where $f(x_1)$ is the constriction shape function already mentioned.

West: $x_1 = -8$, $\partial x_2/\partial \xi_1 = 0$.

East: $x_1 = 20$, $\partial x_2/\partial \xi_1 = 0$.

To maintain clarity, the generated grids are shown in Figure 8.6 for $-6 < x_1 < 5$ only. Three values of constants (1.0, 0.75, and 0.5) are used and are indicated in the figure. For the largest value, the ξ_2 grid lines are more evenly spaced in the range $0.25 < x_2 < 0.8$. For smaller values, the grid nodes are attracted more towards the north and the south boundaries, yielding fewer nodes in the middle range of x_2.

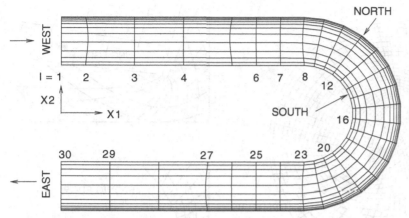

Figure 8.7. Example of C – grid.

C Grid

Figure 8.7 is an example of the C grid. The figure shows a channel with a $180°$ bend. The inner radius of the bend is $R_i = 1$ and the outer radius is $R_o = 2$. The flow enters the west boundary and exits from the east boundary. There are 30 nodes in the I (or, ξ_1) direction and 12 nodes in the J (or, ξ_2) direction. The grids are generated using the following specifications:

West: $x_1 = 0$, $\partial x_2/\partial \xi_1 = 0$, $x_2(1, 1) = 1$, and $x_2(1, JN) = 2$.

East: $x_1 = 0$, $\partial x_2/\partial \xi_1 = 0$, $x_2(1, 1) = -1$, and $x_2(1, JN) = -2$.

South: $x_2(i, 1) = 1$ for $i = 1$ to 8, $x_1(8, 1) = x_1(8, JN) = 5$, $x_1(i, 1) = x_1(8, 1) + R_i(\cos\theta - 1)$, $x_2(i, 1) = R_i \sin\theta$ for $i = 9$ to 23, $x_2(i, 1) = -1$ for $i = 24$ to IN, and $x_1(24, 1) = x_1(24, JN) = 5$.

North: $x_2(i, JN) = 2$ for $i = 1$ to 8, $x_1(i, JN) = x_1(8, JN) + R_o(\cos\theta - 1)$, $x_2(i, JN) = R_o \sin\theta$ for $i = 9$ to 23, and $x_2(i, JN) = -2$ for $i = 24$ to IN.

In these specifications, θ varies from $0°$ to $180°$. The grids are generated with $\Delta s_0 = \Delta s_{max} = 0.05$ and $a = b = c = d = 0.7$. The ξ_2 grid lines show much closer spacings near the north boundary than near the south boundary.

O Grid

Figure 8.8 shows 74 (ξ_1 or circumferential) $\times$ 25 (ξ_2 or radial) grids around the GE90 gas-turbine blade whose surface (south boundary) coordinates are known.[3] The outer circle (radius $= 3\times$ the axial chord) forms the north boundary. The west and east boundaries are periodic and, therefore, x_1 and x_2 coordinates at $i = 1$ and $i = IN$ coincide. The figure also shows details of the grid structure near the trailing and leading edges of the blade.

It must be remembered that grid generation is somewhat of an art because different choices of node locations on the boundaries and the constants in the

---

[3] Although a more practical situation involves a cascade of blades, here the blade is treated as an isolated airfoil.

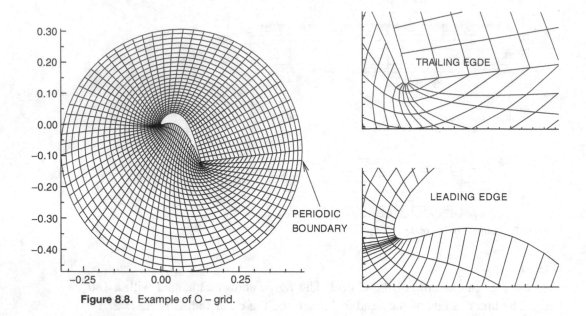

Figure 8.8. Example of O – grid.

stretching functions can produce different grid spacings and stretchings inside the domain. One needs to make a few trials before accepting the generated grid. A graphics package such as TECPLOT for mesh visualisation is therefore necessary. The package also has a *zooming* facility to permit visualisation of dense-grid regions.

8.5 Unstructured Mesh Generation

8.5.1 Main Task

Unstructured mesh generation essentially involves two tasks:

1. locating vertices in the domain and
2. creating vertex and element files (as mentioned in Chapter 6).

These tasks can be carried out in a variety of ways. The two most commonly used are the following:

1. Locating vertices by curvilinear grid generation so that a regular (i, j) structure is readily available for vertex numbering.
2. Locating vertices according to *rules* that yield arbitrary vertices without (i, j) structure. In this *automatic grid generation* method, node numbering requires care.

These alternatives are considered next for further explanation.

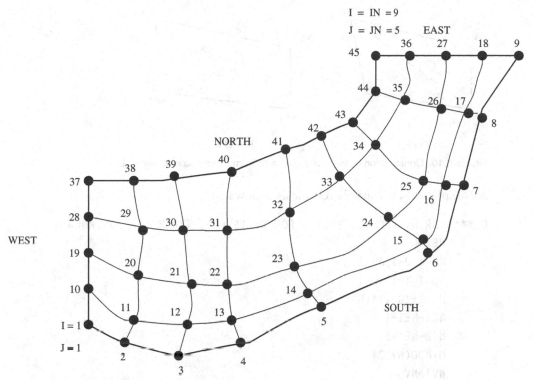

Figure 8.9. Linear numbering of a structured grid.

8.5.2 Domains with (i, j) Structure

Consider the complex domain shown in Figure 8.9. The domain is laid with a curvilinear structured grid. A typical vertex (i, j), therefore, will have eight immediate neighbours: $(i + 1, j), (i + 1, j + 1), (i, j + 1), (i - 1, j + 1), (i - 1, j), (i - 1, j - 1), (i, j - 1),$ and $(i + 1, j - 1)$. We now designate each vertex by a *one-dimensional* address system rather than a two-dimensional one. Thus, vertex (i, j) can be referred to by vertex number NV (say), where

$$NV = i + (j - 1) \times IN. \tag{8.51}$$

In Figure 8.9, nodes are linearly numbered for a grid with $IN = 9$ and $JN = 5$. According to Equation 8.51, vertex (IN, JN) will be referred to by NVMAX $= IN \times JN$, whereas for vertex $(1, 1)$, $NV = 1$. Now, since coordinates of vertices are known, one can readily form the vertex file.

With this linear numbering, one can construct a minimum of two triangular elements out of each quadrilateral element. This formation can be of two types as shown in Figure 8.10. In each case, elements must be numbered along with the associated three vertex numbers to form the element file. This task can be

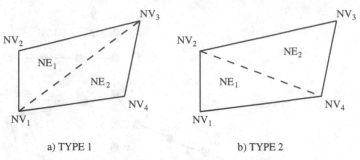

a) TYPE 1   b) TYPE 2

Figure 8.10. Construction of triangular elements from a quadrilateral element.

accomplished by a simple routine as follows:

```
C *** FOR NV (ODD), TYPE1 , FOR NV (EVEN), TYPE2 (IN, JN KNOWN)
      NE1=0
      DO 1 J=1,JN-1
      DO 1 I=1,IN-1
      NV=I+(J-1)*IN
      NE1=NE1+1
      NE2=NE1+1
      M=MOD(NV,2)
      NV1=NV
      NV2=NV1+IN
      NV3=NV2+1
      NV4=NV1+1
      IF(M.EQ.1)THEN
      WRITE(6,*)NE1,NV1,NV3,NV2
      WRITE(6,*)NE2,NV1,NV4,NV3
      ELSE IF(M.EQ.0)THEN
      WRITE(6,*)NE1,NV1,NV4,NV2
      WRITE(6,*)NE2,NV4,NV3,NV2
      ENDIF
      NE1=NE2
1     CONTINUE
```

Figure 8.11 shows the element numbering for the grid shown in Figure 8.9. The numbering is carried out using the routine given here.

8.5.3 Automatic Grid Generation

Automatic grid generation (AGG) is used to generate elements having desired properties and desired density (i.e., clustering). For example, when 2D triangular elements are generated, one may desire that each element has a prespecified area or

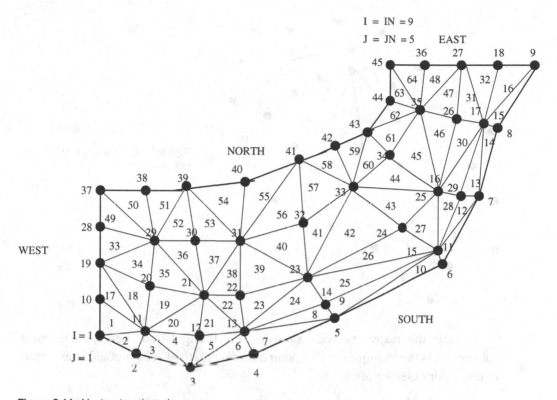

Figure 8.11. Unstructured mesh.

that no included angle shall exceed 90°. There are several ways in which this may be achieved and the subject matter is as much an art as it is a science. Fortunately, useful reviews of methods for AGG are published from time to time and the reader is referred to one such review [27] by way of an example.

Methods for AGG can be classified based on element type, element shape, mesh density control, and time efficiency. The most popular mesh-generation methods first *create* all vertices (boundary and interior) and then connect them by lines to form triangles. The question then arises as to what is the *best* triangulation on a given set of points. The most popular principle for triangulating is called *Delaunay* triangulation.

To understand the scheme, consider a set of vertices on a domain as shown in Figure 8.12. In this figure, triangle A represents a Delaunay triangle because the *circumcircle* passing through the three vertices encloses *no other vertices*. This, however, is not true for triangle B, which is therefore *not* a Delaunay triangle. It is obvious that if the set of vertices were arbitrarily chosen, and their locations were fixed, then it would be difficult to meet the requirement of Delaunay triangulation. Without proof, we state that Delaunay triangulation is achieved in such a way that *thin* elements are avoided [27] whenever possible.

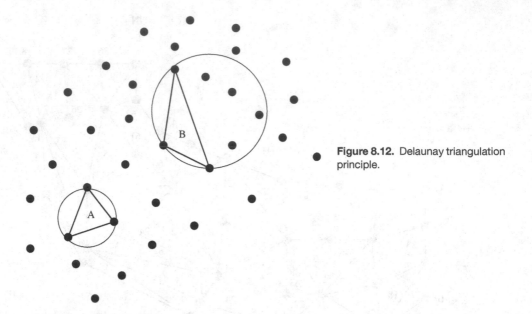

Figure 8.12. Delaunay triangulation principle.

Among the many methods available for triangulation, perhaps the most convenient is the *triangulation by point insertion* method. The method is executed in three steps (see Figure 8.13):

1. Define and discretise domain boundaries. Straight boundaries can be discretised by employing formulas such as Equation 8.2 or by a cubic-spline technique [63]. Curved boundaries, however, require further care.
2. Triangulate the boundary points using the Delaunay triangulation principle. This creates new vertices interior to the domain.
3. Triangulate the remaining interior domain by point insertion. Starting from an existing pair of vertices (1 and 2, say), a third vertex can be searched under a variety of constraints. One such constraint is the *aspect ratio* $AR = r_i / (2r_c)$, where r_i is the radius of *inscribed circle* and r_c is the radius of *circumscribed circle*.[4] The new *inserted vertex* is now placed at the *circumcentre* of the triangle 1–2–3 with *minimum* AR.

One can thus complete the triangulation of the entire domain. These three steps can be cast in the form of an algorithm and a computer program can be written for its implementation. A computer program based on a method by Watson [87] is available in [67]. The next task is to create the *data structure*. This refers to creation of vertex and element numbering to prepare the required vertex and element files. Several commercial packages for AGG are available that can create *mixed* elements and three-dimensional polyhedra. Using these packages, meshes can be generated

---

[4] Here, $r_i = A/s, r_c = 0.25 \times a \times b \times c/A$, semiperimeter $s = (a + b + c)/2$, and area $A = \sqrt{s(s-a)(s-b)(s-c)}$. a, b, and c are lengths of sides of the triangle 1-2-3.

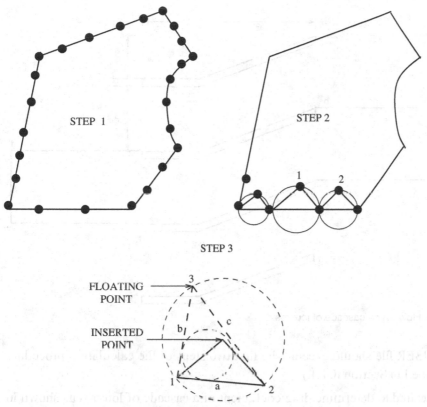

Figure 8.13. Point insertion technique.

to describe flow over an entire aircraft or over and through a car (including the engine space below the bonnet). In such applications, millions of elements are needed and the question of the efficiency with which an AGG algorithm is devised becomes important. The task of AGG has thus assumed considerable significance to be recognised as a specialised branch of CFD.

EXERCISES

1. Derive formulas analogous to Equations 8.2 and 8.3 to determine the distribution of $x_c(i)$.

2. Generate x_1 and x_2 coordinates of an ellipsoidal duct by algebraic grid generation, exploiting symmetry.

3. Starting with Equation 8.19, derive Equations 8.26 and 8.27.

4. Discretise Equation 8.45.

5. Develop a generalised computer program to solve Equation 8.45 for $\Phi = x_1$ and x_2. (Hint: You will need to develop a USER file and a LIBRARY file.

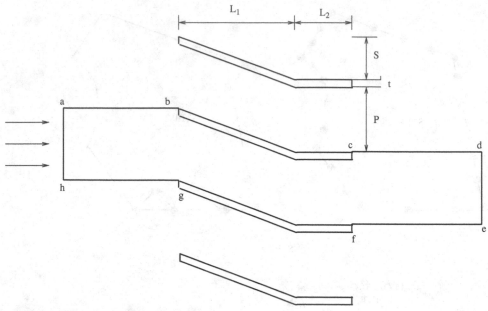

Figure 8.14. Flow over a cascade of louvres.

The USER file should execute the first two steps of the calculation procedure described in Section 8.4.4.)

6. It is desired to determine drag coefficient of a cascade of louvres as shown in Figure 8.14. For this purpose, an analyst selects the domain a–b–c–d–e–f–g–h. Use the computer program developed in Exercise 5 to generate curvilinear grids and provide boundary conditions for x_1 and x_2. Take $ab = 1$, $L_1 = 1.5$, $L_2 = 1.0$, $S = 0.25$, $P = 0.5$, $t = 0.05$, and $cd = 1.5$.

7. Repeat Exercise 6 for the GE90 gas-turbine blade cascade shown in Figure 8.15. The coordinates[5] of the suction and pressure surface of the blade are given in Table 8.2 (30 points on the suction surface and 46 points on the pressure surface). The other dimensions are as follows: axial chord $C_{ax} = 12.964$ cm, pitch $P = 13.811$ cm, blade inlet angle $\beta_1 = 35°$, and blade outlet angle $\beta_2 = -72.49°$. (Hint: If more points are required on the blade surface, their coordinates can be generated using spline interpolation [63].)

8. Consider flow in a duct of square cross section in which a twisted tape has been inserted as shown in Figure 8.16. The width of the tape equals the duct-side length D. This three-dimensional flow can be analysed by generating 2D grids at several cross sections along the axis at different angles Φ from the vertical. One such section A–A at angle $\Phi = 22.5°$ is shown in the figure. The thickness

[5] The author is grateful to Prof. R. J. Goldstein of the University of Minnesota for providing the coordinate data.

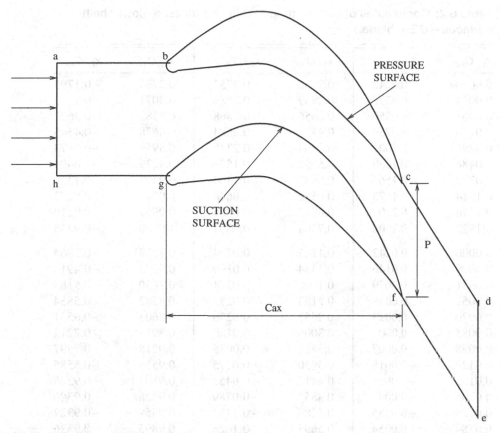

Figure 8.16. Schematic of a gas-turbine blade cascade

of the tape $\delta/D = 0.04$. The flow is symmetric about the tape with secondary flow being transferred through the gaps c–d and e–f. Therefore, curvilinear grids may be generated over only half of the duct cross section. Select west, north, east, and south boundaries and adapt the computer program of Exercise 5 to generate the curvilinear grid. Also specify the boundary conditions for the velocity components u_i, $i = 1, 2$, and 3. (Hint: For the purpose of generating the curvilinear grid, assume $\delta = 0$ to avoid any sharp protrusion into the domain.)

9. The vertex file for the domain of Figure 8.9 is given in Table 8.3. Reading this file, prepare an element file using the routine given in Section 8.5.2 to generate a triangular mesh as shown in Figure 8.11. Now, with reference to Chapter 6, develop a computer program to do the following:

   (a) Identify neighboring element numbers of each element. Store this information in array NHERE (N, K).

   (b) Define boundary nodes B and assign node numbers to them.

Table 8.2: Coordinates of suction (upper half) and pressure (lower half) surfaces – GE90 blade.

x_1/C_{ax}	x_2/C_{ax}	x_1/C_{ax}	x_2/C_{ax}	x_1/C_{ax}	x_2/C_{ax}
0.0000	0.0242	0.2365	0.2752	0.7735	−0.1793
0.0014	0.0377	0.2989	0.2886	0.8071	−0.2703
0.0063	0.0550	0.3656	0.2868	0.8383	−0.3621
0.0155	0.0759	0.4328	0.2684	0.8678	−0.4545
0.0296	0.1001	0.4967	0.2348	0.8959	−0.5473
0.0484	0.1269	0.5556	0.1878	0.9229	−0.6404
0.0722	0.1565	0.6083	0.1304	0.9491	−0.7338
0.1014	0.1878	0.6552	0.0646	0.9747	−0.8273
0.1376	0.2200	0.6942	−0.0025	0.9997	−0.9210
0.1822	0.2506	0.7364	−0.0897	1.0000	−0.9235
0.0000	0.0242	0.1147	0.0124	0.7238	−0.3864
0.0009	0.0146	0.1434	0.0190	0.7603	−0.4915
0.0031	0.0079	0.1760	0.0244	0.7950	−0.5183
0.0052	0.0038	0.2133	0.0273	0.8282	−0.5854
0.0070	0.0013	0.2551	0.0256	0.8603	−0.6531
0.0085	0.0000	0.3006	0.0180	0.8914	−0.7212
0.0098	−0.0007	0.3478	0.0035	0.9218	−0.7897
0.0120	−0.0018	0.3950	−0.0175	0.9515	−0.8585
0.0153	−0.0031	0.4412	−0.0452	0.9807	−0.9274
0.0205	−0.0046	0.4857	−0.0789	0.9828	−0.9306
0.0279	−0.0055	0.5286	−0.1184	0.9859	−0.9327
0.0384	−0.0054	0.5695	−0.1626	0.9895	−0.9336
0.0522	−0.0035	0.6088	−0.2112	0.9932	−0.9330
0.0694	0.0003	0.6441	−0.2596	0.9968	−0.9309
0.0903	0.0058	0.6853	−0.3222	0.9992	−0.9276
–	–	–	–	1.0000	−0.9235

(c) Calculate geometric coefficients B11 (N , K) and B21 (N , K); cell-face area ACF (N , K); lengths LP2E2 (N , K), LX1 (N , K), LX2 (N , K), DX1 (N , K), and DX2 (N , K); and weighting factor FM (N , K).

(d) Calculate the cell volume VOL (N) of each element.
Including the boundary nodes, what is the total number of nodes, NMAX?

10. To dispel the idea that unstructured meshes must necessarily be triangular or polygonal, an analyst maps a complex domain with essentially a *Cartesian* mesh, as shown in Figure 8.17. Now, it is seen that cells with more or less than four faces occur near an irregular boundary (see the enlarged view) and the dimensions of such cells can be determined from the known coordinates of the irregular boundary. Essentially, therefore, the mesh can be generated by algebraic specification. It is also possible to obtain any desired cell density. Of course, to do this automatically, a computer program must be written. Further,

Table 8.3: Vertex file data.

NV	x_1	x_2	NV	x_1	x_2	NV	x_1	x_2	NV	x_1	x_2
1	0	0	13	33	1	25	78	27	37	0	27
2	8	−3	14	50	6	26	82	41	38	10	27
3	10	−6	15	78	15	27	84	51	39	20	28
4	35	−4	16	83	26	28	0	20	40	33	29
5	54	4	17	88	40	29	12	18	41	46	33
6	79	13	18	92	51	30	22	18	42	55	36
7	87	26	19	0	14	31	32	18	43	62	39
8	91	40	20	12	10	32	48	21	44	66	45
9	100	51	21	23	8	33	58	27	45	66	51
10	0	6	22	32	8	34	67	35			
11	10	0	23	48	11	35	74	43			
12	22	0	24	70	20	36	75	51			

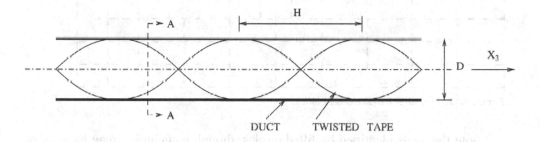

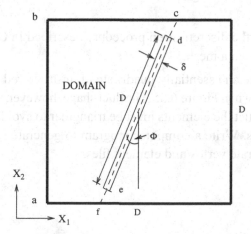

Figure 8.16. Flow in a duct of square cross section containing a twisted tape.

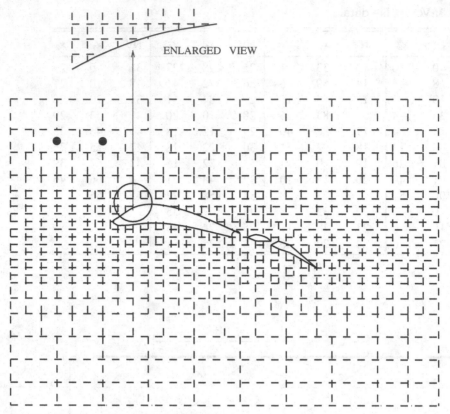

Figure 8.17. Flow over a multielement airfoil.

note that cells identified by filled circles, though rectangular, may have more than four neighboring cells.

(a) Identify the number of neighbouring cells for the two cells marked with filled circles.

(b) Examine whether the discretisation procedure described in Chapter 6 can be employed for such a mesh.

11. It is desired to generate an essentially quadrilateral unstructured mesh for the moon-shaped duct shown in Figure 6.35. The duct shape, however, is such that in some portions of the duct the elements must be triangular to avoid unnecessary concentration of nodes. Write a computer program to generate such a *mixed-element* grid and generate vertex and element files.

9 Convergence Enhancement

9.1 Convergence Rate

In all the preceding chapters it was shown that discretising the differential transport equations results in a set of algebraic equations of the following form:

$$AP\Phi_P = \sum A_k \Phi_k + S, \qquad (9.1)$$

where suffix k refers to appropriate neighbouring nodes of node P. In pure conduction problems ($\Phi = T$), A_k and S may be functions of T. In the general problem of convective–diffusive transport, Φ may stand for any transported variable and A_k and S may again be functions of the Φ under consideration or any other Φ relevant to the system. In curvilinear grid generation, $\Phi = x_1, x_2$, and A_k and S are again functions of x_1 and x_2. In all such cases, if there are N *interior* nodes, we need to solve N equations for each variable Φ in a prespecified sequence. An *iterative* solution is particularly attractive when the algebraic equations for different Φs are strongly coupled through coefficients and sources.

In an iterative procedure, *convergence* implies *numerical* satisfaction of Equation 9.1 at each interior node for each Φ. This satisfaction is checked by the *residual* in Equation 9.1 at each iteration level l (say). Thus

$$R_P^\Phi = AP\,\Phi_P^l - \sum A_k\,\Phi_k^l - S. \qquad (9.2)$$

The whole-field convergence is declared when

$$R^\Phi = \frac{\left[\sum_{\text{all nodes}} \{R_P^\Phi\}^2\right]^{0.5}}{R_{\text{norm}}} < CC, \qquad (9.3)$$

where CC stands for the *convergence criterion* and R_{norm} is a dimensionally correct normalising quantity defined by the CFD analyst. For example, in a problem with total inflow $\dot{m}_{\text{in}}$ and average property $\overline{\Phi}_{\text{in}}$, $R_{\text{norm}} = \dot{m}_{\text{in}} \times \overline{\Phi}_{\text{in}}$ (say). If no such representative quantity is found then $R_{\text{norm}} = 1$. Ideally, CC must be as small as the machine accuracy will permit but typically $CC = 10^{-5}$ (say) suffices for most engineering applications.

The *convergence rate* CR may be defined as

$$CR = -\frac{dR^\Phi}{dl},\tag{9.4}$$

where l is the iteration level. Economic computations will require that CR must be as high as possible. Algebraic equation solvers such as the GS, the TDMA, and the ADI introduced in Chapter 5, however, demonstrate the following convergence rate properties:

1. Overall CR is higher when A_k and S are constants rather than when they are dependent on Φ.
2. The initial (small l) CR is high but progressively decreases as convergence is approached.
3. CR is higher when the A_k are small (for example, coarse grids) than when they are large (fine grids).
4. CR is higher when Dirichlet boundary conditions are specified at all boundaries than when Neumann (or gradient) boundary conditions are specified. This is one reason why the pressure-correction equation is slow to converge.
5. The convergence history (i.e., $R^\Phi \sim l$ relationship) is typically monotonic when A_k and S are constants but can be highly nonmonotonic (or oscillatory) when the equations are strongly coupled.

This last point is concerned with the *stability* of the iterative procedure. The reader may wish to relate this phenomenon with *damping of waves* discussed in Chapter 3.

The CR of the basic iterative methods (GS and ADI for 2D problems) can be enhanced by several techniques. Here, a few of them that have the facility of being incorporated in a generalised computer code will be considered. It is important to note, however, that all convergence enhancement techniques essentially take ever greater account of the *implicitness* embodied in the equation set (9.1). Thus, it is recognised that Φ_P is implicitly related not only to its immediate neighbours but also to its distant neighbours. The objective, therefore, is to strengthen this relationship with the distant neighbours.

The merit of this observation has already been sensed in Chapter 2, where convergence rates of GS (point-by-point) and TDMA (line-by-line) procedures were compared for a 1D problem. In this chapter, the main interest is to consider 2D problems. The enhancement techniques considered can also be extended to 3D problems.

9.2 Block Correction

The block-correction technique is used to enhance the convergence rate of the ADI method. Thus, we rewrite Equation 9.1 as

$$AP_{i,j}\,\Phi_{i,j} = AE_{i,j}\,\Phi_{i+1,j} + AW_{i,j}\,\Phi_{i-1,j}$$
$$+ AN_{i,j}\,\Phi_{i,j+1} + AS_{i,j}\,\Phi_{i,j-1} + Su_{i,j}, \tag{9.5}$$

where

$$AP_{i,j} = AE_{i,j} + AW_{i,j} + AN_{i,j} + AS_{i,j} + Sp_{i,j}. \tag{9.6}$$

Equation 9.5 is written such that the boundary coefficients of the near-boundary nodes are zero and the boundary conditions are absorbed through Su and Sp, as explained in Chapter 5. Thus,

$$AW_{2,j} = AE_{IN-1,j} = AS_{i,2} = AN_{i,JN-1} = 0. \tag{9.7}$$

The central idea of the block-correction technique is that an unconverged field $\Phi_{i,j}^l$ is corrected by adding *uniform* correction $\overline{\Phi}_i$ along lines of constant i. Thus, let

$$\Phi_{i,j} = \Phi_{i,j}^l + \overline{\Phi}_i. \tag{9.8}$$

Now, the correction $\overline{\Phi}_i$ is chosen such that the *integral* conservation over all control-volumes on a constant-i strip is exactly satisfied. The equation governing $\overline{\Phi}_i$ is thus obtained by a two-step procedure. First, Equation 9.8 is substituted in Equation 9.5 so that

$$AP_{i,j}\left(\Phi_{i,j}^l + \overline{\Phi}_i\right) = AE_{i,j}\left(\Phi_{i+1,j}^l + \overline{\Phi}_{i+1}\right) + AW_{i,j}\left(\Phi_{i-1,j}^l + \overline{\Phi}_{i-1}\right)$$
$$+ AN_{i,j}\left(\Phi_{i,j+1}^l + \overline{\Phi}_i\right) + AS_{i,j}\left(\Phi_{i,j-1}^l + \overline{\Phi}_i\right)$$
$$+ Su_{i,j}. \tag{9.9}$$

Then all such equations for $j = 2, 3, \ldots, JN - 1$ are added. Thus, one obtains

$$BP_i\,\overline{\Phi}_i = BE_i\,\overline{\Phi}_{i+1} + BW_i\,\overline{\Phi}_{i-1} + BS_i, \quad i = 2, \ldots, IN - 1, \tag{9.10}$$

where

$$BP_i = \sum_{j=2}^{JN-1} (AP_{i,j} - AN_{i,j} - AS_{i,j}),$$

$$BE_i = \sum_{j=2}^{JN-1} AE_{i,j},$$

$$BW_i = \sum_{j=2}^{JN-1} AW_{i,j}, \tag{9.11}$$

and

$$BS_i = \sum_{j=2}^{JN-1} \left[AE_{i,j}\, \Phi_{i+1,j}^l + AW_{i,j}\, \Phi_{i-1,j}^l + AN_{i,j}\, \Phi_{i,j+1}^l \right.$$
$$\left. + AS_{i,j}\, \Phi_{i,j-1}^l + Su_{i,j} - AP_{i,j}\, \Phi_{i,j}^l \right]. \tag{9.12}$$

It will be recognized that the quantity inside the summation in Equation 9.12 is simply $-R_{i,j}^{\Phi}$ (see Equation 9.2) at iteration level l. Further, Equation 9.10 can be easily solved by TDMA. In this equation, $BE_{IN-1} = BW_2 = 0$ (see Equation 9.7) and hence $\overline{\Phi}_{IN}$ and $\overline{\Phi}_1$ are *not needed*. A similar exercise in the j direction will result in an equation for $\overline{\Phi}_j$.

The overall procedure is as follows:

1. Solve Equation 9.5 once using ADI to arrive at the $\Phi_{i,j}^l$ field.
2. Form the B coefficients in Equation 9.10 and solve this equation by TDMA to yield $\overline{\Phi}_i$ corrections. Reset $\Phi_{i,j}$ according to Equation 9.8.
3. Repeat step 2 to yield $\overline{\Phi}_j$ corrections and reset $\Phi_{i,j}$ again.
4. Return to step 1 if the convergence criterion is not satisfied.

The block-correction procedure generally produces considerably faster convergence than the ADI method but, in certain circumstances, it may produce an erroneous solution or even divergence. Such a circumstance may arise when Φ is highly nonuniform and $\overline{\Phi}_i$ or $\overline{\Phi}_j$ may produce over- or undercorrections. Therefore, the block-correction procedure may be treated as an optional convergence enhancement device.

9.3   Method of Two Lines

In the ADI method, two sweeps are alternately executed in i and j directions (see Chapter 5). Within each sweep, however, the TDMA is executed only along a *single* line so that Φ values of that line are updated simultaneously. To enhance the convergence rate, it is possible to devise a TDMA procedure for two, three, or multiple lines. By way of illustration, we consider the method of two lines [56, 21] in which the following definition is introduced:

$$\Phi_{i,j+1}^* = \Phi_{i,j}. \tag{9.13}$$

Consider lines j and $j + 1$ for the sweep in the i direction. The discretised equations along these lines will read as

$$AP_{i,j}\, \Phi_{i,j+1}^* = AE_{i,j}\, \Phi_{i+1,j+1}^* + AW_{i,j}\, \Phi_{i-1,j+1}^*$$
$$+ AN_{i,j}\, \Phi_{i,j+1} + AS_{i,j}\, \Phi_{i,j-1} + Su_{i,j}, \tag{9.14}$$

$$AP_{i,j+1}\, \Phi_{i,j+1} = AE_{i,j+1}\, \Phi_{i+1,j+1} + AW_{i,j+1}\, \Phi_{i-1,j+1}$$
$$+ AN_{i,j+1}\, \Phi_{i,j+2} + AS_{i,j+1}\, \Phi_{i,j+1}^* + Su_{i,j+1}. \tag{9.15}$$

In writing these equations, it is again assumed that Equation 9.7 holds. Now let

$$ae_i = \frac{AE_{i,j}}{AP_{i,j}}, \qquad ae_i^* = \frac{AE_{i,j+1}}{AP_{i,j+1}},$$

$$aw_i = \frac{AW_{i,j}}{AP_{i,j}}, \qquad aw_i^* = \frac{AW_{i,j+1}}{AP_{i,j+1}},$$

$$an_i = \frac{AN_{i,j}}{AP_{i,j}}, \qquad an_i^* = \frac{AN_{i,j+1}}{AP_{i,j+1}},$$

$$as_i = \frac{AS_{i,j}}{AP_{i,j}}, \qquad as_i^* = \frac{AS_{i,j+1}}{AP_{i,j+1}},$$

$$d_i = \frac{Su_{i,j}}{AP_{i,j}}, \qquad d_i^* = \frac{Su_{i,j+1}}{AP_{i,j+1}}. \tag{9.16}$$

Using these definitions, Equations 9.14 and 9.15 can be written as

$$\Phi_{i,j^*}^* = ae_i\,\Phi_{i+1,j^*}^* + aw_i\,\Phi_{i-1,j^*}^* + an_i\,\Phi_{i,j^*} + b_i, \tag{9.17}$$

$$\Phi_{i,j^*} = ae_i^*\,\Phi_{i+1,j^*} + aw_i^*\,\Phi_{i-1,j^*} + as_i^*\,\Phi_{i,j^*}^* + b_i^*, \tag{9.18}$$

where

$$j^* = j + 1, \tag{9.19}$$

$$b_i = as_i\,\Phi_{i,j-1} + d_i, \tag{9.20}$$

$$b_i^* = an_i^*\,\Phi_{i,j+2} + d_i^*. \tag{9.21}$$

Equations 9.17 and 9.18 represent two equations with suffix j^*. Our interest is to solve them simultaneously. To do this, let

$$\Phi_i^* = A_i^*\,\Phi_{i+1}^* + B_i^*\,\Phi_{i+1} + C_i^*, \tag{9.22}$$

$$\Phi_i = A_i\,\Phi_{i+1} + B_i\,\Phi_{i+1}^* + C_i, \tag{9.23}$$

where suffix j^* is dropped for convenience. We now evaluate Φ_{i-1}^* from Equation 9.22 and substitute this into Equation 9.17. After some algebra, it can be shown that

$$\Phi_i^* = \alpha_{1i}\,\Phi_{i+1}^* + \alpha_{2i}\,\Phi_i + \alpha_{3i}, \tag{9.24}$$

where

$$\alpha_{1i} = \frac{ae_i}{1 - aw_i\,A_{i-1}^*},$$

$$\alpha_{2i} = \frac{aw_i\,B_{i-1}^* + an_i}{1 - aw_i\,A_{i-1}^*},$$

$$\alpha_{3i} = \frac{aw_i\,C_{i-1}^* + b_i}{1 - aw_i\,A_{i-1}^*}. \tag{9.25}$$

Similarly, evaluating Φ_{i-1} from Equation 9.23 and substituting in Equation 9.18, we have

$$\Phi_i = \beta_{1i}\,\Phi_{i+1} + \beta_{2i}\,\Phi_i^* + \beta_{3i}, \qquad (9.26)$$

where

$$\beta_{1i} = \frac{ae_i^*}{1 - aw_i^*\,A_{i-1}},$$

$$\beta_{2i} = \frac{aw_i^*\,B_{i-1} + as_i^*}{1 - aw_i^*\,A_{i-1}},$$

$$\beta_{3i} = \frac{aw_i^*\,C_{i-1} + b_i^*}{1 - aw_i^*\,A_{i-1}}. \qquad (9.27)$$

If we now substitute Equation 9.26 in Equation 9.24, then comparison with Equation 9.22 will show that

$$A_i^* = \frac{\alpha_{1i}}{1 - \alpha_{2i}\,\beta_{2i}},$$

$$B_i^* = \frac{\alpha_{2i}\,\beta_{1i}}{1 - \alpha_{2i}\,\beta_{2i}},$$

$$C_i^* = \frac{\alpha_{2i}\,\beta_{3i} + \alpha_{3i}}{1 - \alpha_{2i}\,\beta_{2i}}. \qquad (9.28)$$

Similarly, substituting Equation 9.24 in Equation 9.26 and comparison with Equation 9.23 will show that

$$A_i = \frac{\beta_{1i}}{1 - \alpha_{2i}\,\beta_{2i}},$$

$$B_i = \frac{\beta_{2i}\,\alpha_{1i}}{1 - \alpha_{2i}\,\beta_{2i}},$$

$$C_i = \frac{\beta_{2i}\,\alpha_{3i} + \beta_{3i}}{1 - \alpha_{2i}\,\beta_{2i}}. \qquad (9.29)$$

The overall two-line TDMA procedure is thus as follows:

1. Consider j and $j^* = j + 1$ lines.
2. Form as, a^*s, d, and d^* according to Equation 9.16 for $i = 2, 3, \ldots, IN - 1$.
3. Form b_i and b_i^* from Equations 9.20 and 9.21 for $i = 2, 3, \ldots, IN - 1$.
4. Evaluate αs, βs, As, Bs, and Cs for $i = 2, 3, \ldots, IN - 1$ by recurrence. Note that $A_1^* = B_1^* = C_1^* = A_1 = B_1 = C_1 = 0$.
5. Hence solve Equations 9.22 and 9.23 by back substitution (i.e., $i = IN - 1$ to 2).
6. Set $\Phi_{i,j} = \Phi_i^*$ and $\Phi_{i,j+1} = \Phi_i$.

7. Go to step 1 with the next value of j (i.e., $j = j + 1$).
8. Repeat steps 1–7 until $j = JN - 2$.

A similar procedure can be executed for sweep on the j direction. Finally, we note that a procedure for simultaneous solution for three, four, or more consecutive lines can also be devised but the associated algebra is very tedious. It will be realised that if a simultaneous solution procedure is devised for *all lines* in a given direction, one will have a procedure that is equivalent to the matrix inversion method for the whole field.

9.4 Stone's Method

As mentioned in Section 9.1, the convergence rate is sensitive to the structure of the coefficient matrix. Stone [79] devised a whole-field procedure that reduces this sensitivity. To apply the method, it is first necessary to change the 2D node address (i, j) to the 1D address N. Thus,

$$N = i + (j - 1) \times IN, \tag{9.30}$$

where $N = 1, \ldots, N_{max}$ and $N_{max} = IN \times JN$. Equation 9.5 therefore can be written as

$$AP_N \, \Phi_N = AE_N \, \Phi_{N+1} + AW_N \, \Phi_{N-1}$$
$$+ AN_N \, \Phi_{N+IN} + AS_N \, \Phi_{N-IN} + Su_N. \tag{9.31}$$

In matrix form, this equation can be written as

$$|A||\Phi| = |Su|. \tag{9.32}$$

Figure 9.1 shows the fully expanded form of Equation 9.32. Note that matrix A has a maximum of five nonzero elements in each row.

The main idea in Stone's method is to represent matrix A as a product of two matrices, U and L. Thus, the L matrix (or lower matrix) is formed in such a way that all entries above the diagonal are zero. The diagonal element is occupied by 1, and positions of $-AW$ and $-AS$ in the A matrix are now taken by BW and BS (say). Similarly, in the U matrix (or upper matrix) all elements below the diagonal are set to zero; the diagonal elements are occupied by BP and elements occupying positions $-AE$ and $-AN$ are replaced by BE and BN, respectively. The L and U matrices are shown in Figure 9.2. Note that the size of L and U matrices is again $N_{max} \times N_{max}$.

Unfortunately, the product matrix $|U| \times |L|$ does not produce the A matrix exactly. Instead, a matrix shown in Figure 9.3 is produced. This matrix has two additional nonzero entries that occupy positions $NW(i - 1, j + 1)$ and $SE(i + 1, j - 1)$. In terms of elements of the U and L matrices, the elements of

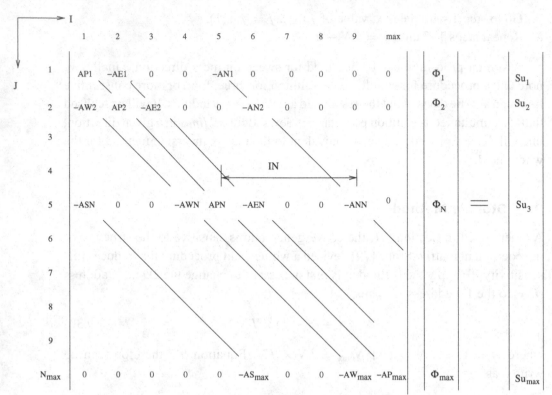

Figure 9.1. Matrix representation of Equation 9.31.

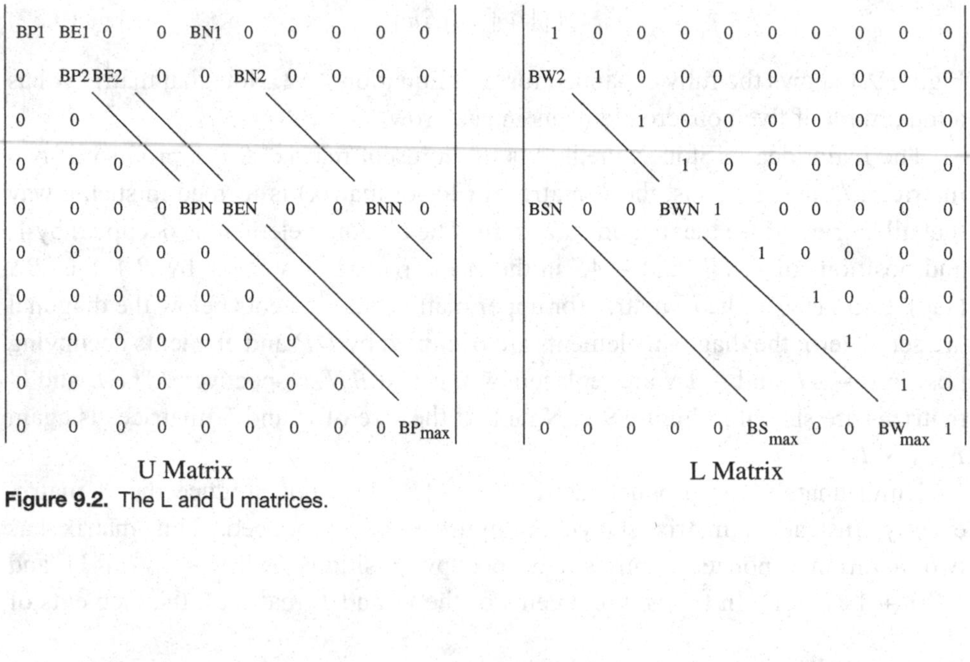

U Matrix   L Matrix

Figure 9.2. The L and U matrices.

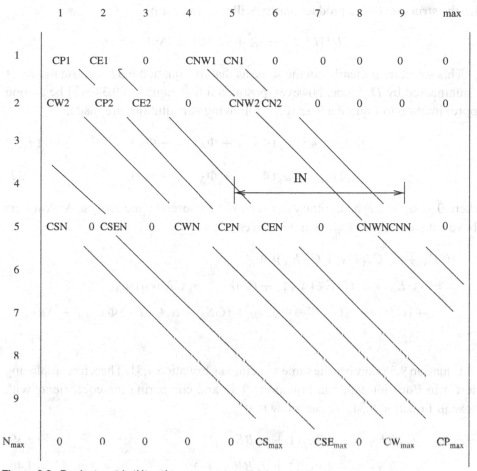

Figure 9.3. Product matrix $|U| \times |L|$.

the product matrix are given by

$$CP_N = BP_N + BE_N\, BW_{N+1} + BN_N\, BS_{N+IN},$$

$$-CE_N = BE_N,$$

$$-CN_N = BN_N,$$

$$-CW_N = BW_N\, BP_N,$$

$$-CS_N = BS_N\, BP_N,$$

$$-CSE_N = BE_N\, BS_{N+1},$$

$$-CNW_N = BN_N\, BW_{N+IN}. \tag{9.33}$$

Thus, the product matrix equation $|U| \times |L| \times |\Phi| = |Su|$ will imply

$$CP_N\, \Phi_N = CE_N\, \Phi_{N+1} + CW_N\, \Phi_{N-1} + CN_N\, \Phi_{N+IN} + CS_N\, \Phi_{N-IN}$$

$$+ CSE_N\, \Phi_{N+1-IN} + CNW_N\, \Phi_{N-1+IN} + Su_N, \tag{9.34}$$

and the structure of the product matrix will take the form

$$|U||L||\Phi| = |A + D||\Phi| = |Su|. \tag{9.35}$$

This structure is clearly not the same as that of Equation 9.32 because matrix A is augmented by D. Stone, however, postulated that Equation 9.34 will be a good approximation to Equation 9.32 if the following substitutions are made:

$$\Phi_{N-1+IN} = \alpha_s (\Phi_{N-1} + \Phi_{N+IN} - \Phi_N), \tag{9.36}$$

$$\Phi_{N+1-IN} = \alpha_s (\Phi_{N+1} + \Phi_{N-IN} - \Phi_N), \tag{9.37}$$

where $0 < \alpha_s < 1$ is an arbitrary constant to be chosen by the analyst. Making the above substitutions in Equation 9.34 gives

$$[CP_N + \alpha_s (CNW_N + CSE_N)] \, \Phi_N$$
$$= (CE_N + \alpha_s CSE_N) \, \Phi_{N+1} + (CW_N + \alpha_s CNW_N) \, \Phi_{N-1}$$
$$+ (CN_N + \alpha_s CNW_N) \, \Phi_{N+IN} + (CS_N + \alpha_s CSE_N) \, \Phi_{N-IN} + Su_N.$$
$$\tag{9.38}$$

Equation 9.38 now has the same structure as Equation 9.31. Therefore, replacing the Cs in Equation 9.38 via Equations 9.33 and comparing the coefficients with those in Equation 9.31, we can show that

$$BE_N = -AE_N / (1 + \alpha_s BS_{N+1}), \tag{9.39}$$

$$BN_N = -AN_N / (1 + \alpha_s BW_{N+IN}), \tag{9.40}$$

$$BP_N = AP_N + \alpha_s (BN_N BW_{N+IN} + BE_N BS_{N+IN})$$
$$- (BE_N BW_{N+1} + BN_N BS_{N+IN}), \tag{9.41}$$

$$BW_N = -(AW_N + \alpha_s BN_N BW_{N+IN}) / BP_N, \tag{9.42}$$

$$BS_N = -(AS_N + \alpha_s BE_N BS_{N+1}) / BP_N. \tag{9.43}$$

Now, it is expected that the product matrix will be a close approximation to the A matrix (i.e., $D \to 0$). In actual solving, therefore, the product matrix equation is written as

$$|A + D||\Phi^{l+1}| = |A + D||\Phi^l| + |Su| - |A||\Phi^l|. \tag{9.44}$$

We now define

$$|\delta| = |\Phi^{l+1}| - |\Phi^l|, \tag{9.45}$$

$$|R| = -[|A||\Phi^l| - |Su|], \tag{9.46}$$

where δ is the change in Φ over one iteration and R is the *negative* of the nodal residual. Therefore, Equation 9.44 can be written as

$$|A + D||\delta| = |R| = |U||L||\delta|. \tag{9.47}$$

The overall procedure is thus as follows:

1. Form elements of the residual R_N matrix from AP, AE, AW, AN, AS, and Su.
2. Form BW_N, BS_N, BE_N, BN_N, and BP_N by recurrence (i.e., from $N = N_{max}$ to 1) using Equations 9.39–9.43. Store BW_N and BS_N.
3. Form $|V| = |L||\delta| = |R||U|^{-1}$. This implies that

$$V_N = (R_N - BE_N V_{N+1} - BN_N V_{N+IN})/BP_N \tag{9.48}$$

for $N = N_{max}, \ldots, 1$.
4. Hence, determine $|\delta| = |V||L|^{-1}$, which implies

$$\delta_N = V_N - BS_N \delta_{N-IN} - BW_N \delta_{N-1} \tag{9.49}$$

for $N = 1, \ldots, N_{max}$.
5. Update $\Phi_N^{l+1} = \Psi_N^l + \delta_N$.

In Stone's method, α_s turns out to be problem dependent. However, advice on the choice of $\alpha_{s,max}$ is available in [29].

9.5 Applications

In this section, convergence enhancement procedures described in the previous sections will be tested against four problems. In each problem, convergence rate and computation times for different grid sizes are recorded. A depiction of typical convergence history in Problem 4 is also provided.

Consider a rectangular domain $0 \le X \le a$ and $0 \le Y \le b$. Assume steady-state heat conduction with the following boundary conditions:

Problem 1: $T(0, Y) = T(a, Y) = T(X, 0) = 0$, $T(X, b) = T_b = 1$, $a = 2$, and $b = 1$.

Problem 2: $T(0, Y) = T(a, Y) = T(X, 0) = 0$, $T(X, b) = T_b = 1$, $a = 5$, and $b = 1$.

Problem 3: $T(0, Y) = T(a, Y) = T(X, b) = 0$, $h(X, 0) = 5$, $T_\infty = 20$, $a = 2$, and $b = 1$.

Problem 4: Same as Problem 3 but with temperature-dependent conductivity $k = k_{ref}(1.0 + 0.1\,T + 0.001\,T^2)$.

In each problem, the residual (see Equation 9.3) is reduced to 10^{-5} and no underrelaxation is employed.

Table 9.1: Problem 1 ($IN = 33$, $JN = 17$).

Procedure	Iterations	CPU (s)
GS	403	121
ADI	104	44
Block correction	30	11
Two-line TDMA	37	22
Stone ($\alpha_s = 0.8$)	48	22
Stone ($\alpha_s = 0.9$)	31	16

Table 9.2: Problem 2 ($JN = 17$).

Procedure	$IN = 33$		$IN = 53$	
	Iterations	CPU (s)	Iterations	CPU (s)
GS	299	93	366	138
ADI	43	22	83	43
Block correction	24	11	34	22
Two-line TDMA	17	11	30	17
Stone ($\alpha_s = 0.9$)	18	11	27	17

The exact solution for Problems 1 and 2 is given by

$$\frac{T}{T_b} = \frac{2}{\pi} \sum_{n=1}^{\infty} \frac{[1 - \cos(n\pi)]}{n \sinh(n\pi b/a)} \sin(n\pi x/a) \sinh(n\pi y/a). \qquad (9.50)$$

Table 9.1 shows results for Problem 1. The results show the expected trend in that the ADI procedure is faster[1] than the GS procedure. The block correction, two-line TDMA, and Stone's procedures are considerably faster. On this relatively coarse grid (though sufficient for obtaining accurate solutions) Stone's procedure is faster when $\alpha_s = 0.9$ than when $\alpha_s = 0.8$.

Table 9.2 shows results for Problem 2. Here, the a dimension is increased but IN still equals 33. The AE and AW coefficients become smaller than those in Problem 1. This results in faster convergence in all methods. When $IN = 53$, the AE and AW coefficients again become bigger and the convergence rate decreases.

The exact solution to Problem 3 is given by

$$\frac{T}{T_\infty} = \sum_{n=1}^{\infty} A_n \sin(n\pi x/a)[e^{-n\pi y/a} - (e^{-2n\pi b/a} e^{n\pi y/a})],$$

$$A_n = \frac{2h}{k} \left[\frac{1 - \cos(n\pi)}{n\pi} \right] \left[\frac{h}{k}(1 - e^{-2n\pi b/a}) + \frac{n\pi}{a}(1 + e^{-2n\pi b/a}) \right]^{-1}.$$

---

[1] Note that the CPU times mentioned in the table depend on the processor used. The quoted times thus have no intrinsic relevance; they are mentioned for the purpose of comparison between different methods.

Table 9.3: Problem 3 h boundary condition.

Procedure	$IN = 33$, $JN = 17$		$IN = 81$, $JN = 41$	
	Iterations	CPU (s)	Iterations	CPU (s)
GS	514	160	3,259	3,433
ADI	129	44	847	1,115
Block correction	209	77	159	242
Two-line TDMA	63	27	288	472
Stone ($\alpha_s = 0.9$)	107	39	213	286

Table 9.4: Problem 4 variable conductivity ($IN = 81$, $JN = 41$).

Procedure	Iterations	CPU (s)
GS	3,546	4,100
ADI	893	1,256
Block correction	133	208
Two-line TDMA	299	550
Stone ($\alpha_s = 0.9$)	236	337

The results are shown in Table 9.3. Here, owing to heat transfer coefficient boundary condition at $Y = 0$, both T_0 and q_0 are not a priori known. Therefore, in this problem with a nonlinear boundary condition, the computer times are greater than in Problem 1 for the $IN = 33$ and $JN = 17$ grid. However, despite the nonlinear boundary condition, GS and ADI showed monotonic convergence (not shown here) whereas the block correction, two-line TDMA, and Stone's methods showed mildly oscillatory convergence. On both grids, Stone's method is attractively fast. Incidentally, for such problems, Patankar [53] recommends that convergence may be checked by overall domain heat balance rather than by the magnitude of the residual. In the present problem, the overall heat balance was satisfied within 0.0025%.

Table 9.4 shows results for Problem 4. In this problem, conductivity varies with temperature so that coefficients AE, AW, AN, and AS change with iterations. Computations are carried out for a very fine grid. The convergence rate now slows down compared with the rates mentioned for Problem 3. For this problem, the convergence history (R_l / R_1) is plotted in Figure 9.4. It is seen that, in all methods, the initial CR is high but decreases with increase in l. For the block-correction procedure, however, the initial rate is almost maintained throughout the iterative process, yielding the overall fastest convergence rate . The overall heat balance was satisfied within 0.025%.

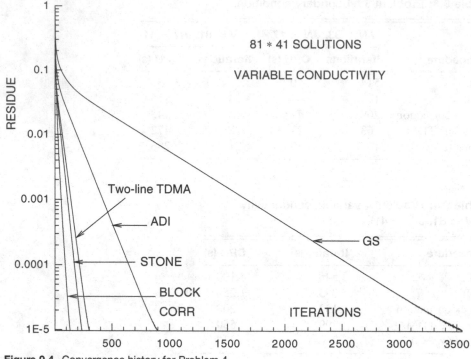

Figure 9.4. Convergence history for Problem 4.

EXERCISES

1. Derive appropriate block-correction equations for lines of constant j.

2. Starting with Equation 9.16, derive Equations 9.28 and 9.29.

3. Derive equations of two-line TDMA for lines of constant i and $i + 1$.

4. Starting with Equation 9.34, derive Equations 9.39–9.43.

5. Using the notation of the program LIB2D.FOR in Appendix C, write subroutines to implement block-correction, two-line TDMA, and Stone's procedures.

APPENDIX A

Derivation of Transport Equations

A.1 Introduction

In the study of transport phenomena in moving fluids, the fundamental laws of motion (conservation of mass and Newton's second law) and energy (first law of thermodynamics) are applied to an elemental fluid. Two approaches are possible:

1. a particle approach or
2. a continuum approach.

In the *particle approach*, the fluid is assumed to consist of particles (molecules, atoms, etc.) and the laws are applied to study particle motion. Fluid motion is then described by the statistically averaged motion of a group of particles. For most applications arising in engineering and the environment, however, this approach is too cumbersome[1] because the significant dimensions of the flow are considerably bigger than the mean-free-path length between molecules. In the *continuum approach*, therefore, statistical averaging is assumed to have been *already performed* and the fundamental laws are applied to portions of fluid (or *control volumes*) that contain a large number of particles. The information lost in averaging must however be recovered. This is done by invoking some further *auxiliary laws* and by empirical specifications of *transport properties* such as viscosity μ, thermal conductivity k, and mass diffusivity D. The transport properties are typically determined from experiments. Notionally, the continuum approach is very attractive because one can now speak of temperature, pressure, or velocity *at a point* and relate them to what is *measured* by most practical instruments.

Guidance for deciding whether the particle or continuum approach is to be used can be obtained from the *Knudsen number* $Kn = l/L$, where l is the mean-free-path length between molecules and L is a characteristic dimension (say, the radius of

---

[1] This can be appreciated from Avogadro's number, which specifies that, at normal temperature and pressure, a gas will contain 6.022×10^{26} molecules per kmol. Thus in air, for example, there will be 10^{16} molecules/mm^3.

a pipe) of the flow. When Kn is very small ($<10^{-5}$), the continuum approach is considered valid. In engineering and environmental flows, therefore, the continuum approach is adopted.

Control Volume

The notion of a controlvolume (CV) is very important in the continuum approach. The CV may be defined as a region in space across the boundaries of which matter, energy, and momentum may flow; it is a region within which source or sink of the same quantities may prevail. Further, it is a region on which external forces may act.

In general, a CV may be large or infinitesimally small. However, consistent with the idea of a *differential* in a continuum, an infinitesimally small CV is considered. Thus, when the laws are to be expressed through differential equations, the CV is located within a moving fluid. Again, two approaches are possible:

1. a Lagrangian approach or
2. a Eulerian approach.

In the *Lagrangian approach*, the CV is considered to be moving with the fluid as a whole. In the *Eulerian approach*, in contrast, the CV is assumed *fixed* in space and the fluid is assumed to flow through and past the CV. Except when dealing with certain types of unsteady flows (waves, for example), the Eulerian approach is generally used for its notional simplicity. Also, measurements made using stationary instruments can be directly compared with the solutions of differential equations obtained using the Eulerian approach.

Finally, it is important to note that the fundamental laws define *total flows* of mass, momentum, and energy not only in terms of *magnitude* but also in terms of *direction*. In a general problem of convection, neither magnitude nor direction is known a priori at different positions in the flowing fluid. The problem of ignorance of direction is circumvented by resolving velocity, force, and scalar fluxes in three directions that define the space.

In the derivations to follow, the three chosen directions will be along Cartesian coordinates. The derivations are carried out using the continuum approach within a Eulerian specification of the CV. Figure A.1 shows the considered CV of dimensions Δx_1, Δx_2, and Δx_3 located at (x_1, x_2, x_3) from a fixed origin.

A.2   Mass Conservation – Fluid Mixture

The law of conservation of mass states that

$$\text{Rate of accumulation of mass } (\dot{M}_{ac}) = \text{Rate of mass in } (\dot{M}_{in})$$
$$- \text{Rate of mass out } (\dot{M}_{out}).$$

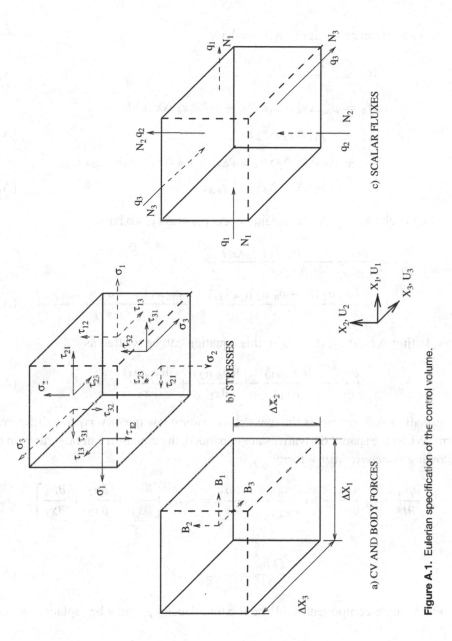

Figure A.1. Eulerian specification of the control volume.

a) CV AND BODY FORCES

b) STRESSES

c) SCALAR FLUXES

Thus, with reference to Figure A.1, we have

$$\dot{M}_{ac} = \frac{\partial(\rho_m \, \Delta V)}{\partial t}, \tag{A.1}$$

$$\dot{M}_{in} = \rho_m \, \Delta x_2 \, \Delta x_3 \, u_1 \, |_{x_1} + \rho_m \, \Delta x_3 \, \Delta x_1 \, u_2 \, |_{x_2}$$
$$+ \rho_m \, \Delta x_1 \, \Delta x_2 \, u_3 \, |_{x_3}, \tag{A.2}$$

$$\dot{M}_{out} = \rho_m \, \Delta x_2 \, \Delta x_3 \, u_1 \, |_{x_1+\Delta x_1} + \rho_m \, \Delta x_3 \, \Delta x_1 \, u_2 \, |_{x_2+\Delta x_2}$$
$$+ \rho_m \, \Delta x_1 \, \Delta x_2 \, u_3 \, |_{x_3+\Delta x_3}. \tag{A.3}$$

Dividing each term by $\Delta V (\text{constant}) = \Delta x_1 \, \Delta x_2 \, \Delta x_3$, we have

$$\frac{\partial \rho_m}{\partial t} = \frac{(\rho_m u_1 \, |_{x_1} - \rho_m u_1 \, |_{x_1+\Delta x_1})}{\Delta x_1}$$
$$+ \frac{(\rho_m u_2 \, |_{x_2} - \rho_m u_2 \, |_{x_2+\Delta x_2})}{\Delta x_2} + \frac{(\rho_m u_3 \, |_{x_3} - \rho_m u_3 \, |_{x_3+\Delta x_3})}{\Delta x_3}. \tag{A.4}$$

Now, letting $\Delta x_1, \Delta x_2, \Delta x_3 \rightarrow 0$, this equation can be written as

$$\frac{\partial \rho_m}{\partial t} + \frac{\partial(\rho_m u_1)}{\partial x_1} + \frac{\partial(\rho_m u_2)}{\partial x_2} + \frac{\partial(\rho_m u_3)}{\partial x_3} = 0. \tag{A.5}$$

Equation A.5 represents the mass conservation law in *conservative* differential form. When the spatial derivatives are expanded, the equation can be written in the following *nonconservative* form:

$$\frac{\partial \rho_m}{\partial t} + u_1 \frac{\partial \rho_m}{\partial x_1} + u_2 \frac{\partial \rho_m}{\partial x_2} + u_3 \frac{\partial \rho_m}{\partial x_3} = -\rho_m \left[\frac{\partial u_1}{\partial x_1} + \frac{\partial u_2}{\partial x_2} + \frac{\partial u_3}{\partial x_3} \right], \tag{A.6}$$

or

$$\frac{D \rho_m}{D t} = -\rho_m \, \nabla \cdot V. \tag{A.7}$$

For a single-component fluid, the mixture density ρ_m may be replaced by ρ.

A.3 Momentum Equations

Newton's second law of motion states that for a *given direction*

Rate of accumulation of momentum (Mom_{ac})

= Rate of momentum in (Mom_{in})

− Rate of momentum out (Mom_{out})

+ Sum of forces acting on the CV (F_{cv}).

Thus, with respect to Figure A.1, we can write the contributions in the x_1 direction as

$$Mom_{ac} = \frac{\partial(\rho_m \Delta V u_1)}{\partial t}, \tag{A.8}$$

$$Mom_{in} = (\rho_m \Delta x_2 \Delta x_3 u_1) u_1 |_{x_1} + (\rho_m \Delta x_3 \Delta x_1 u_2) u_1 |_{x_2}$$
$$+ (\rho_m \Delta x_1 \Delta x_2 u_3) u_1 |_{x_3}, \tag{A.9}$$

$$Mom_{out} = (\rho_m \Delta x_2 \Delta x_3 u_1) u_1 |_{x_1 + \Delta x_1} + (\rho_m \Delta x_3 \Delta x_1 u_2) u_1 |_{x_2 + \Delta x_2}$$
$$+ (\rho_m \Delta x_1 \Delta x_2 u_3) u_1 |_{x_3 + \Delta x_3},$$

$$F_{cv} = -(\sigma_1 |_{x_1} - \sigma_1 |_{x_1 + \Delta x_1}) \Delta x_2 \Delta x_3 + (\tau_{21} |_{x_2 + \Delta x_2} - \tau_{21} |_{x_2}) \Delta x_3 \Delta x_1$$
$$+ (\tau_{31} |_{x_3 + \Delta x_3} - \tau_{31} |_{x_3}) \Delta x_1 \Delta x_2 + \rho_m B_1 \Delta V, \tag{A.10}$$

where B_1 is the body force per unit mass, the σs are tensile normal stresses, and τs are shear stresses. Now, dividing by ΔV and letting $\Delta x_1, \Delta x_2, \Delta x_3 \to 0$, it can be shown that

x_1 Direction Momentum Equation

$$\frac{\partial(\rho_m u_1)}{\partial t} + \frac{\partial(\rho_m u_1 u_1)}{\partial x_1} + \frac{\partial(\rho_m u_2 u_1)}{\partial x_2} + \frac{\partial(\rho_m u_3 u_1)}{\partial x_3}$$
$$= \frac{\partial(\sigma_1)}{\partial x_1} + \frac{\partial(\tau_{21})}{\partial x_2} + \frac{\partial(\tau_{31})}{\partial x_3} + \rho_m B_1. \tag{A.11}$$

A similar exercise in the x_2 and x_3 directions will yield

x_2 Direction Momentum Equation

$$\frac{\partial(\rho_m u_2)}{\partial t} + \frac{\partial(\rho_m u_1 u_2)}{\partial x_1} + \frac{\partial(\rho_m u_2 u_2)}{\partial x_2} + \frac{\partial(\rho_m u_3 u_2)}{\partial x_3}$$
$$= \frac{\partial(\tau_{12})}{\partial x_1} + \frac{\partial(\sigma_2)}{\partial x_2} + \frac{\partial(\tau_{32})}{\partial x_3} + \rho_m B_2. \tag{A.12}$$

x_3 Direction Momentum Equation

$$\frac{\partial(\rho_m u_3)}{\partial t} + \frac{\partial(\rho_m u_1 u_3)}{\partial x_1} + \frac{\partial(\rho_m u_2 u_3)}{\partial x_2} + \frac{\partial(\rho_m u_3 u_3)}{\partial x_3}$$
$$= \frac{\partial(\tau_{13})}{\partial x_1} + \frac{\partial(\tau_{23})}{\partial x_2} + \frac{\partial(\sigma_3)}{\partial x_3} + \rho_m B_3. \tag{A.13}$$

A few comments on these equations are now in order:

1. By making use of Equation A.5, the left-hand sides of Equations A.11, A.12, and A.13 can be replaced by $\rho_m D(u_1)/Dt$, $\rho_m D(u_2)/Dt$, and $\rho_m D(u_3)/Dt$,

respectively. Such equations are called *nonconservative* forms of momentum equations.

2. Equations A.5, A.11, A.12, and A.13 define the fluid motion completely. However, they contain twelve unknowns (three velocity components and nine stresses). By invoking the rule of complementarity of stresses (i.e., $\tau_{ij} = \tau_{ji}$, $i \neq j$), the unknowns can be reduced to nine. Still, the number of unknowns exceeds the number of available equations (four).

3. A solvable system must have the same number of unknowns and equations. To do this, Stokes's stress laws are invoked:

Stress Laws

$$\tau_{ij} = \mu \left(\frac{\partial u_i}{\partial x_j} + \frac{\partial u_j}{\partial x_i} \right),$$ (A.14)

$$\sigma_i = -p + \sigma_i' = -p + 2\mu \left(\frac{\partial u_i}{\partial x_i} \right) \quad \text{(no summation)}, \quad \text{(A.15)}$$

where σ_i' is called the deviatoric stress,[2] p is pressure (compressive), and μ is the viscosity of the fluid.[3]

4. When Equations A.14 and A.15 are substituted in Equations A.11, A.12, and A.13, the new equations can be compactly written in tensor notation as

Momentum Equations u_i (i = 1, 2, 3)

$$\frac{\partial(\rho_m u_i)}{\partial t} + \frac{\partial(\rho_m u_j u_i)}{\partial x_j} = \frac{\partial}{\partial x_j} \left[\mu_{\text{eff}} \frac{\partial u_i}{\partial x_j} \right] - \frac{\partial p}{\partial x_i} + \rho_m B_i + S_{u_i}. \quad \text{(A.16)}$$

This equation is the same as Equation 1.3 in Chapter 1. The three equations (A.16) now contain only four unknowns (u_1, u_2, u_3, and p). Along with Equation A.5, therefore, there are as many unknowns as there are equations.

---

[2] In Chapter 1, the deviatoric stress is expressed as

$$\sigma_i' = 2\mu \left(\frac{\partial u_i}{\partial x_i} \right) + q$$

and significance of q is explained in Section 1.5.

[3] In turbulent flows, the total stress comprises additive contributions of laminar and turbulent components. The turbulent stress $\tau_{ij}^t = -\rho_m \overline{u_i' u_j'}$ is again represented in the manner of Equation A.14 by invoking turbulent viscosity μ_t. This is known as the Boussinesq approximation. Then the total stress τ_{ij}^{tot} in a turbulent flow is given by

$$\tau_{ij}^{\text{tot}} = \tau_{ij} + \tau_{ij}^t = (\mu + \mu_t) \left(\frac{\partial u_i}{\partial x_j} + \frac{\partial u_j}{\partial x_i} \right) - \frac{2}{3} \rho_m e \, \delta_{ij},$$

where δ_{ij} is the Kronecker delta and $e = \sum_{i=1}^{3} \overline{u_i' u_i'}/2$ is the kinetic energy of velocity fluctuations.

5. In incompressible flows, the density ρ_m is externally specified as a constant or as a function of temperature and the sum of partial densities of mixture components. In compressible flow, however, the density is recovered from an equation of state. Thus, according to the law of corresponding states, for reduced[4] pressure $p_r < 0.5$ and reduced temperature $T_r > 1.5$, the density is calculated from the perfect gas relation

$$\rho_m = \frac{p}{R_g\,T} = \frac{p\,M_g}{R_u\,T},\tag{A.17}$$

where M_g is the molecular weight of the gas and R_u is the universal gas constant.

A.4 Equation of Mass Transfer

The conservation of mass for species k of the mixture is stated as

Rate of accumulation of mass $(\dot{M}_{k,\mathrm{ac}})$ = Rate of mass in $(\dot{M}_{k,\mathrm{in}})$

$-$ Rate of mass out $(\dot{M}_{k,\mathrm{out}})$

$+$ Rate of generation within CV (R_k)

To apply this principle, let ρ_k be the density of the species k in a fluid mixture of density ρ_m. Similarly, let $N_{i,k}$ be the *mass transfer flux* (kg/m²-s) of species k in the i direction. Then

$$\dot{M}_{k,\mathrm{ac}} = \frac{\partial(\rho_k\,\Delta V)}{\partial t},$$

$$\dot{M}_{k,\mathrm{in}} = N_{1,k}\,\Delta x_2\,\Delta x_3\,|_{x_1} + N_{2,k}\,\Delta x_3\,\Delta x_1\,|_{x_2} + N_{3,k}\,\Delta x_1\,\Delta x_2\,|_{x_3},$$

$$\dot{M}_{k,\mathrm{out}} = N_{1,k}\,\Delta x_2\,\Delta x_3\,|_{x_1+\Delta x_1} + N_{2,k}\,\Delta x_3\,\Delta x_1\,|_{x_2+\Delta x_2} + N_{3,k}\,\Delta x_1\,\Delta x_2\,|_{x_3+\Delta x_3}.$$

Dividing each term by ΔV and letting $\Delta x_1, \Delta x_2, \Delta x_3 \to 0$, we get

$$\frac{\partial(\rho_k)}{\partial t} + \frac{\partial(N_{1,k})}{\partial x_1} + \frac{\partial(N_{2,k})}{\partial x_2} + \frac{\partial(N_{3,k})}{\partial x_3} = R_k.\tag{A.18}$$

Now, the total mass transfer flux $N_{i,k}$ is the sum of *convective* flux due to bulk fluid motion (with each species having the same velocity as the bulk fluid) and *diffusion* flux $(m''_{i,k})$. Thus,

$$N_{i,k} = \rho_k\,u_i + m''_{i,k}.\tag{A.19}$$

---

[4] Reduced pressure and temperature are defined as $p_r = p/p_{cr}$ and $T_r = T/T_{cr}$, where the suffix cr stands for the critical point.

Under certain restricted circumstances of interest in this book, the diffusion flux is given by Fick's law of mass diffusion

$$m''_{i,k} = -D \frac{\partial \rho_k}{\partial x_i}, \tag{A.20}$$

where D (m^2/s) is the mass diffusivity.[5] Substituting Equations A.19 and A.20 in Equation A.18, we can show that

$$\frac{\partial(\rho_k)}{\partial t} + \frac{\partial(\rho_k u_1)}{\partial x_1} + \frac{\partial(\rho_k u_2)}{\partial x_2} + \frac{\partial(\rho_k u_3)}{\partial x_3} = \frac{\partial}{\partial x_1}\left(D\frac{\partial \rho_k}{\partial x_1}\right) + \frac{\partial}{\partial x_2}\left(D\frac{\partial \rho_k}{\partial x_2}\right)$$

$$+ \frac{\partial}{\partial x_3}\left(D\frac{\partial \rho_k}{\partial x_3}\right) + R_k. \tag{A.21}$$

It is a common practise to refer to species k via its *mass fraction* ω_k defined as

$$\omega_k = \frac{\rho_k}{\rho_m} \qquad \sum_{\text{all species}} \omega_k = 1. \tag{A.22}$$

Using this definition, Equation A.21 can be compactly written as

$$\frac{\partial(\rho_m \omega_k)}{\partial t} + \frac{\partial(\rho_m u_j \omega_k)}{\partial x_j} = \frac{\partial}{\partial x_j}\left(\rho_m D \frac{\partial \omega_k}{\partial x_j}\right) + R_k. \tag{A.23}$$

Note that when the mass transfer equation is summed over all species of the mixture, the mass conservation equation for the bulk fluid (Equation A.5) is retrieved. This is because $\sum R_k = 0$. That is, when some species are generated by a chemical reaction, others are destroyed so that there is no net mass generation in the bulk fluid.

A.5　Energy Equation

The first law of thermodynamics, when considered in *rate* form (W/m^3), can be written as

$$\dot{E} = \dot{Q}_{\text{conv}} + \dot{Q}_{\text{cond}} + \dot{Q}_{\text{gen}} - \dot{W}_s - \dot{W}_b, \tag{A.24}$$

where

$$\dot{E} = \text{Rate of change of energy of the CV,}$$

$$\dot{Q}_{\text{conv}} = \text{Net rate of energy transferred by convection,}$$

$$\dot{Q}_{\text{cond}} = \text{Net rate of energy transferred by conduction,}$$

---

[5] The mass diffusivity is defined only for a binary mixture of two fluids 1 and 2 as D_{12}. In multicomponent gaseous mixtures, however, diffusivities for pairs of species are nearly equal and a single symbol D suffices for all species. Incidentally, in turbulent flows, this assumption of equal (effective) diffusivities has even greater validity.

$$\dot{Q}_{gen} = \text{Net volumetric heat generation within the CV,}$$

$$\dot{W}_s = \text{Net rate of work done by surface forces, and}$$

$$\dot{W}_b = \text{Net rate of work done by body forces.}$$

Each term will now be represented by a mathematical expression.

Rate of Change

The equation for the rate of change is

$$\dot{E} = \frac{\partial(\rho_m e^o)}{\partial t}, \quad e^o = e + \frac{V^2}{2} = h - \frac{p}{\rho_m} + \frac{V^2}{2}, \tag{A.25}$$

where e represents specific energy (J/kg), h is specific enthalpy (J/kg), and $V^2 = u_1^2 + u_2^2 + u_3^2$. In the expression for e^o, contributions from other forms of energy (potential, chemical, electromagnetic, etc.) are neglected.

Convection and Conduction

Following the convention that heat energy flowing *into* the CV is positive (and vice versa), it can be shown that

$$\dot{Q}_{conv} = -\frac{\partial \sum (N_{j,k} e_k^o)}{\partial x_j}, \tag{A.26}$$

where $N_{j,k}$ is given by Equation A.19. Now, since all species have the same velocity,

$$\dot{Q}_{conv} = -\frac{\partial}{\partial x_j} \sum \left[N_{j,k} (h_k - p_k/\rho_k + V^2/2) \right], \tag{A.27}$$

where p_k is the partial pressure of species k. After some algebra, it can be shown that

$$\dot{Q}_{conv} = -\frac{\partial(\rho_m u_j e^o)}{\partial x_j} - \frac{\partial \left(\sum m''_{j,k} h_k \right)}{\partial x_j}. \tag{A.28}$$

The conduction contribution is given by Fourier's law of heat conduction, so that

$$\dot{Q}_{cond} = -\frac{\partial q_j}{\partial x_j} = \frac{\partial}{\partial x_j} \left[k_m \frac{\partial T}{\partial x_j} \right]. \tag{A.29}$$

Volumetric Generation

Two principal components of volumetric energy generation are chemical energy ($\dot{Q}_{chem}$) and radiative transfer ($\dot{Q}_{rad}$). Thus,

$$\dot{Q}_{gen} = \dot{Q}_{chem} + \dot{Q}_{rad}. \tag{A.30}$$

The chemical energy is positive for exothermic reactions and negative for endothermic reactions. Evaluation of $\dot{Q}_{\text{chem}}$ depends on the chemical reaction model employed in a particular situation. The $\dot{Q}_{\text{rad}}$ term represents the net radiation exchange between the control volume and its surroundings. Evaluation of this term, in general, requires solution of integro-differential equations [48]. However, in certain restrictive circumstances, the term may be represented analogous to $\dot{Q}_{\text{cond}}$ with k replaced by *radiation conductivity* k_{rad} as

$$k_{\text{rad}} = \frac{16\,\sigma\,T^3}{a+s},\tag{A.31}$$

where σ is the Stefan–Boltzmann constant and a and s are absorption and scattering coefficients, respectively.

Work Done by Surface and Body Forces

Following the convention that the work done *on* the CV is negative, it can be shown that

$$-\dot{W}_{\text{s}} = \frac{\partial}{\partial x_1}\left[\sigma_1 u_1 + \tau_{12} u_2 + \tau_{13} u_3\right] + \frac{\partial}{\partial x_2}\left[\tau_{21} u_1 + \sigma_2 u_2 + \tau_{23} u_3\right]$$

$$+ \frac{\partial}{\partial x_3}\left[\tau_{31} u_1 + \tau_{32} u_2 + \sigma_3 u_3\right],\tag{A.32}$$

$$-\dot{W}_{\text{b}} = \rho_{\text{m}}\left(B_1 u_1 + B_2 u_2 + B_3 u_3\right).\tag{A.33}$$

Adding these two equations and making use of Equations A.11–A.14 can show that

$$-(\dot{W}_{\text{s}} + \dot{W}_{\text{b}}) = \rho_{\text{m}} \frac{D}{Dt}\left[\frac{V^2}{2}\right] + \mu\,\Phi_{\text{v}} - p\,\nabla \cdot V,\tag{A.34}$$

where $V^2/2$ is the mean kinetic energy and the *viscous dissipation* function is given by

$$\Phi_{\text{v}} = 2\left[\left(\frac{\partial u_1}{\partial x_1}\right)^2 + \left(\frac{\partial u_2}{\partial x_2}\right)^2 + \left(\frac{\partial u_3}{\partial x_3}\right)^2\right]$$

$$+ \left(\frac{\partial u_1}{\partial x_2} + \frac{\partial u_2}{\partial x_1}\right)^2 + \left(\frac{\partial u_1}{\partial x_3} + \frac{\partial u_3}{\partial x_1}\right)^2 + \left(\frac{\partial u_3}{\partial x_2} + \frac{\partial u_2}{\partial x_3}\right)^2. \tag{A.35}$$

Combining Equations A.24–A.35 therefore leads to

$$\frac{\partial \rho_{\text{m}} e^o}{\partial t} + \frac{\partial(\rho_{\text{m}} u_j e^o)}{\partial x_j} = \frac{\partial}{\partial x_j}\left[k_{\text{m}} \frac{\partial T}{\partial x_j}\right] - \frac{\partial\left(\sum m''_{j,k} h_k\right)}{\partial x_j}$$

$$+ \frac{D}{Dt}\left[\frac{V^2}{2}\right] - p\,\nabla \cdot V + \mu\,\Phi_{\text{v}} + \dot{Q}_{\text{chem}} + \dot{Q}_{\text{rad}}.$$

$$\tag{A.36}$$

By using Equation A.5, the left-hand side of this equation can be replaced by $\rho_m \, De^o/Dt$. Further, if e^o is replaced by enthalpy h (see Equation A.25), Equation A.36 can also be written as

$$\rho_m \frac{Dh}{Dt} = \frac{\partial}{\partial x_j}\left[k_m \frac{\partial T}{\partial x_j}\right] - \frac{\partial\left(\sum m''_{j,k} h_k\right)}{\partial x_j} + \mu \, \Phi_v + \frac{Dp}{Dt} + \dot{Q}_{chem} + \dot{Q}_{rad}.$$

(A.37)

For reacting or nonreacting mixtures and under various assumptions listed in [33], it is possible to combine energy transfer by conduction and mass diffusion so that Equation A.37 may also be written as

$$\rho_m \frac{Dh}{Dt} = \frac{\partial}{\partial x_j}\left[\frac{k_m}{Cp_m} \frac{\partial h}{\partial x_j}\right] + \mu \, \Phi_v + \frac{Dp}{Dt} + \dot{Q}_{chem} + \dot{Q}_{rad}. \quad \text{(A.38)}$$

APPENDIX B

1D Conduction Code

B.1 Structure of the Code

The 1D conduction code is divided into two parts:

1. a *user part* containing files COM1D.FOR and USER1D.FOR and
2. a *library part* containing file LIB1D.FOR.

The user part is *problem dependent*. Therefore, the two files in this part are used to specify the problem to be solved. In contrast, the library part is *problem independent*. Thus, the LIB1D.FOR file remains unaltered for *all* problems. In this sense, the library part may be called the *solver* whereas the user part may be called the *pre-* and *postprocessor*.

This structure is central to creation of a generalised code. To *execute* the code, USER1D.FOR and LIB1D.FOR files are compiled separately and then *linked* before execution. The COM1D.FOR is common to both parts and its contents are brought into each subroutine or function via the "INCLUDE" statement in FORTRAN. Variable names starting with I, J, K, L, M, and N are *integers* whereas all others are *real* by default. The list of variable names with their meanings is given in Table B.1. The listings of each file are given at the end of this appendix.

B.2 File COM1D.FOR

In this file, logical, real, and integer variables are included. The PARAMETER statement is used to specify the maximum array dimension IT and values of π, GREAT, and SMALL. The latter are frequently required for generalised coding. The variable names are given in a *labelled* COMMON as in COMMON/BOUND/..., where BOUND is the label. Here, variables of relevance to boundary conditions are included. If required, the user may add more variable names or arrays for the specific problem at hand as shown at the bottom of the file.

Table B.1: List of variables 1D for conduction code.

Variable	Meaning
ACF	Array containing cross-sectional area (m^2) at cell face w
AE, AW	Array containing east and west coefficients
AL	Domain length (m)
AP	Array containing coefficient of variable Φ_P
COND	Array containing conductivity (W/m-K) at node P
CONDREF	Reference conductivity
CC	Convergence criterion
DELT	Time step (s)
DUM1,DUM2	Dummy arrays
FCMX	Maximum absolute fractional change
GAUSS	Logical – refers to Gauss–Seidel method
GREAT	Parameter having a large value 10^{30}
H1SPEC	Logical – refers to h-boundary condition at node 1
HB1	Heat transfer coefficient (W/m^2-K) at node 1
HB1O	Heat transfer coefficient at node 1 at old time
HBN	Heat transfer coefficient at node N
HBNO	Heat transfer coefficient at node N at old time
HNSPEC	Logical – refers to h-boundary condition at node N
HPREF	Heat transfer coefficient at any x
HPREFO	Heat transfer coefficient at any x at old time
ISTOP	STOP index – used in unsteady problems
IT	Parameter containing array size
ITER	Iteration counter
ITERMX	Maximum number of allowable iterations
N	Total number of nodes
NTIME	Current time counter
PERIM	Array containing perimeter (m) at any x
PI	Value of π
PSI	Variable Ψ for choosing explicit/implicit scheme
Q1SPEC	Logical – refers to q-boundary condition at node 1
QB1	Heat flux (W/m^2) at node 1
QB1O	Heat flux at node 1 at old time
QBN	Heat flux at node N
QBNO	Heat flux at node N at old time
QNSPEC	Logical – refers to q-boundary condition at node N
RHO	Array for density (kg/m^3)
RP	Relaxation parameter α
SMALL	Parameter having a small value 10^{-30}
SP	Array containing Sp
SPH	Array containing specific heat (J/kg-K) at node P
SPHREF	Reference specific heat
STAB	Array for storing boundary coefficients
STEADY	Logical – refers to steady-state calculation
SU	Array containing Su

(continued)

Table B.1 (*continued*)

Variable	Meaning
T	Array containing temperature (°C or K)
T1	Temperature at node 1
T1O	Temperature at node 1 at old time
T1SPEC	Logical – refers to T-boundary condition at node 1
THOMAS	Logical – refers to TDMA
TIMEMX	Maximum allowable time
TINF	Temperature T_∞
TINFO	Temperature T_∞ at old time
TINF1	Temperature T_∞ near node 1
TINFN	Temperature T_∞ near node N
TINF1O	Temperature T_∞ near node 1 at old time
TINFNO	Temperature T_∞ near node N at old time
TN	Temperature at node N
TNO	Temperature at node N at old time
TNSPEC	Logical – refers to T-boundary condition at node N
TO	Array containing temperature at old time
TTIME	Total current time
UNSTEADY	Logical – refers to unsteady-state calculation
VOL	Array containing cell volume (m^3)
X	Coordinate of node P (m)
XCELL	Logical – refers to cell-face coordinate specification
XCF	Coordinate of cell face at w
XNODE	Logical – refers to node coordinate specification

B.3   File USER1D.FOR

This is the main control file at the command of the user. The first routine PROGRAM ONED is the command routine from where subroutine MAIN is called. The latter is the first subroutine of the LIB1D.FOR file. When all operations are completed, PROGRAM ONED calls the RESULT subroutine, which is a part of the USER1D.FOR file.

Following the listing of the COM1D.FOR file, listings of two USER1D.FOR files are given. They correspond to the two solved problems in Chapter 2. The reader is advised to refer to these files as well as to Table B.1 to understand the description of each routine in USER1D.FOR file.

BLOCK DATA This routine at the end of the USER1D.FOR file specifies all the problem-dependent data such as properties, boundary conditions, and other control parameters. It is assumed that all data are given in consistent units. Here, SI units are used except for the grid data XCF or X, which are dimensionless. The physical coordinates in meters are then evaluated by multiplying by AL (the domain length) in PROGRAM ONED. Dimensionless specification provides better appreciation of nonuniformity (if any) in the specified grid. When a nonuniform grid is specified, it is advisable to ensure that the ratio of two consecutive cell sizes does not exceed 2.

Subroutine INIT In this routine, an initial guess for T at ITER $= 0$ in a steady-state problem or at $t = 0$ in an unsteady-state problem is given. In a steady-state problem, the number of iterations (and hence the computer time) greatly depends on how close the initial guess is to the final converged solution. In the fin problem (Problem 2, Chapter 2), a linear temperature profile is given with $T1 = 225$ (given) and TN $= 205$ (which is guessed) although the converged solution is nonlinear.

Subroutine NEWVAL In this routine, boundary conditions at a new time (if different from the initial time) are specified.

Subroutine PROPS In this routine, thermal conductivity and specific heat are given. They may be functions of x, t, or T. The density is of course constant in our formulation (see Chapter 2).

Subroutine SORCE A problem-dependent source ($q''' \Delta V$) is given in this routine. It may be a function of T, x, and/or Ψ.

Subroutine INTPRI This routine prints the converged solution at the current time step. The routine can also be used to store current values in dummy arrays DUM1 and DUM2 for later printing or plotting. Here, the STOP condition may be given.

Functions HPERI, AREA, and PERI These function routines calculate heat transfer coefficient at node I and area and perimeter at location X or XCF as per the specifications in their arguments. Note that heat transfer coefficients may be functions of T, x, and/or t.

Subroutine RESULT In this last routine, the converged solution is printed along with evaluation and printing of derived parameters. For example, in Problem 2 of Chapter 2, it is of interest to calculate heat loss from the fin as well as fin effectiveness and compare them with the exact solutions. This routine can also be used to create files containing results for postprocessing using graphics packages such as GNUPLOT or GRAPHER.

B.4 File LIB1D.FOR

Subroutine MAIN All subroutines in the code are called from this subroutine. First, subroutines GRID and INIT are called. Then, starting with TTIME $= 0$, an *outer* DO loop (3000) is initiated to begin calculations at a time step NTIME and TTIME is incremented by DELT. Subroutine NEWVAL is called to set boundary conditions at a new time step. Then, iterations are carried out in an *inner* loop (1000) in which subroutines PROPS, COEF, SORCE, BOUND, and SOLVE are called in turn. The SOLVE routine returns the value of FCMX. If this value is less than 10^{-4}, the inner loop is exited; otherwise a further iteration is carried out by returning to "1000 ITER = ITER + 1." In a steady-state problem, a minimum of two iterations are performed. If the problem is steady, there is no need to carry out calculations at a new time step and, therefore, the outer loop is also now exited and control is transferred to statement "5000 CONTINUE." If the problem is unsteady, subroutines UPDATE and INTPRI are called and the outer loop continues.

Subroutine GRID In this routine, depending on logical XCELL or XNODE, coordinates XCF or X are set and area, perimeter, and cell volume are calculated and printed. It is always desirable to check these specifications in the output file OO (see PROGRAM ONED).

Subroutine COEF In this routine, coefficients AE and AW are evaluated. Note that cell-face conductivities are evaluated by harmonic mean.

Subroutine BOUND This routine implements specified boundary conditions at I = 1 and I = N. The implementation is carried out by updating Su and Sp at near-boundary nodes as explained in Chapter 2.

Subroutine SOLVE In this routine, Su and Sp are further updated if the problem is unsteady. Also, if the stability criterion is violated, a warning message is printed. AP and Su are further augmented to take account of the underrelaxation factor. Thus, all coefficients are ready to solve the discretised equations. This is done by GS or by TDMA depending on the user choice specified in the BLOCK DATA routine.

Subroutine UPDATE This routine sets all new variables to their "OLD" counterparts.

Subroutine PRINT The arguments of this general routine carry the variable F and its logical name "HEADER" specified from point-of-call. The routine is written to print six variables on a line. If N > 6, the next six variables are printed on the next line, and so on. The values are printed in E-format but the user may change to F-format, if desired.

COMMON BLOCK COM1D.FOR

```
C *** THIS IS COMMON BLOCK FOR 1-D CONDUCTION PROGRAM
        PARAMETER(IT=50,PI=3.1415927,SMALL=1E-30,GREAT=1E30)
        LOGICAL T1SPEC,H1SPEC,Q1SPEC,TNSPEC,HNSPEC,QNSPEC
        LOGICAL STEADY,UNSTEADY,GAUSS,THOMAS,XCELL,XNODE
        COMMON/BOUNDS/T1SPEC,H1SPEC,Q1SPEC,TNSPEC,HNSPEC,QNSPEC
        COMMON/STATE/STEADY,UNSTEADY,GAUSS,THOMAS,XCELL,XNODE
        COMMON/CVAR/T(IT),TO(IT),SPH(IT),COND(IT),RHO(IT)
        COMMON/COORDS/X(IT),XCF(IT),ACF(IT),PERIM(IT),VOL(IT),AL
        COMMON/COEFF/AP(IT),AE(IT),AW(IT),SU(IT),SP(IT),STAB(IT)
        COMMON/CONTRO/ITERMX,N,RP,RSU,FCMX,CC,ISTOP
```

```
      COMMON/CTRAN/DELT,TIMEMX,MXSTEP,PSI,ITER,NTIME,TTIME
      COMMON/CPROPS/CONDREF,RHOREF,SPHREF
      COMMON/CDAT1/T1,TN,QB1,QBN,HB1,HBN,TINF1,TINFN,HPREF,TINF
      COMMON/CDAT1/QB10,QBNO,HB10,HBNO,TINF10,TINFNO,HPREFO,TINFO
      COMMON/CDUM/DUM1(5000),DUM2(5000),DUM3(5000)
C ADDITIONAL PROBLEM-DEPENDENT VARIABLES
C VARIABLES FOR PROB2
      COMMON/CP2/BREADTH,THICK
C VARIABLES FOR PROB3
      COMMON/CRADS/R1,R2,R3
```

USER File for Problem 1 – Chapter 2

```
C ***************************************************
      PROGRAM ONED
      INCLUDE 'COM1D.FOR'
C ***************************************************
      OPEN(6,FILE='00')
      WRITE(6,*)' ***********************************************'
      WRITE(6,*)' ADHESION OF PLASTIC SHEETS - PROB1-CHAPTER2'
      WRITE(6,*)' ***********************************************'
      DO 1 I=1,N
1     XCF(I)=XCF(I)*AL
      CALL MAIN
      CALL RESULT
      STOP
      END
C ***************************************************
      SUBROUTINE INIT
      INCLUDE 'COM1D.FOR'
C ***************************************************
C GIVE INITIAL GUESS AT TIME=0.0  OR AT ITER=0 FOR STEADY STATE
      TIN=30
      DO 1 I=1,N
      T(I)=30
      IF(I.EQ.1.OR.I.EQ.N)T(I)=250
1     CONTINUE
      RETURN
      END
C ***************************************************
      SUBROUTINE NEWVAL
      INCLUDE 'COM1D.FOR'
C ***************************************************
```

```
C SET NEW VALUES OF HB1,HBN,QB1,QBN,TINF1,TINFN OR SOURCES
      RETURN
      END
C *************************************************
      SUBROUTINE PROPS
      INCLUDE 'COM1D.FOR'
C *************************************************
C COND(I) AND SPH(I) ARE DEFINED AT NODE P
      DO 1 I=1,N
      RHO(I)=RHOREF
      COND(I)=CONDREF
1     SPH(I)=SPHREF
      RETURN
      END
C *************************************************
      SUBROUTINE SORCE
      INCLUDE 'COM1D.FOR'
C *************************************************
C FORM PROBLEM DEPENDENT SOURCE TERM INCLUDING SU AND SP
      DO 1 I=2,N-1
      SU(I)=SU(I)+0.0
1     CONTINUE
      RETURN
      END
C *************************************************
      SUBROUTINE INTPRI
      INCLUDE 'COM1D.FOR'
      CHARACTER*20 HEADER
C *************************************************
      WRITE(6,*)' TIMESTEP = ',NTIME,' TOTAL TIME = ',TTIME
C PRINT TEMPERATURES AT THE CURRENT STEP
      HEADER=' TEMP '
      CALL PRINT(T,HEADER)
C STORE MID-POINT TEMPERATURE
      DUM1(NTIME)=T(4)
C GIVE STOP CONDITION
      IMID=4
      IF(T(IMID).GT.140)ISTOP=1
      RETURN
      END
```

```
C ****************************************************
C      FUNCTION ROUTINES
C ****************************************************
      FUNCTION HPERI(II)
      INCLUDE 'COM1D.FOR'
C H AT PERIMETER
      I=II
      HPERI=HPREF*0.0+X(I)*0.0+T(I)*0.0
      RETURN
      END
C -----------------------------------------------
      FUNCTION AREA(XX)
      INCLUDE 'COM1D.FOR'
C AREA OF CROSS-SECTION
      AREA=1.0+0.0*XX
      RETURN
      END
C -----------------------------------------------
      FUNCTION PERI(XX)
      INCLUDE 'COM1D.FOR'
C PERIMETER
      PERI=0*XX
      RETURN
      END
C ****************************************************
      SUBROUTINE RESULT
      INCLUDE 'COM1D.FOR'
      CHARACTER*20 HEADER
C ****************************************************
      HEADER=' FINAL-TEMP '
      CALL PRINT(T,HEADER)
      HEADER=' X(I) '
      CALL PRINT(X,HEADER)
      HEADER=' XCF(I) '
      CALL PRINT(XCF,HEADER)
C EXTRACT PROBLEM DEPENDENT PARAMETERS IF ANY
      WRITE(6,*)' PRINT MID-POINT TEMPERATURE'
      DO 1 I=1,NTIME
      TT=FLOAT(I)*DELT
1     WRITE(6,*)TT,DUM1(I)
      TNOW=DUM1(NTIME)
      TOLD=DUM1(NTIME-1)
```

```
      TT=FLOAT(NTIME-1)*DELT
      TIME=(140-TOLD)/(TNOW-TOLD)*DELT+TT
      WRITE(6,*)' TIME FOR ADHESION = ',TIME
      RETURN
      END
C ************************************************
      BLOCK DATA
      INCLUDE 'COM1D.FOR'
C ************************************************
C LOGICAL DECLARATIONS
      DATA STEADY,UNSTEADY,GAUSS,THOMAS/.FALSE.,.TRUE.,.TRUE.,.FALSE./
C ----------------------------------------------
C CONTROL PARAMETERS
C FULLY IMPLICIT (PSI=1),FULLY EXPLICIT (PSI=0),SEMI IMPLICIT (0<PSI<1)
      DATA PSI,DELT,MXSTEP,ITERMX,RP,CC/0.0,10,10000,500,1.0,1E-5/
C ----------------------------------------------
C BOUNDARY  SPECIFICATION
      DATA T1SPEC,Q1SPEC,H1SPEC/.TRUE.,2*.FALSE./
      DATA TNSPEC,QNSPEC,HNSPEC/.TRUE.,2*.FALSE./
      DATA T1,TN,QB1,QBN,HB1,HBN/250.0,250.0,0.0,0.0,0.0,0.0/
C     DATA TINF,TINF1,TINFN,HPREF/25,150,250,12.0/
      DATA CONDREF,RHOREF,SPHREF/0.25,1300,2000.0/
C ----------------------------------------------
C GRID SPECIFICATION
      DATA XCELL,XNODE/.TRUE.,.FALSE./
      DATA N,AL/7,0.01/
      DATA XCF/0.0,0.0,0.2,0.4,0.6,0.8,1.0,43*1.0/
      END
```

USER File for Problem 2 – Chapter 2

```
C ************************************************
      PROGRAM ONED
      INCLUDE 'COM1D.FOR'
C ************************************************
      OPEN(6,FILE='OO')
      WRITE(6,*)' **********************************************'
      WRITE(6,*)' RECTANGULAR FIN - PROB2-CHAPTER2'
      WRITE(6,*)' SOLVE BY GS AND TDMA'
      WRITE(6,*)' **********************************************'
      DO 1 I=1,N
```

```
1     XCF(I)=XCF(I)*AL
      CALL MAIN
      CALL RESULT
      STOP
      END
C ************************************************
      SUBROUTINE INIT
      INCLUDE 'COM1D.FOR'
C ************************************************
C GIVE INITIAL GUESS AT TIME=0.0  OR AT ITER=0 FOR STEADY STATE
      BB=(TN-T1)/AL
      DO 1 I=1,N
1     T(I)=T1+BB*X(I)
      RETURN
      END
C ************************************************
      SUBROUTINE NEWVAL
      INCLUDE 'COM1D.FOR'
C ************************************************
C SET NEW VALUES OF HB1,HBN,QB1,QBN,TINF1,TINFN OR SOURCES
      RETURN
      END
C ************************************************
      SUBROUTINE PROPS
      INCLUDE 'COM1D.FOR'
C ************************************************
C COND(I) AND SPH(I) ARE DEFINED AT NODE P
      DO 1 I=1,N
      COND(I)=CONDREF
1     SPH(I)=SPHREF
      RETURN
      END
C ************************************************
      SUBROUTINE SORCE
      INCLUDE 'COM1D.FOR'
C ************************************************
C FORM PROBLEM DEPENDENT SOURCE TERM INCLUDING SU AND SP
      DO 1 I=2,N-1
      TERM=HPERI(I)*PERIM(I)*(XCF(I+1)-XCF(I))
      SU(I)=SU(I)+TERM*TINF
      SP(I)=SP(I)+TERM
```

```
      1     CONTINUE
            RETURN
            END
C ****************************************************
            SUBROUTINE INTPRI
            INCLUDE 'COM1D.FOR'
C ****************************************************
            RETURN
            END
C ****************************************************
C        FUNCTION ROUTINES
C ****************************************************
            FUNCTION HPERI(II)
            INCLUDE 'COM1D.FOR'
C H AT PERIMETER
            I=II
            HPERI=HPREF+X(I)*0.0+T(I)*0.0
            RETURN
            END
C -----------------------------------------------
            FUNCTION AREA(XX)
            INCLUDE 'COM1D.FOR'
C AREA OF CROSS-SECTION
            AREA=BREADTH*THICK+0.0*XX
            RETURN
            END
C -----------------------------------------------
            FUNCTION PERI(XX)
            INCLUDE 'COM1D.FOR'
C PERIMETER
            PERI=2*BREADTH+0.0*XX
            RETURN
            END
C ****************************************************
            SUBROUTINE RESULT
            INCLUDE 'COM1D.FOR'
            CHARACTER*20 HEADER
C ****************************************************
            HEADER=' FINAL-TEMP '
            CALL PRINT(T,HEADER)
C EXTRACT PROBLEM DEPENDENT PARAMETERS IF ANY
```

```
C EXACT SOLUTION
      AM=SQRT(HPREF*PERIM(2)/CONDREF/ACF(2))
      QLOSS=SQRT(HPREF*PERIM(2)*CONDREF*ACF(2))*(T1-TINF)*TANH(AM*AL)
      EFF=TANH(AM*AL)/(AM*AL)
      WRITE(6,*)' EXACT SOLUTION          '
      WRITE(6,*)' QLOSS = ',QLOSS,' EFF = ',EFF
C NUMERICAL SOLUTION
      QLOSS=ACF(2)*CONDREF*(T(1)-T(2))/(X(2)-X(1))
      QMAX=2*AL*BREADTH*HPREF*(T(1)-TINF)
      EFF=QLOSS/QMAX
      WRITE(6,*)' NUMERICAL SOLUTION       '
      WRITE(6,*)' QLOSS = ',QLOSS,' EFF = ',EFF
      RETURN
      END
C **************************************************
      BLOCK DATA
      INCLUDE 'COM1D.FOR'
C **************************************************
C LOGICAL DECLARATIONS
C *** DECLARE STEADY OR UNSTEADY AND SOLUTION METHOD
      DATA STEADY,UNSTEADY,GAUSS,THOMAS/.TRUE.,.FALSE.,.TRUE.,.FALSE./
C -----------------------------------------
C CONTROL PARAMETERS
C FULLY IMPLICIT(PSI=1),FULLY EXPLICIT(PSI=0),SEMI IMPLICIT (0<PSI<1)
      DATA PSI,DELT,MXSTEP,ITERMX,RP,CC/1.0,5,100,500,1.0,1E-5/
C -----------------------------------------
C BOUNDARY  SPECIFICATION
      DATA T1SPEC,Q1SPEC,H1SPEC/.TRUE.,2*.FALSE./
      DATA TNSPEC,QNSPEC,HNSPEC/.FALSE.,.TRUE.,.FALSE./
      DATA T1,TN,QB1,QBN,HB1,HBN/225.0,205.0,0.0,0.0,0.0,0.0/
      DATA TINF,TINF1,TINFN,HPREF/25,0.0,0.0,15.0/
      DATA CONDREF,RHOREF,SPHREF/45.0,1.0,1.0/
C -----------------------------------------
C GRID SPECIFICATION
      DATA XCELL,XNODE/.TRUE.,.FALSE./
      DATA N,AL/7,0.02/
      DATA XCF/0.0,0.0,0.2,0.4,0.6,0.8,1.0,43*1.0/
C PROBLEM DEPENDENT PARAMETERS (IF ANY)
      DATA BREADTH,THICK/0.2,0.002/
      END
```

USER File for Problem 3 – Chapter 2

```
C *************************************************
C THIS IS USER FILE USER1D.FOR - A. W. DATE
C *************************************************
      PROGRAM ONED
      INCLUDE 'COM1D.FOR'
C *************************************************
      OPEN(6,FILE='OO')
      WRITE(6,*)' *********************************************'
      WRITE(6,*)' ANNULAR COMPOSITE FIN - PROB3-CHAPTER2'
      WRITE(6,*)' SOLVE BY  TDMA'
      WRITE(6,*)' *********************************************'
      DX=(R3-R1)/FLOAT(N-2)
      XCF(1)=0
      XCF(2)=0.0
      DO 1 I=3,N
1     XCF(I)=XCF(I-1) + DX
      CALL MAIN
      CALL RESULT
      STOP
      END
C *************************************************
      SUBROUTINE INIT
      INCLUDE 'COM1D.FOR'
C *************************************************
C GIVE INITIAL GUESS AT TIME=0.0  OR AT ITER=0 FOR STEADY STATE
      T(1)=T1
      T(N)=TN
      RETURN
      END
C *************************************************
      SUBROUTINE NEWVAL
      INCLUDE 'COM1D.FOR'
C *************************************************
C SET NEW VALUES OF HB1,HBN,QB1,QBN,TINF1,TINFN OR SOURCES
      RETURN
      END
C *************************************************
      SUBROUTINE PROPS
      INCLUDE 'COM1D.FOR'
C *************************************************
```

```
C COND(I) AND SPH(I) ARE DEFINED AT NODE P
      RR=R2-R1
      DO 1 I=1,N
      IF(X(I).LT. RR)COND(I)=200
      IF(X(I).GT. RR)COND(I)=40
1     SPH(I)=SPHREF
      RETURN
      END
C ************************************************
      SUBROUTINE SORCE
      INCLUDE 'COM1D.FOR'
C ************************************************
C FORM PROBLEM DEPENDENT SOURCE TERM INCLUDING SU AND SP
      DO 1 I=2,N-1
      TERM=HPERI(I)*PERIM(I)*(XCF(I+1)-XCF(I))
      SU(I)=SU(I)+TERM*TINF
      SP(I)=SP(I)+TERM
1     CONTINUE
      RETURN
      END
C ************************************************
      SUBROUTINE INTPRI
      INCLUDE 'COM1D.FOR'
C ************************************************
      RETURN
      END
C ************************************************
C        FUNCTION ROUTINES
C ************************************************
      FUNCTION HPERI(II)
      INCLUDE 'COM1D.FOR'
C H AT PERIMETER
      I=II
      HPERI=HPREF+X(I)*0.0+T(I)*0.0
      RETURN
      END
C ---------------------------------------------
      FUNCTION AREA(XX)
      INCLUDE 'COM1D.FOR'
C AREA OF CROSS-SECTION
      AREA=2*PI*(R1+XX)*THICK
```

```
      RETURN
      END
C ----------------------------------------------
      FUNCTION PERI(XX)
      INCLUDE 'COM1D.FOR'
C PERIMETER
      PERI=4*PI*(R1+XX)
      RETURN
      END
C ************************************************
      SUBROUTINE RESULT
      INCLUDE 'COM1D.FOR'
      CHARACTER*20 HEADER
C ************************************************
      HEADER=' FINAL-TEMP '
      CALL PRINT(T,HEADER)
C EXTRACT PROBLEM DEPENDENT PARAMETERS IF ANY
      QLOSS=ACF(2)*COND(1)*(T(1)-T(2))/(X(2)-X(1))
      QMAX=2*PI*(R3**2-R1**2)*HPREF*(T(1)-TINF)
      EFF=QLOSS/QMAX
      WRITE(6,*)' NUMERICAL SOLUTION            '
      WRITE(6,*)' QLOSS = ',QLOSS,' EFF = ',EFF
C PLOT TEMP PROFILE
       OPEN(12,FILE='TEXT3.DAT')
       WRITE(12,*)'TITLE  =  ANNULAR FIN'
       WRITE(12,*)'VARIABLES = XX TT '
       WRITE(12,*)'ZONE T = ZONE1, I = ',N,' ,F = POINT'
       DO 51 J=1,N
51     WRITE(12,*)X(J),T(J)
       CLOSE(12)
      RETURN
      END
C ************************************************
      BLOCK DATA
      INCLUDE 'COM1D.FOR'
C ************************************************
C LOGICAL DECLARATIONS
C *** DECLARE STEADY OR UNSTEADY AND SOLUTION METHOD
      DATA STEADY,UNSTEADY,GAUSS,THOMAS/.TRUE.,.FALSE.,.FALSE.,.TRUE./
C ----------------------------------------------
C CONTROL PARAMETERS
C FULLY IMPLICIT (PSI=1),FULLY EXPLICIT (PSI=0),SEMI IMPLICIT (0<PSI<1)
```

```
      DATA PSI,DELT,MXSTEP,ITERMX,RP,CC/1.0,5,100,500,1.0,1E-5/
C --------------------------------------------------
C BOUNDARY  SPECIFICATION
      DATA T1SPEC,Q1SPEC,H1SPEC/.TRUE.,2*.FALSE./
      DATA TNSPEC,QNSPEC,HNSPEC/.FALSE.,.TRUE.,.FALSE./
      DATA T1,TN,QB1,QBN,HB1,HBN/200.0,150.0,0.0,0.0,0.0,0.0/
      DATA TINF,TINF1,TINFN,HPREF/25,0.0,0.0,20.0/
      DATA CONDREF,RHOREF,SPHREF/1.0,1.0,1.0/
C --------------------------------------------------
C GRID SPECIFICATION
      DATA XCELL,XNODE/.TRUE.,.FALSE./
      DATA N/8/
C PROBLEM DEPENDENT PARAMETERS (IF ANY)
      DATA THICK/0.001/
      DATA R1,R2,R3/0.0125,0.025,0.0375/
      END
```

Library File LIB1D.FOR

```
C ***********************************************
C   THIS IS LIBRARY LIB1D.FOR - A. W. DATE
C ***********************************************
      SUBROUTINE MAIN
      INCLUDE 'COM1D.FOR'
C ***************************************************
      WRITE(6,*)' ****************************************'
      IF(THOMAS)WRITE(6,*)' SOLUTION BY TDMA'
      IF(GAUSS)WRITE(6,*)' SOLUTION BY GAUSS SIEDEL'
      WRITE(6,*)' ****************************************'
C***   CALCULATE CELL FACE COORDINATES, AREA AND VOLUME.
      CALL GRID
C***   SPECIFY INITIAL TEMPERATURE DISTRIBUTION (USER FILE)
      CALL INIT
      ISTOP=0
      IF(STEADY)PSI=1.0
      IF(UNSTEADY)THEN
      DO 101 I=1,N
101   TO(I)=T(I)
      IF(PSI.EQ.0.0)ITERMX=0
      ENDIF
      TTIME=0.0
C*** BEGIN TIME STEP
      TIMEMX=MXSTEP*DELT
```

```
          DO 3000 NTIME=1,MXSTEP
          TTIME=TTIME+DELT
C SET NEW VALUES AT THE BOUNDARY OR SOURCES (USER FILE)
          IF(UNSTEADY)CALL NEWVAL
C***      BEGIN ITERATIONS AT A TIME STEP
          IF(PSI.NE.0.0)WRITE(6,*)'  ITER     FCMX '
          ITER=0
1000      ITER=ITER+1
C CALL PROPERTIES ROUTINE (USER FILE)
          CALL PROPS
C***      CALCUALTE THE COEFFICIENTS AW AND AE
          CALL COEF
C***      CALCULATE THE SOURCE TERMS SU AND SP (USER FILE)
          CALL SORCE
C***      SPECIFY THE BOUNDARY CONDITIONS
          CALL BOUND
C***      SOLVE THE DISCRETISED EQUATION
          CALL SOLVE
C***      WRITE RESIDUAL, CHECK CONVERGENCE
          WRITE(6,500)ITER,FCMX
          IF(ITER.GT.ITERMX) GO TO 2000
          IF(STEADY.AND.ITER.EQ.1)GO TO 1000
          IF(FCMX.GT.CC) GO TO 1000
2000      CONTINUE
          IF(STEADY)GO TO 5000
C END OF TIME STEP
C UPDATE OLD TEMPERATURES AND PRINT OUT VARIABLES (USER FILE)
          CALL INTPRI
          CALL UPDATE
          IF(ISTOP.EQ.1)GO TO 5000
          IF(TTIME.GT.TIMEMX)GO TO 5000
3000      CONTINUE
5000      CONTINUE
500       FORMAT(I5,6X,E10.3)
          RETURN
          END
C ************************************************
          SUBROUTINE GRID
          INCLUDE 'COM1D.FOR'
          CHARACTER*20 HEADER
C ************************************************
C GRID DATA ARE GIVEN IN BLOCK DATA (USER FILE)
```

```
          IF(XCELL)THEN
          XCF(2)=XCF(1)
          X(1)=XCF(1)
          DO 1 I=2,N-1
1         X(I)=0.5*(XCF(I)+XCF(I+1))
          X(N)=XCF(N)
          ELSE
          XCF(1)=X(1)
          XCF(2)=X(1)
          DO 2 I=3,N-1
2         XCF(I)=0.5*(X(I)+X(I-1))
          XCF(N)=X(N)
          ENDIF
C CALCULATE PERIMETER,CELL-FACE AREA AND CELL VOLUME
C AREA AND PERI ARE FUNCTION ROUTINES (USER FILE)
          DO 3 I=1,N
          ACF(I)=AREA(XCF(I))
          PERIM(I)=PERI(X(I))
3         CONTINUE
          DO 4 I=2,N-1
4         VOL(I)=AREA(X(I))*(XCF(I+1)-XCF(I))
          HEADER=' X(I) '
          CALL PRINT(X,HEADER)
          HEADER=' XCF(I) '
          CALL PRINT(XCF,HEADER)
          HEADER=' CELL FACE AREA '
          CALL PRINT(ACF,HEADER)
          HEADER=' PERIMETER '
          CALL PRINT(PERIM,HEADER)
          HEADER=' CELL-VOLUME '
          CALL PRINT(VOL,HEADER)
          RETURN
          END
C ***********************************************
          SUBROUTINE COEF
          INCLUDE 'COM1D.FOR'
C ***********************************************
          DO 1 I=2,N-1
C INITIALISE SU ANS SP
          STAB(I)=0.0
          SU(I)=0.0
          SP(I)=0.0
```

```
            LW=0
            LE=0
            IF(I.EQ.2)LW=1
            IF(I.EQ.N-1)LE=1
            DXE=X(I+1)-X(I)
            DXEP=X(I+1)-XCF(I+1)
            DXEM=XCF(I+1)-X(I)
            DXW=X(I)-X(I-1)
            DXWP=X(I)-XCF(I)
            DXWM=XCF(I)-X(I-1)
C***    CALCULATE CELL FACE CONDUCTIVITY BY HARMONIC MEAN.
            CONDSME=DXE/(DXEM/COND(I)+DXEP/COND(I+1))*(1-LE)+LE*COND(I+1)
            CONDSMW=DXW/(DXWP/COND(I)+DXWM/COND(I-1))*(1-LW)+LW*COND(I-1)
            AW(I)=CONDSMW*ACF(I)/DXW
            AE(I)=CONDSME*ACF(I+1)/DXE
1       CONTINUE
            RETURN
            END
C ***************************************************
            SUBROUTINE BOUND
            INCLUDE 'COM1D.FOR'
C ***************************************************
            STAB(2)=AW(2)
            STAB(N-1)=AE(N-1)
C***    FOR I=1 BOUNDARY
            IF(T1SPEC) THEN
            SU(2)=SU(2)+AW(2)*(PSI*T(1)+(1-PSI)*(T0(1)-T0(2)))
            SP(2)=SP(2)+AW(2)*PSI
            AW(2)=0.0
            ELSE IF(Q1SPEC) THEN
            SU(2)=SU(2)+ACF(2)*(PSI*QB1+(1-PSI)*QB10)
            T(1)=QB1*ACF(2)/(AW(2)+SMALL)+T(2)
            AW(2)=0.0
            ELSE IF (H1SPEC) THEN
            TERM1=HB1*ACF(2)+SMALL
            TERM2=AW(2)+SMALL
            TERM=1/(1/TERM1+ 1/TERM2)
            SU(2)=SU(2)+PSI*TERM*TINF1+TERM1*(1-PSI)*(TINF10-T0(1))
            SP(2)=SP(2)+PSI*TERM
            T(1)=(T(2)+TERM1/TERM2*TINF1)/(1+TERM1/TERM2)
            AW(2)=0.0
            ENDIF
```

```
C***    FOR I=N BOUNDARY
        IF(TNSPEC) THEN
        SU(N-1)=SU(N-1)+AE(N-1)*(PSI*T(N)+(1-PSI)*(TO(N)-TO(N-1)))
        SP(N-1)=SP(N-1)+AE(N-1)*PSI
        AE(N-1)=0.0
        ELSE IF(QNSPEC)THEN
        SU(N-1)=SU(N-1)+ACF(N)*(PSI*QBN+(1-PSI)*QBNO)
        T(N)=QBN*ACF(N)/(AE(N-1)+SMALL)+T(N-1)
        AE(N-1)=0.0
        ELSE IF(HNSPEC) THEN
        TERM1=HBN*ACF(N)+SMALL
        TERM2=AE(N-1)+SMALL
        TERM=1/(1/TERM1+ 1/TERM2)
        SU(N-1)=SU(N-1)+PSI*TERM*TINFN+TERM1*(1-PSI)*(TINFNO-TO(N))
        SP(N-1)=SP(N-1)+PSI*TERM
        T(N)=(T(N-1)+TERM1/TERM2*TINFN)/(1+TERM1/TERM2)
        AE(N-1)=0.0
        ENDIF
        RETURN
        END
C ***************************************************
        SUBROUTINE SOLVE
        INCLUDE 'COM1D.FOR'
C ***************************************************
        DIMENSION AA(IT),BB(IT)
C***    ASSEMBLE SU AND SP TERMS
        DO 1 I=2,N-1
        IF(UNSTEADY)THEN
        BP=RHO(I)*SPH(I)/DELT*VOL(I)
        SP(I)=SP(I)+BP
        SU(I)=SU(I)+(1-PSI)*(AE(I)*TO(I+1)+AW(I)*TO(I-1))
        SU(I)=SU(I)+(BP-(1-PSI)*(AE(I)+AW(I)))*TO(I)
C CHECK FOR STABILITY CONDITION
        TERM=BP-(1-PSI)*(AE(I)+AW(I)+STAB(I))
        IF(TERM.LT.0.0)WRITE(*,*)' COEF OF TPOLD IS NEGATIVE AT I = ',I
        ENDIF
        AP(I)=PSI*(AE(I)+AW(I))+SP(I)
C UNDER-RELAX
        E=(1.-RP)/RP*AP(I)
        AP(I)=AP(I)+E
        SU(I)=SU(I)+E*T(I)
1       CONTINUE
```

```fortran
        FCMX=0.0
C -----------------------------------------------------
C***    SOLVE BY GAUSS-SIEDEL METHOD
C -----------------------------------------------------
        IF(GAUSS)THEN
        DO 2 I=2,N-1
        TL=T(I)
        ANUM=PSI*(AE(I)*T(I+1)+AW(I)*T(I-1))+SU(I)
        T(I)=ANUM/AP(I)
        DIFF=(T(I)-TL)/(TL+SMALL)
        IF(ABS(DIFF).GT.FCMX)FCMX=ABS(DIFF)
2       CONTINUE
        ENDIF
C -----------------------------------------------------
C***    SOLVE BY TDMA
C -----------------------------------------------------
        IF(THOMAS)THEN
C CALCULATE  COEFFICIENTS BY RECURRENCE
        AA(2)=PSI*AE(2)/AP(2)
        BB(2)=SU(2)/AP(2)
        DO 3 I=3,N-1
        DEN=1.0-PSI*AW(I)/AP(I)*AA(I-1)
        AA(I)=PSI*AE(I)/AP(I)/(DEN+SMALL)
3       BB(I)=(PSI*AW(I)*BB(I-1)+SU(I))/AP(I)/(DEN+SMALL)
C BACK SUBSTITUTION
        DO 4 I=N-1,2,-1
        TL=T(I)
        T(I)=AA(I)*T(I+1)+BB(I)
        DIFF=(T(I)-TL)/(TL+SMALL)
        IF(ABS(DIFF).GT.FCMX)FCMX=ABS(DIFF)
4       CONTINUE
        ENDIF
        RETURN
        END
C ***************************************************
        SUBROUTINE UPDATE
        INCLUDE 'COM1D.FOR'
C ***************************************************
C RESET OLD VALUES
        DO 200 I=1,N
200     TO(I)=T(I)
        QB1O=QB1
```

```
            QBNO=QBN
            HB1O=HB1
            HBNO=HBN
            TINF1O=TINF1
            TINFNO=TINFN
            HPREFO=HPREF
            TINFO=TINF
            RETURN
            END
C ************************************************
      SUBROUTINE PRINT(F,HEADER)
      INCLUDE 'COM1D.FOR'
       CHARACTER*20 HEADER
C ************************************************
      DIMENSION F(IT)
      WRITE(6,*)'****************************************'
      WRITE(6,*)'DISTRIBUTION OF ',HEADER
      IB=1
      IE=IB+6
      IF(IE.GT.N)IE=N
100   CONTINUE
      WRITE(6,500)(F(I),I=IB,IE)
      WRITE(6,600)(I,I=IB,IE)
      IF(IE.LT.N) THEN
      IB=IE+1
      IE=IB+6
      IF(IE.GT.N)IE=N
      GO TO 100
      ENDIF
      WRITE(6,*)'****************************************'
500   FORMAT(7E10.3)
600   FORMAT(4X,I3,6I10)
      RETURN
      END
```

APPENDIX C

2D Cartesian Code

C.1 Structure of the Code

The structure of the 2D Cartesian code is similar to that of the 1D conduction code. The code is again divided into two parts: The *problem-dependent* user part containing files COM2D.FOR and USER2D.FOR and the *problem independent* library part that contains the LIB2D.FOR file.[1] The listings of each file are given at the end of this appendix. The list of variable names with their meanings is given in Table C.1.

C.2 File COM2D.FOR

In this file, again logical, real, and integer variables are included. All other contents of this file bear the same description as the COM1D.FOR file.

C.3 File USER2D.FOR

This is the main control file at the command of the user. The first routine PROGRAM MAIN is the command routine from where subroutine MAINPR is called. The latter is the first subroutine of the LIB2D.FOR file. Here, listings of USER files are given for three problems solved in Chapter 5. These are (a) 1D porous body flow, (b) turbulent flow in an axisymmetric pipe expansion, and (c) natural convection evaporation.

BLOCK DATA This routine at the end of the USER2D.FOR file specifies all the problem-dependent data such as control parameters, relaxation parameters, Prandtl numbers, flow conditions, equations to be solved, and convection scheme used. The

---

[1] The library file does not contain two features that can be generalised. These are (a) modifications to coefficients of the pressure-correction equation for a compressible flow and (b) modifications for a fixed pressure boundary condition. However, these can be incorporated by the user via the ADSORB.FOR routine in the USER file.

Table C.1: List of variables for 2D Cartesian code.

Variable	Meaning
AE, AW	Array containing east and west coefficients
AN, AS	Array containing north and south coefficients
AMW	Mass influx at the wall (kg/m^2-s)
AP	Array containing coefficient of Φ_P
AP1	Array containing coefficient of p'_P
APU, APV	Array containing coefficient of u_P and v_P
AXISYMM	Logical – =.TRUE. refers to axisymmetric case
CAPPA	Constant in log law of the wall
CC	Convergence criterion
CCTM	Stop condition for unsteady problem
CD1	Constant in e–ϵ turbulence model
CD2	Constant in e–ϵ turbulence model
CONMAS	Logical – refers to imposition of mass balance at exit plane
D	Array for turbulent energy dissipation rate ϵ
DELT	Time step (s)
DENSIT	Reference density (kg/m^3)
DO	Array for ϵ at old time
DP1	Periodic pressure change in I direction
DP2	Periodic pressure change in J direction
DXMI	Array containing increment $X(I) - X(I-1)$
DXP	Array containing increment $XC(I+1) - XC(I)$
DYMI	Array containing increment $Y(J) - Y(J-1)$
DYP	Array containing increment $YC(J+1) - YC(J)$
E	Array for turbulent kinetic energy e
ELOG	Constant in log law of the wall
EO	Array for e at old time
FDIF	Array for storing $\Phi^{l+1} - \Phi^l$ or $\Phi - \Phi^o$
FTRAN	Logical – refers to false transient solution
GAMMA	Multiplier of $p - \overline{p}$
GRCELL	Logical – refers to specification of cell-face coordinates
GRNODE	Logical – refers to specification of node coordinates
GREAT	Parameter having a large value 10^{30}
HH	Array containing enthalpy variable (J/kg)
HYBRID	Logical – refers to hybrid convection scheme
IN	Maximum number of nodes in I direction
INM	IN–1
IPERIOD	Index for periodicity in I direction – see BLOCK DATA
IPREF	I index of reference point of pressure
IREAD	= .TRUE. when file NSIN is to be read
IT	Parameter containing array size in I direction
IWRITE	= .TRUE. when file NSOUT is to be written
JN	Maximum number of nodes in J direction
JNM	JN–1
JPERIOD	Index for periodicity in J direction – see BLOCK DATA

(continued)

Table C.1 (*continued*)

Variable	Meaning
JPREF	J index of reference point of pressure
JT	Parameter containing array size in J direction
MFREQ	NSOUT file written after every MFREQ iteration
MXGR	Parameter containing bigger of IT and JT
MXIT	Maximum number of allowable iterations
MXSTEP	Maximum number of allowable time steps
NITER	Iteration count
NPERIOD	= 1 for periodic boundary condition, = 0 otherwise
NSWEEP	Array containing maximum sweeps per iteration
NTAG	Array for identifying interior nodes
NTAGE	Array for identifying east near-boundary nodes
NTAGN	Array for identifying north near-boundary nodes
NTAGS	Array for identifying south near-boundary nodes
NTAGW	Array for identifying west near-boundary nodes
NTIME	Current time step number
NVAR	Maximum number of variables solved
O	Array for mass fraction ω
P	Array for pressure (N/m^2)
PI	Parameter π
PO	Array for pressure at old time
POWER	Logical – refers to power-law scheme
PP	Array for pressure correction
PR	Array for fluid Prandtl number
PRT	Array for turbulent Prandtl number
PSM	Array for smoothing pressure correction
QW	Array for wall-heat in flux (W/m^2)
R	Array for storing radius at the node
RC	Array for storing radius at the south cell face
RHO	Array for density (kg/m^3)
RNORM	Array for storing residual normalising factors
RP	Array relaxation parameter α
RSDU	Array for storing maximum residual
SMALL	Parameter having a small value 10^{-30}
SLVE	Logical array for specifying variable to be solved
SP	Array containing Sp
SPH	Array for specific heat (J/kg-K)
STEADY	Logical – refers to steady-state calculation
STIME	Time at the start of a transient
TTIME	Total time after NTIME steps
SU	Array containing Su
T	Array containing temperature (°C or K)
TAUW	Array containing shear stress at the wall (N/m^2)
TO	Array containing temperature at old time
TURBUL	Logical – refers to turbulent flow
U	Array containing u_1 velocity

Variable	Meaning
UNSTDY	Logical – refers to unsteady-state calculation
UO	Array containing u_1 velocity at old time
UPWIND	Logical – refers to upwind difference scheme
V	Array containing u_2 velocity
VISCOS	Reference viscosity (N-s/m^2)
VIS	Array containing laminar viscosity
VIST	Array containing turbulent viscosity
VO	Array containing u_2 velocity at old time
VOL	Array containing cell volume (m^3)
W	Array containing u_3 velocity
WO	Array containing u_3 velocity at old time
X	Coordinate of node I (m)
XC	Coordinate of cell face at w
Y	Coordinate of node J (m)
YC	Coordinate of cell face at s

user may introduce additional problem-dependent indices such as IB1, IB2, etc. The grid coordinate data may be normalised or real.

Subroutine TITLE In this routine, the problem-specific title is inserted.

Subroutine INIT In this routine, the best known initial guesses for all relevant variables are given. Also, known INFLOW conditions are specified.

Subroutine BSPEC Here, boundary types are specified. Identifiers for boundary type and boundary condition are declared by logical variables such as WEST or SYMM. WALLT and WALLQ stand for, respectively, temperature and heat influx specified at wall boundaries. EXIT1 and EXIT2 stand for exit boundary conditions. When EXIT1, the first normal derivative at the boundary is set to zero, for EXIT2, the second normal derivative is zero. BLOCK identifies blocked regions of the domain. These specifications must be made carefully and a hand sketch of the domain will assist correct specifications. Also, reference may be made to the node-tagging section in Chapter 5.

Subroutine ADSORB This routine is used to add any special source terms for each variable solved. The standard source terms are included in the SORCE routine in LIB2D.FOR file. The routine is also used to overwrite USER-defined specifications for density, specific heat, and viscosity. The routine is also used to specify a fixed-pressure boundary condition. Further, the routine is used to give periodic boundary conditions.

Subroutine RESULT In this routine, the final converged solutions are printed. The routine is also used to extract useful parameters such as friction factor or Nusselt number from the converged solution. This routine is also used to create output files for contour and vector plots using graphics packages such as TECPLOT.

Subroutine OMEGA In this routine, the user sets out solution of the mass transfer equation(s) as well as the enthalpy equation by defining new variables O (I, J) and HH (I, J). If the problem is unsteady then OO (I, J) and HHO (I, J) must also be defined. If there are several mass fractions solved (as in a combustion problem), then additional variable names must be defined and declared in the COM2D.FOR file. For all these variables, subroutine COEF is first called. Then, source terms and boundary conditions are specified for each variable with appropriate updates of Su and Sp. Finally, the variable under consideration is solved by calling the SOLVE routine. The USER file for the evaporation problem shows how this is done by solving for the vapor mass fraction.

C.4   File LIB2D.FOR

Subroutine MAINPR This is the main routine from which all other routines are called for program execution. The sequence of calling is important.

Subroutine INITIA Here, all variables are initialised.

Subroutine TAG In this routine node tagging is accomplished. The routine is called from subroutine BSPEC in the USER2D.FOR file. CHAR1 and CHAR2 carry logical variables whereas IB, IL and JB, JL carry specification limits in the I and J directions. Note that NTAG (I, J) = 1 for the blocked region; it is already set to zero in routine INITIA otherwise.

Subroutine BOUND This routine implements the boundary conditions. The routine can be written more compactly. Here, boundary conditions for west, east, south, and north are written explicitly for ease of understanding. Note that periodic boundaries are treated as inflow boundaries. Therefore, velocities at such boundaries must be provided by the user in the ADSORB routine in the USER2D.FOR file. The routine provides boundary conditions for only six variables: u, v, w, e, ϵ, and T. For all other variables such as mass fractions or enthalpy, the boundary conditions are given in the routine OMEGA in the USER2D.FOR file.

Subroutine GRID In this routine, node and/or cell-face coordinates are evaluated depending on logical specifications GRCELL and GRNODE. Also, repeatedly used incremental distances and cell volume are calculated and stored. The sum of cell volumes must equal the domain volume. Hence, the latter is printed via SUMVOL. The USER should always check SUMVOL in output file OO opened in PROGRAM MAIN. Note that R (J) and RC (J) are set to 1 for the plane case but are equated to Y (J) and YC (J), respectively, for the axisymmetric case.

Subroutine COEF In this routine, coefficients AE, AW, AN, and AS are calculated for transport equations and for the pressure-correction equation. Note that, in evaluation of transport equation coefficients, the cell-face viscosities are evaluated by harmonic mean. Similarly, care is exercised in evaluation of periodic boundary coefficients in the pressure-correction equation.

Subroutine SORCE This routine includes standard source terms for all variables. Thus, for u and v velocities, pressure-gradient terms are included but body force terms are excluded. In the turbulent kinetic energy source term, the near-wall boundary node is *excluded* as required for implementation of the high Reynolds number (HRE) turbulence model (see Chapter 5). For the LRE model, the USER will have to modify entries in BOUND and SORCE routines. For temperature, the standard source term is set to zero. To include effects of viscous dissipation or heat generation due to chemical reaction or radiation, the ADSORB routine in the USER file must be used.

Subroutine APCOF In this routine the coefficient of Φ_P is assembled by adding SP (I , J) and dividing by α. The APU and APV coefficients store the AP coefficient of the two momentum equations whereas AP1 stores the coefficient of p'_P.

Subroutine PROPS Here, density, specific heat, and viscosity variables are specified. If BSOR (8) is .TRUE. then the default specifications can be overwritten in the ADSORB routine in the USER2D.FOR file.

Subroutine UNST In this routine, Su and Sp are appropriately updated for truly unsteady or false-transient calculation.

Subroutine UPDATE In this routine, all new time values are set in old time values.

Subroutine INFLUX In this routine values of $\dot{m}\Phi$ at inlet boundaries are evaluated for all variables to form the RNORM array. The latter is used to normalise the residual calculation in subroutine SOLVE.

Subroutine MASBAL This routine calculates the domain exit mass flow rate based on specifications of velocity boundary conditions in subroutine BOUND. However, this mass flow rate must be the same as the sum of all mass flow rates specified at the inlet boundaries. During an iterative solution, this balance is rarely maintained. Therefore, *before* solving the pressure-correction equation, the prevailing exit-plane velocities are uniformly corrected in this routine by the ratio of inlet to exit mass flow rates.

Subroutine PVCOR In this routine, the mass-conserving pressure-correction p'_m is first recovered and then pressure and velocities are corrected. The smoothing pressure-correction p'_{sm} is stored for printing when desired. The routine also calculates the mass residual R_m, as explained in Chapter 5.

Subroutine BOUNDP In this routine, boundary pressures are extrapolated from near-boundary values as explained in Chapter 5. Note that p'_m values are also extrapolated to effect correct velocity corrections at the near-boundary nodes in routine PVCOR. Further, care is exercised to effect correct interpolation of pressure at the periodic boundaries. When BSOR (9) is .TRUE., fixed-pressure conditions are given in the ADSORB routine of the USER2D.FOR file.

Subroutine INDATA This routine simply writes out input data given in BLOCK DATA in the OO file for verification.

Subroutine SOLVE This routine solves the discretised equations by the ADI method. At the start of this routine (i.e., at iteration level l) residuals are calculated at each node and the root-sum-square value of residuals is stored in RSUM.

Subroutine SOLP This routine is the same as the SOLVE routine but is exclusively used for the pressure-correction equation. Note that subroutine BOUNDP is called at each sweep when a periodic boundary condition is specified.

Subroutine EQN In this routine the outer DO loop (2000) is initiated for an unsteady calculation and the inner loop (1000) carries out iterative calculations. For each chosen variable (specified by logical SLVE), subroutines COEF, SORCE, BOUND, APCOF, and SOLVE are called in turn. To carry out appropriate updates of Su and Sp, UNST and ADSORB routines are also called. When all relevant variables are covered, subroutine PROPS is called to update the properties. Then, the maximum residual among all variables is stored in RSTOP. In a *steady-state problem*, if the convergence criterion is satisfied, the inner loop is exited and control is returned to the subroutine MAINPR. In an *unsteady problem*, upon exiting the inner loop, subroutine UPDATE is called to reset the values and the STOP condition is based on the maximum value of FDIF. The outer loop is continued until the maximum number of steps specified in BLOCK DATA is executed. At each time step, however, the inner loop is executed for the MXIT number of iterations. In many problems, because of the impossibility of specifying good initial guesses, the number of iterations required may run into the thousands. For this reason, variables are written out in file NSOUT at every MFREQ iteration, where MFREQ is set in the BLOCK DATA routine.[2]

Subroutine TDMA This routine calculates the recurrence coefficients and carries out back substitution as required in TDMA execution.

Subroutine OPT In this routine, all variables are written out in *binary* form in file NSOUT.

Subroutine IPT In this routine, all variables are read in *binary* form from file NSIN. Therefore, before execution of the program in a continuation can commence, file NSOUT must be copied to file NSIN.

Subroutines PRINTK This routine is used to print out 2D variables.

---

[2] At every iteration, the three main residuals for u, v, and p' are stored in arrays RESIU(NITER), RESIV(NITER), and RESIM(NITER). Their evaluation has been commented on. However, the USER may activate this evaluation to enable printing of residual history in the RESULT routine when required.

Subroutines PR1D This routine is used to print out 1D variables.

Function STAN This function routine is called from subroutine BOUND to implement a wall-function boundary condition for temperature and mass-fraction variables in a turbulent flow. The routine evaluates the Stanton number based on specification of PF, which the USER may change if required. For mass-fraction variables, the function is called from routine OMEGA.

Functions FINTW, FINTE, FINTS, and FINTN These function routines evaluate variable Φ at cell faces, w, e, s, and n, respectively, using linear interpolation.

COMMON BLOCK COM2D.FOR

```
C*******************************
C THIS IS COM2D.FOR
C *******************************
C IT AND JT CHANGE WITH THE PROBLEM
        PARAMETER(IT=37,JT=37,MXGR=37)
        PARAMETER(GREAT=1.0E+20,SMALL=1.0E-20,PI=3.1415926)
        LOGICAL TURBUL,STEADY,UNSTDY,FTRAN,CONMAS,AXISYMM,BSOR
        LOGICAL UPWIND,HYBRID,POWER,SLVE,IREAD,IWRITE
        LOGICAL GRCELL,GRNODE
C
        COMMON/CFLOW/TURBUL,STEADY,UNSTDY,FTRAN,CONMAS,AXISYMM
        COMMON/SCHEME/UPWIND,HYBRID,POWER,SLVE(7),IREAD,IWRITE
        COMMON/CGRID/IN,JN,INM,JNM,IPREF,JPREF,CORP,NPERIOD,MFREQ
     1        ,GRCELL,GRNODE
        COMMON/CONTR1/CC,MXIT,CCTM,MXSTEP,DELT,STIME,TTIME
        COMMON/CONTR2/RP(9),NSWEEP(7),NITER,RSDU(7),FDIF(7),RNORM(7)
        COMMON/CPROP/DENSIT,VISCOS,PR(7),PRT(7),RHO(IT,JT)
     1        ,SPHEAT,SPH(IT,JT),VIS(IT,JT),VIST(IT,JT),GAMMA
        COMMON/CTURB/CD1,CD2,CMU,ELOG,CAPPA
        COMMON/CTAG/NTAG(IT,JT),NTAGW(IT,JT),NTAGE(IT,JT)
     1           ,NTAGS(IT,JT),NTAGN(IT,JT)
        COMMON/COFV/AW(IT,JT),AE(IT,JT),AS(IT,JT),AN(IT,JT)
        COMMON/CSOR/SU(IT,JT),SP(IT,JT),BSOR(9)
        COMMON/CAP/AP1(IT,JT),AP(IT,JT),APU(IT,JT),APV(IT,JT)
        COMMON/CVAR/U(IT,JT),V(IT,JT),W(IT,JT),P(IT,JT)
     1           ,E(IT,JT),D(IT,JT),T(IT,JT),PP(IT,JT),PSM(IT,JT)
        COMMON/CVAO/UO(IT,JT),VO(IT,JT),WO(IT,JT),PO(IT,JT)
     1           ,EO(IT,JT),DO(IT,JT),TO(IT,JT),RHOO(IT,JT)
        COMMON/CORD/X(IT),Y(JT),XC(IT),YC(JT),R(JT),RC(JT)
     1           ,DXMI(IT),DYMI(JT),DXP(IT),DYP(JT),VOL(IT,JT)
        COMMON/CHEAT/QW(IT,JT),TAUW(IT,JT),AMW(IT,JT)
```

```
      COMMON/CPERIOD/DP1,DP2,IPERIOD,JPERIOD
      COMMON/CDUMT/DUM1(IT,JT),DUM2(IT,JT),DUM3(IT,JT)
      COMMON/CRES/RESIU(5000),RESIV(5000),RESIM(5000)
C   ADDITIONAL PROBLEM-DEPENDENT COMMON STATEMENTS
      COMMON/CEVAP/IB1,IB2,JB1,GRM,SC,OBR,OWT,O(IT,JT),HH(IT,JT)
      COMMON/CKRAL/D2,D1,U2,U1
      COMMON/CPOROS/UU(IT,50),PRES(IT,50),PPP(IT,50),PPS(IT,50)
     1         ,EPSI,RESIST,PIN
```

1D Porous Body Problem – Chapter 5

The USER file that follows shows how a *fixed-pressure* boundary condition (i.e., $p' = 0$) is implemented in the ADSORB subroutine.

```
C ****************************************
C THIS IS USER FILE POROS.FOR - PFIX BOUNDARY CONDITION
C ****************************************
      PROGRAM MAIN
      INCLUDE 'COM2D.FOR'
      OPEN(UNIT=6,FILE='OO')
      WRITE(*,*)'-------- output is in OO file --------------'
C **** INITIAL DATA
      WRITE(*,*)'GIVE ----- MXIT,IREAD,GAMMA'
      READ(*,*)MXIT,IREAD,GAMMA
      DX=1/FLOAT(IN-1)
      X(1)=0.0
      DO 1 I=2,IN
1     X(I)=X(I-1)+DX
      Y(1)=0
      Y(2)=0.5
      Y(3)=1.0
      INM=IN-1
      JNM=JN-1
      VISCOS=VISCOS/EPSI
      DENSIT=DENSIT/EPSI**2
C
      CALL MAINPR
      STOP
      END
C ****************************************
      SUBROUTINE TITLE
      INCLUDE 'COM2D.FOR'
C ****************************************
```

```
      WRITE(6,*)'*****************************************'
      WRITE(6,*)' PROGRAM TO CALCULATE POROS MEDIUM FLOW '
      WRITE(6,*)'*****************************************'
      RETURN
      END
C *************************************
      SUBROUTINE INIT
      INCLUDE 'COM2D.FOR'
C *************************************
C INITIAL GUESS
      PIN=RESIST
      DO 1 I=1,IN
      DO 1 J=1,JN
C     P(I,J)=PIN*(1-X(I)/X(IN))
1     CONTINUE
      P(1,2)=PIN
      RETURN
      END
C *************************************
      SUBROUTINE BSPEC
      INCLUDE 'COM2D.FOR'
C *************************************
C **** PROVIDE BOUNDARY & BLOCKED REGIONS
C
      CHARACTER*10 BLOCK,WEST,EAST,SOUTH,NORTH
      CHARACTER*10 INFLOW,EXIT1,SYMM,EXIT2,WALLT,WALLQ,PERIOD
      DATA  BLOCK,WEST,EAST,SOUTH,NORTH
     1 /'BLOCK','WEST','EAST','SOUTH','NORTH'/
      DATA   INFLOW,EXIT1,SYMM,EXIT2,WALLT,WALLQ,PERIOD
     1 /'INFLOW','EXIT1','SYMM','EXIT2','WALLT','WALLQ','PERIOD'/
C ***** BLOCKED REGIONS
C      CALL TAG(BLOCK,BLOCK,2,IB1-1,JB2,JNM)
C      CALL TAG(BLOCK,BLOCK,IB1,INM,2,JB1-1)
C ***** DEFINES W & E BOUNDARIES
      CALL TAG(WEST,EXIT1,2,2, 2,JNM)
      CALL TAG(EAST,EXIT1,INM,INM,2,JNM)
C ***** DEFINES N&S BOUNDARIES
      CALL TAG(NORTH,SYMM,2,INM,JNM,JNM)
      CALL TAG(SOUTH,SYMM,2,INM,2,2)
      RETURN
      END
```

```
C  ***************************************
       SUBROUTINE RESULT
       INCLUDE 'COM2D.FOR'
C  ***************************************
       CHARACTER*20 HEADER
       JSTEP=-1
       WRITE(6,*)' NITER=',NITER
       HEADER=' U-VEL'
       CALL PRINTK(U,1,IN,2,JNM,HEADER,JSTEP)
       HEADER=' PRESS'
       CALL PRINTK(P,1,IN,2,JNM,HEADER,JSTEP)
       OPEN(12,FILE='PORU.DAT')
       WRITE(12,*)'IN = ',IN
       WRITE(12,*)' X(I) = '
       WRITE(12,500)(X(I),I=1,IN)
       WRITE(12,*)' U- VELOCITY '
       DO 11 NN=1,NITER-1
11     WRITE(12,500)NN,(UU(I,NN),I=1,IN)
       CLOSE(12)
       OPEN(13,FILE='PORP.DAT')
       WRITE(13,*)'IN = ',IN
       WRITE(13,*)' X(I) = '
       WRITE(13,500)(X(I),I=1,IN)
       WRITE(13,*)' PRESSURE'
       DO 12 NN=1,NITER-1
12     WRITE(13,500)NN,(PRES(I,NN),I=1,IN)
       CLOSE(13)
       OPEN(14,FILE='PORPP.DAT')
       WRITE(14,*)'IN = ',IN
       WRITE(14,*)' X(I) = '
       WRITE(14,500)(X(I),I=1,IN)
       WRITE(14,*)' PPM'
       DO 122 NN=1,NITER-1
122     WRITE(14,500)NN,(PPP(I,NN),I=1,IN)
       CLOSE(14)
       OPEN(15,FILE='PORPS.DAT')
       WRITE(15,*)'IN = ',IN
       WRITE(15,*)' X(I) = '
       WRITE(15,500)(X(I),I=1,IN)
       WRITE(15,*)' PPS'
       DO 222 NN=1,NITER-1
222     WRITE(15,500)NN,(PPS(I,NN),I=1,IN)
```

```
        CLOSE(15)
600     FORMAT(2X,7E10.3)
500     FORMAT(I4,2X,7E10.3)
        RETURN
        END
C ***************************************
        SUBROUTINE ADSORB(NN)
        INCLUDE 'COM2D.FOR'
C ***************************************
        N=NN
        GO TO (10,20,30,40,50,60,70,80,90),N
C *** FOR PRESSURE CORRECTION - PFIX CONDITION
10      CONTINUE
        SP(2,2)=SP(2,2)+RHO(1,2)*(R(2)*DYP(2))**2/APU(2,2)
        SP(INM,2)=SP(INM,2)+RHO(IN,2)*(R(2)*DYP(2))**2/APU(INM,2)
        GO TO 1000
C *** FOR U-VEL
20      CONTINUE
        DO 21 I=2,INM
        TERM=U(I+1,2)/DXMI(I+1)*DYP(2)+U(I-1,2)/DXMI(I)*DYP(2)
        SU(I,2)=SU(I,2)+TERM*VISCOS
        SP(I,2)=SP(I,2)+VISCOS*DYP(2)*(1./DXMI(I+1)+1./DXMI(I))
21      SP(I,2)=SP(I,2)+VISCOS*RESIST*EPSI*VOL(I,2)
        GO TO 1000
C *** FOR V-VEL
30      GO TO 1000
C *** FOR W-VEL
40      GO TO 1000
C *** FOR K. ENERGY
50      GO TO 1000
C *** FOR DISSIPATION
60      GO TO 1000
C *** FOR TEMPERATURE
70      GO TO 1000
C *** FOR FLUID PROPERTIES
80      GO TO 1000
C *** CALLED FROM BOUNDP
90      P(1,2)=PIN
        P(IN,2)=0
        PP(IN,2)=0
        PP(1,2)=0
C DUMMY VARABLES FOR UU,PRES,PPP,PPS FOR PRINTING
```

```
            DO 91 I=1,IN
            UU(I,NITER)=U(I,2)
            PPS(I,NITER)=PSM(I,2)/(P(I,2)+SMALL)
            PRES(I,NITER)=P(I,2)
            PPP(I,NITER)=PP(I,2)/(P(I,2)+SMALL)
91          CONTINUE
1000        CONTINUE
            RETURN
            END
C ***************************************
            SUBROUTINE OMEGA
            INCLUDE 'COM2D.FOR'
C ***************************************
C SPECIES EQUATION
            RETURN
            END
C ***************************************
            BLOCK DATA
            INCLUDE 'COM2D.FOR'
C ***************************************
C **** INITIAL DATA(make sure that IN,JN equal IT,JT)
            DATA CC,IPREF,JPREF,MXIT,GAMMA/1.0E-06,3,2,55,0.5/
            DATA CCTM,MXSTEP,DELT,STIME,MFREQ/1.0E-06,65,1.00,0.0,20/
C                 PP  U    V    W    E    D    T   VIS  P
            DATA   RP/1.0,0.95,0.5,0.5,0.5,1.0,1.0,1.0,0.95/
            DATA NSWEEP/ 1, 1 , 1 , 1 , 1 , 1 , 1          /
            DATA   PR/1.0,1.0,1.0,1.0,1.0,1.0,0.7           /
            DATA   PRT/1.0,1.0,1.0,1.0,1.0,1.3,0.9          /
            DATA DENSIT,VISCOS,SPHEAT/1.0,1.0,1.0/
            DATA CD1,CD2,CMU,ELOG,CAPPA/1.44,1.92,0.09,9.793,0.4187/
C **** LOGICAL DATA
            DATA TURBUL, STEADY, UNSTDY,   FTRAN , CONMAS, AXISYMM
     1        /.FALSE.,.TRUE., .FALSE., .FALSE., .FALSE., .FALSE./
            DATA UPWIND,HYBRID,POWER/.TRUE.,.FALSE.,.FALSE./
            DATA SLVE/2*.TRUE.,5*.FALSE./
            DATA BSOR/.TRUE.,.TRUE.,6*.FALSE.,.TRUE./
            DATA IREAD,IWRITE/.FALSE. ,.TRUE./
C PERIODIC BC
            DATA IPERIOD,JPERIOD/0,0/
            DATA DP1,DP2/0.0,0.0/
C **** READ GRID DATA
            DATA GRCELL,GRNODE/.FALSE.,.TRUE./
```

```
      DATA IN,JN/5,3/
      DATA EPSI,RESIST/0.1,4E5/
      END
```

Pipe-Expansion Problem – Chapter 5

In the USER file that follows see how inlet conditions for e and ϵ are given in the INIT subroutine. Also, in the BLOCK DATA routine, see that CONMAS and AXISYMM are set active. The pipe-expansion step is designated by JB1. The file is first executed with NTEMP = 0 and only flow variables are calculated. Then, reading NSIN, the temperature equation is solved with NTEMP = 1. Also, in subroutine RESULT, note how the reattachment length and Nusselt numbers are evaluated.

```
C ****************************************
C THIS IS USER FILE FOR PIPE-EXPANSION
C ****************************************
      PROGRAM MAIN
      INCLUDE 'COM2D.FOR'
      OPEN(UNIT=6,FILE='OO')
      WRITE(*,*)'      --- output is in OO file --------------'
C **** INITIAL DATA
      INM=IN-1
      JNM=JN-1
      D2=2.0
      D1=1.0
      U2=1.0
C
      WRITE(*,*)' IF NTEMP = 1, ONLY TEMP SOLUTION '
      WRITE(*,*)'GIVE ----- MXIT,IREAD,DELT,VISCOS,NTEMP '
      READ(*,*)MXIT,IREAD,DELT,VISCOS,NTEMP
      IF(NTEMP.EQ.1)THEN
      SLVE(1)=.FALSE.
      SLVE(2)=.FALSE.
      SLVE(3)=.FALSE.
      SLVE(5)=.FALSE.
      SLVE(6)=.FALSE.
      SLVE(7)=.TRUE.
      ENDIF
      PRT(7)=0.85+0.0309*(PR(7)+1)/PR(7)
      REY=U2*D2*DENSIT/VISCOS
      WRITE(6,*)' REYNOLDS NO = ',REY
C
      CALL MAINPR
```

```
            STOP
            END
C *************************************
            SUBROUTINE TITLE
            INCLUDE 'COM2D.FOR'
C *************************************

            WRITE(6,*)'*******************************'
            WRITE(6,*)' PROGRAM TO CALCULATE SUDDEN EXPANSION D2/D1 = 2 '
            WRITE(6,*)' KRALL AND SPARROW '
            WRITE(6,*)'*******************************'
            RETURN
            END
C *************************************
            SUBROUTINE INIT
            INCLUDE 'COM2D.FOR'
C *************************************
C INITIAL GUESS
            RATIO=10*0.563E-3/VISCOS
            UBAR=U2
            DO 10 I=1,IN
            QW(I,JN)=1.0
            DO 10 J=1,JNM
            U(I,J)=UBAR
            U(1,J)=(D2/D1)**2*UBAR
            IF(J.GE.JB1)U(1,J)=0.0
            IF(TURBUL)THEN
            E(I,J)=0.1*0.1*U(I,J)**2
            D(I,J)=CMU*DENSIT*E(I,J)**2/VISCOS/RATIO
            ENDIF
10          CONTINUE
            RETURN
            END
C *************************************
            SUBROUTINE BSPEC
            INCLUDE 'COM2D.FOR'
C *************************************
C **** PROVIDE BOUNDARY & BLOCKED REGIONS
        CHARACTER*10 BLOCK,WEST,EAST,SOUTH,NORTH
        CHARACTER*10 INFLOW,EXIT1,SYMM,EXIT2,WALLT,WALLQ,PERIOD
        DATA  BLOCK,WEST,EAST,SOUTH,NORTH
     1    /'BLOCK','WEST','EAST','SOUTH','NORTH'/
```

```
      DATA    INFLOW,EXIT1,SYMM,EXIT2,WALLT,WALLQ,PERIOD
     1   /'INFLOW','EXIT1','SYMM','EXIT2','WALLT','WALLQ','PERIOD'/
C ***** BLOCKED REGIONS
C         CALL TAG(BLOCK,BLOCK,IB1,IB2-1,JB1,JNM)
C ***** DEFINES W & E BOUNDARIES
      CALL TAG(WEST,INFLOW,2,2, 2,JB1-1)
      CALL TAG(WEST,WALLQ,2,2,JB1,JNM)
      CALL TAG(EAST,EXIT2,INM,INM,2,JNM)
C ***** DEFINES N&S BOUNDARIES
      CALL TAG(NORTH,WALLQ,2,INM, JNM,JNM)
      CALL TAG(SOUTH,SYMM,2,INM,2,2)
      RETURN
      END
C **************************************
      SUBROUTINE RESULT
      INCLUDE 'COM2D.FOR'
C **************************************
      CHARACTER*20 HEADER
      DO 1 J=1,JN
      DO 1 I=1,IN
1     VIST(I,J)=VIST(I,J)/VISCOS

      JSTEP=-1
      WRITE(6,*)' NITER=',NITER
      HEADER=' U-VEL'
      CALL PRINTK(U,1,IN,1,JN,HEADER,JSTEP)
      HEADER=' V-VEL'
      CALL PRINTK(V,1,IN,1,JN,HEADER,JSTEP)
      HEADER=' PRESS'
      CALL PRINTK(P,1,IN,1,JN,HEADER,JSTEP)
      IF(TURBUL)THEN
      HEADER=' K ENERGY'
      CALL PRINTK(E,1,IN,1,JN,HEADER,JSTEP)
      HEADER=' DISS'
      CALL PRINTK(D,1,IN,1,JN,HEADER,JSTEP)
      HEADER=' MUT'
      CALL PRINTK(VIST,1,IN,1,JN,HEADER,JSTEP)
      ENDIF
      IF(SLVE(7))THEN
      HEADER=' TEMP'
      CALL PRINTK(T,1,IN,1,JN,HEADER,JSTEP)
      ENDIF
```

```
      C
            STEP=2.0*R(JN)
            XATTCH=0.0
            COND=VISCOS*SPHEAT/PR(7)
            DO 10 I=2,INM
            XX=X(I)-XC(1)
            IF(U(I,JNM).GT.0.0.AND.U(I-1,JNM).LT.0.0)THEN
            BB=(U(I,JNM)-U(I-1,JNM))/DXMI(I)
            AA=U(I,JNM)-BB*X(I)
            XATTCH=-AA/BB-XC(1)
            ENDIF
            FLUX=QW(I,JN)
            SHEAR=ABS(TAUW(I,JN))
            UPLUS=U(I,JNM)/SQRT(SHEAR/DENSIT)
            YPLUS=(Y(JN)-Y(JNM))*SQRT(SHEAR/DENSIT)/VISCOS
            TW=T(I,JN)
            ANUM=0.0
            DEN=0.0
            DEN1=0.0
            DO 11 J=2,JNM
            ANUM=ANUM+T(I,J)*ABS(U(I,J))*R(J)*DYP(J)
            DEN=DEN+ABS(U(I,J))*R(J)*DYP(J)
            DEN1=DEN1+U(I,J)*R(J)*DYP(J)
      11    CONTINUE
            TB=ANUM/DEN
            UBAR=DEN1*2/R(JN)**2
            ANU=FLUX/(TW-TB+SMALL)*2.0*R(JN)/COND
            XX=XX/STEP
            REY=UBAR*DENSIT*2*R(JN)/VISCOS
            ANUTH=0.0123*REY**0.874*PR(7)**0.4
            ANUR=ANU/ANUTH
            WRITE(6,*)XX,ANUR,UPLUS,YPLUS
      10    CONTINUE
            WRITE(6,*)' ANUTH = ', ANUTH,' REY = ',REY

            XATTCH=XATTCH/STEP
            WRITE(6,*)' ATTACHMENT X = ',XATTCH
            IBEG=2
            IEND=35
            ITOT=IEND-IBEG+1
            OPEN(12,FILE='EXPN.DAT')
            WRITE(12,*)'TITLE  =  EXPANSION - KRALL'
```

```
          WRITE(12,*)'VARIABLES = XX YY UU VV TT '
          WRITE(12,*)'ZONE T = ZONE1, I = ',ITOT,', J = ',JN,' ,F = BLOCK'
          DO 111 J=1,JN
111          WRITE(12,*)(X(I),I=IBEG,IEND)
          DO 112 J=1,JN
112          WRITE(12,*)(Y(J),I=IBEG,IEND)
          DO 113 J=1,JN
113          WRITE(12,*)(U(I,J),I=IBEG,IEND)
          DO 114 J=1,JN
114          WRITE(12,*)(V(I,J),I=IBEG,IEND)
          DO 115 J=1,JN
115         WRITE(12,*)(T(I,J),I=IBEG,IEND)
          CLOSE(12)
          RETURN
          END
C ****************************************
          SUBROUTINE ADSORB(NN)
          INCLUDE  'COM2D.FOR'
C ****************************************
          N=NN
          GO TO (10,20,30,40,50,60,70,80,90),N
C *** FOR PRESSURE CORRECTION
10        GO TO 1000
C *** FOR U-VEL
20        GO TO 1000
C *** FOR V-VEL
30        GO TO 1000
C *** FOR W-VEL
40        GO TO 1000
C *** FOR K. ENERGY
50        GO TO 1000
C *** FOR DISSIPATION
60        GO TO 1000
C *** FOR TEMPERATURE
70        GO TO 1000
C *** FOR FLUID PROPERTIES
80        GO TO 1000
C *** CALLED FROM BOUNDP
90     CONTINUE
1000   CONTINUE
          RETURN
          END
```

```
C ****************************************
       SUBROUTINE OMEGA
       INCLUDE 'COM2D.FOR'
C ****************************************
       RETURN
       END
C ****************************************
       BLOCK DATA
       INCLUDE 'COM2D.FOR'
C ****************************************
C **** INITIAL DATA(make sure that IN,JN equal IT,JT)

       DATA CC,IPREF,JPREF,MXIT,GAMMA/1.0E-05,10,10,55,0.5/
       DATA CCTM,MXSTEP,DELT,STIME,MFREQ/1.0E-05,65,1.00,0.0,20/
C                   PP  U   V   W   E   D   T   VIS  P
       DATA     RP/1.0,1.0,1.0,1.0,1.0,1.0,1.0,0.5,0.1/
       DATA NSWEEP/ 10, 1 , 1 , 1 , 1 , 1 , 1           /
       DATA     PR/1.0,1.0,1.0,1.0,1.0,1.0,3.0          /
       DATA    PRT/1.0,1.0,1.0,1.0,1.0,1.3,0.9          /
       DATA DENSIT,VISCOS,SPHEAT/1.0,0.5E-3,1.0/
       DATA CD1,CD2,CMU,ELOG,CAPPA/1.44,1.92,0.09,9.793,0.4187/
C **** LOGICAL DATA
       DATA TURBUL, STEADY, UNSTDY,   FTRAN , CONMAS, AXISYMM
      1    /.TRUE.,.TRUE., .FALSE., .TRUE., .TRUE., .TRUE./
       DATA UPWIND,HYBRID,POWER/.TRUE.,.FALSE.,.FALSE./
       DATA SLVE/3*.TRUE.,.FALSE.,2*.TRUE.,.FALSE./
       DATA BSOR/9*.FALSE./
       DATA IREAD,IWRITE/.FALSE. ,.TRUE./
C PERIODIC BC
       DATA IPERIOD,JPERIOD/0,0/
       DATA DP1,DP2/0.0,0.0/
C **** READ GRID DATA
       DATA GRCELL,GRNODE/.TRUE.,.FALSE./
       DATA IN,JN/67,24/
       DATA JB1/14/
      DATA  YC/0.0,0.0,0.02,0.05,0.08,0.1,0.15,0.2,0.25
      1        ,0.30,0.35,0.40,0.45,0.5
      1        ,0.55,0.6,0.65,0.70,0.75
      1        ,0.8,0.85,0.9,0.95,1.0/
      DATA XC/0.0,0.0,0.03,0.06,0.1,0.15,0.2,0.25,0.3,0.35,0.4,0.45
      1        ,0.50,0.55,0.6,0.65,0.7,0.75,0.8,0.85,0.9,0.95,1.0,1.05
      1        ,1.12,1.2,1.3,1.4,1.5,1.6,1.75,1.9,2.05,2.2,2.4,2.6,2.8
```

```
     1          ,3.0,3.3,3.6,3.9,4.2,4.5,5.0,5.5,6.5,7.0,7.5,8.0,8.5,9.0
     1          ,9.7,10.5,11.5,12.5,14.0,16.0,18.0,21.0,25.0,29.0,33.0
     1          ,38.0,43.0,48.0,53.0,58.0/
          END
```

Natural Convection Evaporation – Chapter 5

The USER file that follows shows implementation of the mass transfer equation in subroutine OMEGA. In this subroutine, first coefficients of the discretised equation (AE, AW, AN, and AS) are evaluated through CALL COEF(0,SC,0.9), where $Pr_t = 0.9$ is inserted though not required in actual calculations because the flow is laminar. Then, since there is no source term (case of inert mass-transfer), no update of Su and Sp is made.[3] Now, boundary conditions are given where the mass transfer flux AMW(I, 1) at the south wall is evaluated. Then, the equation is solved through CALL SOLVE(0,RPO,RSU), where RPO is the underrelaxation factor. In the ADSORB subroutine, the source term in the v equation is added to account for buoyancy. Density is taken to be constant.

```
C *************************************
C THIS IS USER FILE NATURAL CONVECTION MASS TRANSFER
C *************************************
      PROGRAM MAIN
      INCLUDE 'COM2D.FOR'
      OPEN(UNIT=6,FILE='00')
      WRITE(*,*)'-------- output is in 00 file --------------'
C **** INITIAL DATA
      INM=IN-1
      JNM=JN-1
      SC=0.614
      OBR=50
      OWT=50
C
      WRITE(*,*)'GIVE ----- MXIT,IREAD,GRM   '
      READ(*,*)MXIT,IREAD,GRM
      CALL MAINPR
      STOP
      END
C *************************************
      SUBROUTINE TITLE
      INCLUDE 'COM2D.FOR'
C *************************************
```

---

[3] In a combustion problem, source terms must be calculated for each $\Phi = \omega_j$ of interest.

```
        WRITE(6,*)'****************************************'
        WRITE(6,*)' WATER EVAPORATION BY NATURAL CONVECTION  '
        WRITE(6,*)'****************************************'
        RETURN
        END
C ****************************************
        SUBROUTINE INIT
        INCLUDE 'COM2D.FOR'
C ****************************************
C INITIAL GUESS (ONLY HEAT FLUX NEEDS TO BE SPECIFIED)
        DO 10 I=1,IN
        DO 10 J=1,JN
        O(I,J)=0.0
        IF(I.LT.IB1)O(I,1)=1.0
10      CONTINUE
        RETURN
        END
C ****************************************
        SUBROUTINE BSPEC
        INCLUDE 'COM2D.FOR'
C ****************************************
C **** PROVIDE BOUNDARY & BLOCKED REGIONS
C
      CHARACTER*10 BLOCK,WEST,EAST,SOUTH,NORTH
      CHARACTER*10 INFLOW,EXIT1,SYMM,EXIT2,WALLT,WALLQ,PERIOD
      DATA  BLOCK,WEST,EAST,SOUTH,NORTH
     1  /'BLOCK','WEST','EAST','SOUTH','NORTH'/
      DATA   INFLOW,EXIT1,SYMM,EXIT2,WALLT,WALLQ,PERIOD
     1  /'INFLOW','EXIT1','SYMM','EXIT2','WALLT','WALLQ','PERIOD'/
C ***** BLOCKED REGIONS
        CALL TAG(BLOCK,BLOCK,IB1,IB2-1,2,JB1-1)
C ***** DEFINES W & E BOUNDARIES
        CALL TAG(WEST,SYMM,2,2, 2,JNM)
        CALL TAG(WEST,WALLQ,IB2,IB2, 2,JB1-1)
        CALL TAG(EAST,WALLQ,INM,INM,2,JNM)
        CALL TAG(EAST,WALLQ,IB1-1,IB1-1,2,JB1-1)
C ***** DEFINES N&S BOUNDARIES
        CALL TAG(NORTH,WALLQ,2,INM, JNM,JNM)
        CALL TAG(SOUTH,WALLQ,2,IB1-1,2,2)
        CALL TAG(SOUTH,WALLT,IB2,INM,2,2)
        CALL TAG(SOUTH,WALLT,IB1,IB2-1,JB1,JB1)
```

```
          RETURN
          END
C ****************************************
          SUBROUTINE RESULT
          INCLUDE 'COM2D.FOR'
C ****************************************
          CHARACTER*20 HEADER
          JSTEP=-1
          WRITE(6,*)' NITER=',NITER
          HEADER=' U-VEL'
          CALL PRINTK(U,1,IN,1,JN,HEADER,JSTEP)
          HEADER=' V-VEL'
          CALL PRINTK(V,1,IN,1,JN,HEADER,JSTEP)
          HEADER=' PRESS'
          CALL PRINTK(P,1,IN,1,JN,HEADER,JSTEP)
          HEADER=' OMEGA'
          CALL PRINTK(O,1,IN,1,JN,HEADER,JSTEP)
          HEADER=' AMW'
          CALL PRINTK(AMW,1,IN,1,JN,HEADER,JSTEP)
C CALCULATE NORALISED EVAPORATION RATE
          SUMWAT=0.0
          DX=0.0
          DO 1 I=2,IB1-1
          DX=DX+DXP(I)
1         SUMWAT=SUMWAT+AMW(I,1)*DXP(I)
          VBAR=SUMWAT/DX
          SUMBR=0.0
          DO 2 I=IB2,INM
2         SUMBR=SUMBR+AMW(I,1)*DXP(I)
C DIFFUSION LIMIT
          B=(0-1)/(1-OWT)
          DL=VISCOS/SC/YC(JB1)*ALOG(1+B)
        · WRITE(6,*)' DIFFUSION LIMIT = ',DL
          WRITE(6,*)' ACTUAL FLUX = ',VBAR
          RR=VBAR/DL
          WRITE(6,*)' RATIO = ',RR,' GRM = ',GRM,' B = ',B
          WRITE(6,*)' SUMWAT = ',SUMWAT,' SUMBR = ',SUMBR
          WRITE(6,*)' OWT = OBR = ',OWT
          IEND=30
          JEND=33
          OPEN(24,FILE='EVAP.DAT')
          WRITE(24,*)'TITLE  =   EVAPORATION'
```

```
            WRITE(24,*)'VARIABLES = XX YY UU VV OO '
            WRITE(24,*)'ZONE T = ZONE1, I = ',IEND
    1             ,' , J = ',JEND,' ,F = BLOCK'
            DO 11 J=1,JEND
11            WRITE(24,*)(X(I),I=1,IEND)
            DO 12 J=1,JEND
12            WRITE(24,*)(Y(J),I=1,IEND)
            DO 13 J=1,JEND
13            WRITE(24,*)(U(I,J),I=1,IEND)
            DO 14 J=1,JEND
14            WRITE(24,*)(V(I,J),I=1,IEND)
            DO 15 J=1,JEND
15            WRITE(24,*)(O(I,J),I=1,IEND)
            CLOSE(24)
            RETURN
            END
C ***************************************
            SUBROUTINE ADSORB(NN)
            INCLUDE 'COM2D.FOR'
C ***************************************
            N=NN
            GO TO (10,20,30,40,50,60,70,80,90),N
C *** FOR PRESSURE CORRECTION
10          GO TO 1000
C *** FOR U-VEL
20           GO TO 1000
C *** FOR V-VEL
30           DO 31 J=2,JNM
            DO 31 I=2,INM
31            SU(I,J)=SU(I,J)+GRM*O(I,J)*VOL(I,J)*(1-NTAG(I,J))
            GO TO 1000
C *** FOR W-VEL
40           GO TO 1000
C *** FOR K. ENERGY
50           GO TO 1000
C *** FOR DISSIPATION
60           GO TO 1000
C *** FOR TEMPERATURE
70           GO TO 1000
C *** FOR FLUID PROPERTIES
80           GO TO 1000
C *** CALLED FORM BOUNDP - FOR PRESSURE
```

```
90      CONTINUE
1000    CONTINUE
        RETURN
        END
C ***************************************
        SUBROUTINE OMEGA
        INCLUDE 'COM2D.FOR'
C ***************************************
C SOLVE FOR MASSFRACTION
        CALL COEF(0,SC,0.9)
C BOUNDARY CONDITIONS
        DO 2 J=2,JNM
        AW(2,J)=0.0
        O(1,J)=O(2,J)
        AE(INM,J)=0.0
        O(IN,J)=O(INM,J)
        IF(J.LT.JB1)THEN
        AE(IB1-1,J)=0.0
        O(IB1,J)=O(IB1-1,J)
        AW(IB2,J)=0.0
        O(IB2-1,J)=O(IB2,J)
        ENDIF
2       CONTINUE
        DO 3 I=2,INM
C NORTH WALL
        AN(I,JNM)=0.0
        O(I,JN)=O(I,JNM)
C TIP WALL
        IF(I.GT.IB1-1.AND.I.LT.IB2)THEN
        AS(I,JB1)=0.0
        O(I,JB1-1)=O(I,JB1)
        ENDIF
C WALL-WATER AND BRINE
        IF(I.LE.IB1-1.OR.I.GE.IB2)THEN
        DELTA=Y(2)-Y(1)
        TERM=VIS(I,1)/DELTA/SC
        OTT=OWT
        IF(I.GE.IB2)OTT=OBR
        B=(O(I,2)-O(I,1))/(O(I,1)-OTT)
        AMW(I,1)=TERM*ALOG(1+B)
C WALL VELOCITY
        V(I,1)=AMW(I,1)/RHO(I,1)
```

```
        ENDIF
3       CONTINUE
        RPO=1.0
        DO 4 J=2,JNM
        DO 4 I=2,INM
        SUM=AW(I,J)+AE(I,J)+AS(I,J)+AN(I,J)
4       AP(I,J)=(SUM+SP(I,J))/RPO
C SOLVE THE O EQN
        CALL SOLVE(O,RPO,RSU)
        DO 5 J=1,JN
        DO 5 I=1,IN
        IF(O(I,J).GT.1.0)O(I,J)=1.0
5       IF(O(I,J).LT.0.0)O(I,J)=0.0
        RETURN
        END
C *************************************
        BLOCK DATA
        INCLUDE 'COM2D.FOR'
C *************************************
C **** INITIAL DATA(make sure that IN,JN equal IT,JT)

        DATA CC,IPREF,JPREF,MXIT,GAMMA/1.0E-05,10,10,55,0.5/
        DATA CCTM,MXSTEP,DELT,STIME,MFREQ/1.0E-05,65,1.00,0.0,20/
C                 PP  U   V   W   E   D   T  VIS P
        DATA    RP/1.0,0.5,0.5,1.0,1.0,1.0,1.0,1.0,0.1/
        DATA NSWEEP/ 10, 1 , 1 , 1 , 1 , 1 , 1           /
        DATA    PR/1.0,1.0,1.0,1.0,1.0,1.0,0.7           /
        DATA    PRT/1.0,1.0,1.0,1.0,1.0,1.3,0.9          /
        DATA DENSIT,VISCOS,SPHEAT/1.0,1.0,1.0/
        DATA CD1,CD2,CMU,ELOG,CAPPA/1.44,1.92,0.09,9.793,0.4187/
C **** LOGICAL DATA
        DATA TURBUL, STEADY, UNSTDY,   FTRAN , CONMAS, AXISYMM
     1     /.FALSE.,.TRUE., .FALSE., .FALSE., .FALSE., .FALSE./
        DATA UPWIND,HYBRID,POWER/.TRUE.,.FALSE.,.FALSE./
        DATA SLVE/3*.TRUE.,4*.FALSE./
        DATA BSOR/2*.FALSE.,.TRUE.,6*.FALSE./
        DATA IREAD,IWRITE/.FALSE. ,.TRUE./
C PERIODIC BC
        DATA IPERIOD,JPERIOD/0,0/
        DATA DP1,DP2/0.0,0.0/
C **** READ GRID DATA
        DATA GRCELL,GRNODE/.TRUE.,.FALSE./
```

```
      DATA IB1,IB2,JB1,IN,JN/16,21,24,37,37/
C h=2 AND H = 8
      DATA  YC/0.0,0.0,0.02,0.04,0.07,0.12,0.18,0.25,0.35,0.5
     1      ,0.65,0.8,0.95,1.1,1.25,1.4,1.55,1.7,1.8,1.85,1.9
     1      ,1.95,1.98,2.0
     1      ,2.02,2.05,2.1,2.2,2.4,2.8,3.3,4.0,5.0,6.0,7.0,7.5,8.0/
C L/2 = 8, t = 0.1, 1/2=0.5
      DATA XC/0.0,0.0,0.02,0.04,0.07,0.1,0.15,0.2,0.25,0.3,0.35,0.4
     1       ,0.43,0.46,0.48,0.5
     1       ,0.52,0.54,0.56,0.58,0.6
     1       ,0.62,0.65,0.7,0.8,0.9,1.0,1.2,1.5,2.0,3.0,4.0,5.0
     1       ,6.0,7.0,7.5,8.0/
      END
```

Library File LIB2D.FOR

```
C ******************************************
C THIS IS LIBRARY FILE LIB2D.FOR ---- A W DATE
C RESIDUALS ARE STORED FOR PLOTTING IN SUBROUTINE EQN
C ******************************************
      SUBROUTINE MAINPR
      INCLUDE 'COM2D.FOR'
C ******************************************
      NPERIOD=0
      CALL TITLE
      CALL GRID
      CALL INITIA
      CALL BSPEC
      CALL INIT
      CALL PROPS
      CALL INDATA
      CALL INFLUX
      IF(IREAD) CALL IPT
      IF(UNSTDY)CALL UPDATE
      CALL EQN
      CALL BOUNDP
      IF(IWRITE)CALL OPT
      CALL RESULT
      RETURN
      END
C ******************************************
      SUBROUTINE INITIA
      INCLUDE 'COM2D.FOR'
```

```
C ******************************************
      DO 1 J=1,JN
      DO 1 I=1,IN
      PP(I,J)=0.0
      P(I,J)=0.0
      U(I,J)=0.0
      V(I,J)=0.0
      W(I,J)=0.0
      E(I,J)=0.0
      D(I,J)=0.0
      T(I,J)=0.0
      QW(I,J)=0.0
      VIS(I,J)=VISCOS
      VIST(I,J)=0.0
      RHO(I,J)=DENSIT
      RHOO(I,J)=DENSIT
      AW(I,J)=0.0
      AE(I,J)=0.0
      AS(I,J)=0.0
      AN(I,J)=0.0
      APU(I,J)=GREAT
      APV(I,J)=GREAT
      AP1(I,J)=GREAT
      AP(I,J)=GREAT
      NTAG(I,J)=0
      NTAGW(I,J)=0
      NTAGE(I,J)=0
      NTAGS(I,J)=0
1     NTAGN(I,J)=0
      RETURN
      END
C ******************************************
      SUBROUTINE TAG(CHAR1,CHAR2,IB,IL,JB,JL)
      INCLUDE 'COM2D.FOR'
C ******************************************
      CHARACTER*10 CHAR1,CHAR2
      IF(CHAR2.EQ.'PERIOD')NPERIOD=1
      DO 1 J=JB,JL
      DO 1 I=IB,IL
      IF(CHAR1.EQ.'BLOCK')THEN
      NTAG(I,J)=1
      GO TO 1
```

```
        ENDIF
        IF (CHAR1.EQ.'WEST')THEN
        IF(CHAR2.EQ.'INFLOW')NTAGW(I,J)=11
        IF(CHAR2.EQ.'SYMM'  )NTAGW(I,J)=12
        IF(CHAR2.EQ.'EXIT1'  )NTAGW(I,J)=13
        IF(CHAR2.EQ.'EXIT2'  )NTAGW(I,J)=15
        IF(CHAR2.EQ.'WALLT' )NTAGW(I,J)=14
        IF(CHAR2.EQ.'WALLQ' )NTAGW(I,J)=16
        IF(CHAR2.EQ.'PERIOD' )NTAGW(I,J)=17
        ELSE IF(CHAR1.EQ.'EAST')THEN
        IF(CHAR2.EQ.'INFLOW')NTAGE(I,J)=21
        IF(CHAR2.EQ.'SYMM'  )NTAGE(I,J)=22
        IF(CHAR2.EQ.'EXIT1'  )NTAGE(I,J)=23
        IF(CHAR2.EQ.'EXIT2'  )NTAGE(I,J)=25
        IF(CHAR2.EQ.'WALLT' )NTAGE(I,J)=24
        IF(CHAR2.EQ.'WALLQ' )NTAGE(I,J)=26
        IF(CHAR2.EQ.'PERIOD' )NTAGE(I,J)=27
        ELSE IF(CHAR1.EQ.'SOUTH')THEN
        IF(CHAR2.EQ.'INFLOW')NTAGS(I,J)=31
        IF(CHAR2.EQ.'SYMM'  )NTAGS(I,J)=32
        IF(CHAR2.EQ.'EXIT1'  )NTAGS(I,J)=33
        IF(CHAR2.EQ.'EXIT2'  )NTAGS(I,J)=35
        IF(CHAR2.EQ.'WALLT' )NTAGS(I,J)=34
        IF(CHAR2.EQ.'WALLQ' )NTAGS(I,J)=36
        IF(CHAR2.EQ.'PERIOD' )NTAGS(I,J)=37
        ELSE IF(CHAR1.EQ.'NORTH')THEN
        IF(CHAR2.EQ.'INFLOW')NTAGN(I,J)=41
        IF(CHAR2.EQ.'SYMM'  )NTAGN(I,J)=42
        IF(CHAR2.EQ.'EXIT1'  )NTAGN(I,J)=43
        IF(CHAR2.EQ.'EXIT2'  )NTAGN(I,J)=45
        IF(CHAR2.EQ.'WALLT' )NTAGN(I,J)=44
        IF(CHAR2.EQ.'WALLQ' )NTAGN(I,J)=46
        IF(CHAR2.EQ.'PERIOD' )NTAGN(I,J)=47
        ENDIF
1       CONTINUE
        RETURN
        END
C ******************************************
        SUBROUTINE BOUND(NN)
        INCLUDE 'COM2D.FOR'
C ******************************************
        N=NN
```

```
              DO 1 J=2,JNM
              DO 1 I=2,INM
              VOLP=VOL(I,J)
              RHOP=RHO(I,J)
C ***  BLOCKED REGION
              IF(NTAG(I,J).EQ.1)THEN
              IF(N.EQ.2)SU(I,J)=GREAT*U(I,J)
              IF(N.EQ.3)SU(I,J)=GREAT*V(I,J)
              IF(N.EQ.4)SU(I,J)=GREAT*W(I,J)
              IF(N.EQ.5)SU(I,J)=GREAT*E(I,J)
              IF(N.EQ.6)SU(I,J)=GREAT*D(I,J)
              IF(N.EQ.7)SU(I,J)=GREAT*T(I,J)
              SP(I,J)=GREAT
              GO TO 1
              END IF
C *** WEST BOUNDARY
              LW=NTAGW(I,J)
              IF(LW.EQ.0)GO TO 100
              AWNOW=AW(I,J)
C      INLET
              IF(LW.EQ.11.OR.LW.EQ.17)THEN
              AW(I,J)=0.0
              IF(N.EQ.2)SU(I,J)=AWNOW*U(I-1,J)+SU(I,J)
              IF(N.EQ.3)SU(I,J)=AWNOW*V(I-1,J)+SU(I,J)
              IF(N.EQ.4)SU(I,J)=AWNOW*W(I-1,J)+SU(I,J)
              IF(N.EQ.5)SU(I,J)=AWNOW*E(I-1,J)+SU(I,J)
              IF(N.EQ.6)SU(I,J)=AWNOW*D(I-1,J)+SU(I,J)
              IF(N.EQ.7)SU(I,J)=AWNOW*T(I-1,J)+SU(I,J)
              SP(I,J)=AWNOW+SP(I,J)
              ENDIF
C      SYMMETRY
              IF(LW.EQ.12)THEN
              IF(N.EQ.2)SP(I,J)=AWNOW+SP(I,J)
              AW(I,J)=0.0
              IF(N.EQ.2)U(I-1,J)=0.0
              IF(N.EQ.3)V(I-1,J)=V(I,J)
              IF(N.EQ.4)W(I-1,J)=W(I,J)
              IF(N.EQ.5)E(I-1,J)=E(I,J)
              IF(N.EQ.6)D(I-1,J)=D(I,J)
              IF(N.EQ.7)T(I-1,J)=T(I,J)
              ENDIF
C      EXIT
```

```
         IF(LW.EQ.13.OR.LW.EQ.15) THEN
         AW(I,J)=0.0
         RATIO=(X(I)-XC(I))/DXMI(I+1)
         IF(LW.EQ.13)RATIO=0.0
         IF(N.EQ.2)U(I-1,J)=U(I,J)-RATIO*(U(I+1,J)-U(I,J))
         IF(N.EQ.3)V(I-1,J)=V(I,J)-RATIO*(V(I+1,J)-V(I,J))
         IF(N.EQ.4)W(I-1,J)=W(I,J)-RATIO*(W(I+1,J)-W(I,J))
         IF(N.EQ.5)E(I-1,J)=E(I,J)-RATIO*(E(I+1,J)-E(I,J))
         IF(N.EQ.6)D(I-1,J)=D(I,J)-RATIO*(D(I+1,J)-D(I,J))
         IF(N.EQ.7)T(I-1,J)=T(I,J)-RATIO*(T(I+1,J)-T(I,J))
         ENDIF
C        WALL
         IF(LW.EQ.14.OR.LW.EQ.16) THEN
         AW(I,J)=0.0
         DELTA=X(I)-XC(I)
         AREA=R(J)*DYP(J)
         UWAL=U(I-1,J)
         VWAL=V(I-1,J)
         WWAL=W(I-1,J)
         VISWAL=VIS(I-1,J)
         ANG=ATAN(W(I,J)/(V(I,J)+SMALL))
         VT=(V(I,J)-VWAL)*COS(ANG)+(W(I,J)-WWAL)*SIN(ANG)
         VTTAU=CMU**0.25*SQRT(ABS(E(I,J)))
         YPLUS=VTTAU*DELTA*RHOP/VISWAL
         EYPLUS=ELOG*YPLUS
         TMULT=VISWAL*AREA/DELTA
         TAUW(I-1,J)=VISWAL*VT/DELTA
         IF(TURBUL)TAUW(I-1,J)=RHO(I-1,J)*VTTAU**2
         IF(YPLUS.GT.11.6)TMULT=RHOP*CAPPA*VTTAU*AREA/ALOG(EYPLUS)
         IF(N.EQ.2) THEN
         SU(I,J)=AWNOW*UWAL+SU(I,J)
         SP(I,J)=AWNOW+SP(I,J)
         ELSE IF(N.EQ.3) THEN
         SU(I,J)=TMULT*VWAL+SU(I,J)
         SP(I,J)=TMULT+SP(I,J)
         ELSE IF(N.EQ.4) THEN
         SU(I,J)=TMULT*WWAL+SU(I,J)
         SP(I,J)=TMULT+SP(I,J)
         ELSE IF(N.EQ.5) THEN
         GENR=TMULT*VT/AREA*VT/DELTA
         TERM=RHOP**2*CMU*ABS(E(I,J))/VISWAL
         IF(YPLUS.GT.11.6)
```

```
     1   TERM=RHOP*VTTAU*ALOG(EYPLUS)*CMU**0.5/CAPPA/DELTA
         SU(I,J)=GENR*VOLP+SU(I,J)
         SP(I,J)=TERM*VOLP+SP(I,J)
         ELSE IF(N.EQ.6) THEN
         TERM=VTTAU**3/CAPPA/DELTA
         SU(I,J)=GREAT*TERM
         SP(I,J)=GREAT
         ELSE IF(N.EQ.7) THEN
         IF(TURBUL)THEN
         UPLUS=ABS(VT)/VTTAU
         STANTON=STAN(UPLUS,YPLUS,PR(7),PRT(7))
         TERM=RHOP*ABS(VT)*AREA*STANTON
         ELSE
         TERM=VISWAL/(PR(N)*DELTA)*AREA
         ENDIF
         IF(LW.EQ.14)THEN
         SU(I,J)=TERM*T(I-1,J)+SU(I,J)
         SP(I,J)=TERM+SP(I,J)
         QW(I-1,J)=TERM/AREA*(T(I-1,J)-T(I,J))*SPH(I-1,J)
         ELSE IF(LW.EQ.16)THEN
         SU(I,J)=QW(I-1,J)*AREA/SPH(I-1,J)+SU(I,J)
         T(I-1,J)=QW(I-1,J)/TERM*AREA/SPH(I-1,J)+T(I,J)
         ENDIF
         ENDIF
         ENDIF
C *** EAST BOUNDARY
100      LE=NTAGE(I,J)
         IF(LE.EQ.0)GO TO 200
         AENOW=AE(I,J)
C  INLET
         IF(LE.EQ.21.OR.LE.EQ.27)THEN
         AE(I,J)=0.0
         IF(N.EQ.2)SU(I,J)=AENOW*U(I+1,J)+SU(I,J)
         IF(N.EQ.3)SU(I,J)=AENOW*V(I+1,J)+SU(I,J)
         IF(N.EQ.4)SU(I,J)=AENOW*W(I+1,J)+SU(I,J)
         IF(N.EQ.5)SU(I,J)=AENOW*E(I+1,J)+SU(I,J)
         IF(N.EQ.6)SU(I,J)=AENOW*D(I+1,J)+SU(I,J)
         IF(N.EQ.7)SU(I,J)=AENOW*T(I+1,J)+SU(I,J)
         SP(I,J)=AENOW+SP(I,J)
         ENDIF
C  SYMMETRY
         IF(LE.EQ.22)THEN
```

```
          IF(N.EQ.2)SP(I,J)=AENOW+SP(I,J)
          AE(I,J)=0.0
          IF(N.EQ.2)U(I+1,J)=0.0
          IF(N.EQ.3)V(I+1,J)=V(I,J)
          IF(N.EQ.4)W(I+1,J)=W(I,J)
          IF(N.EQ.5)E(I+1,J)=E(I,J)
          IF(N.EQ.6)D(I+1,J)=D(I,J)
          IF(N.EQ.7)T(I+1,J)=T(I,J)
          ENDIF
C   EXIT
          IF(LE.EQ.23.OR.LE.EQ.25) THEN
          AE(I,J)=0.0
          RATIO=(XC(I+1)-X(I))/DXMI(I)
          IF(LE.EQ.23)RATIO=0.0
          IF(N.EQ.2)U(I+1,J)=U(I,J)+RATIO*(U(I,J)-U(I-1,J))
          IF(N.EQ.3)V(I+1,J)=V(I,J)+RATIO*(V(I,J)-V(I-1,J))
          IF(N.EQ.4)W(I+1,J)=W(I,J)+RATIO*(W(I,J)-W(I-1,J))
          IF(N.EQ.5)E(I+1,J)=E(I,J)+RATIO*(E(I,J)-E(I-1,J))
          IF(N.EQ.6)D(I+1,J)=D(I,J)+RATIO*(D(I,J)-D(I-1,J))
          IF(N.EQ.7)T(I+1,J)=T(I,J)+RATIO*(T(I,J)-T(I-1,J))
          ENDIF
C   WALL
          IF(LE.EQ.24.OR.LE.EQ.26) THEN
          AE(I,J)=0.0
          DELTA=XC(I+1)-X(I)
          AREA=R(J)*DYP(J)
          UWAL=U(I+1,J)
          VWAL=V(I+1,J)
          WWAL=W(I+1,J)
          VISWAL=VIS(I+1,J)
          ANG=ATAN(W(I,J)/(V(I,J)+SMALL))
          VT=(V(I,J)-VWAL)*COS(ANG)+(W(I,J)-WWAL)*SIN(ANG)
          VTTAU=CMU**0.25*SQRT(ABS(E(I,J)))
          YPLUS=VTTAU*DELTA*RHOP/VISWAL
          EYPLUS=ELOG*YPLUS
          TMULT=VISWAL*AREA/DELTA
          TAUW(I+1,J)=-VISWAL*VT/DELTA
          IF(TURBUL)TAUW(I+1,J)=RHO(I+1,J)*VTTAU**2
          IF(YPLUS.GT.11.6)TMULT=RHOP*CAPPA*VTTAU*AREA/ALOG(EYPLUS)
          IF(N.EQ.2) THEN
          SU(I,J)=AENOW*UWAL+SU(I,J)
          SP(I,J)=AENOW+SP(I,J)
```

```
          ELSE IF(N.EQ.3) THEN
          SU(I,J)=TMULT*VWAL+SU(I,J)
          SP(I,J)=TMULT+SP(I,J)
          ELSE IF(N.EQ.4) THEN
          SU(I,J)=TMULT*WWAL+SU(I,J)
          SP(I,J)=TMULT+SP(I,J)
          ELSE IF(N.EQ.5) THEN
          GENR=TMULT*VT/AREA*VT/DELTA
          TERM=RHOP**2*CMU*ABS(E(I,J))/VISWAL
          IF(YPLUS.GT.11.6)
     1    TERM=RHOP*VTTAU*ALOG(EYPLUS)*CMU**0.5/CAPPA/DELTA
          SU(I,J)=GENR*VOLP+SU(I,J)
          SP(I,J)=TERM*VOLP+SP(I,J)
          ELSE IF(N.EQ.6) THEN
          TERM=VTTAU**3/CAPPA/DELTA
          SU(I,J)=GREAT*TERM
          SP(I,J)=GREAT
          ELSE IF(N.EQ.7) THEN
          IF(TURBUL)THEN
          UPLUS=ABS(VT)/VTTAU
          STANTON=STAN(UPLUS,YPLUS,PR(7),PRT(7))
          TERM=RHOP*ABS(VT)*AREA*STANTON
          ELSE
          TERM=VISWAL/(PR(N)*DELTA)*AREA
          ENDIF
          IF(LE.EQ.24)THEN
          SU(I,J)=TERM*T(I+1,J)+SU(I,J)
          SP(I,J)=TERM+SP(I,J)
          QW(I+1,J)=TERM/AREA*(T(I+1,J)-T(I,J))*SPH(I+1,J)
          ELSE IF(LE.EQ.26)THEN
          SU(I,J)=QW(I+1,J)*AREA/SPH(I+1,J)+SU(I,J)
          T(I+1,J)=QW(I+1,J)/TERM*AREA/SPH(I+1,J)+T(I,J)
          ENDIF
          ENDIF
          ENDIF
C *** SOUTH BOUNDARY
200       LS=NTAGS(I,J)
          IF(LS.EQ.0)GO TO 300
          ASNOW=AS(I,J)
C  INLET
          IF(LS.EQ.31.OR.LS.EQ.37)THEN
          AS(I,J)=0.0
```

```
        IF(N.EQ.2)SU(I,J)=ASNOW*U(I,J-1)+SU(I,J)
        IF(N.EQ.3)SU(I,J)=ASNOW*V(I,J-1)+SU(I,J)
        IF(N.EQ.4)SU(I,J)=ASNOW*W(I,J-1)+SU(I,J)
        IF(N.EQ.5)SU(I,J)=ASNOW*E(I,J-1)+SU(I,J)
        IF(N.EQ.6)SU(I,J)=ASNOW*D(I,J-1)+SU(I,J)
        IF(N.EQ.7)SU(I,J)=ASNOW*T(I,J-1)+SU(I,J)
        SP(I,J)=ASNOW+SP(I,J)
        ENDIF
C  SYMMETRY
        IF(LS.EQ.32)THEN
        IF(N.EQ.3)SP(I,J)=ASNOW+SP(I,J)
        AS(I,J)=0.0
        IF(N.EQ.3)V(I,J-1)=0.0
        IF(N.EQ.2)U(I,J-1)=U(I,J)
        IF(N.EQ.4)W(I,J-1)=W(I,J)
        IF(N.EQ.5)E(I,J-1)=E(I,J)
        IF(N.EQ.6)D(I,J-1)=D(I,J)
        IF(N.EQ.7)T(I,J-1)=T(I,J)
        ENDIF
C  EXIT
        IF(LS.EQ.33.OR.LS.EQ.35) THEN
        RATIO=(Y(J)-YC(J))/DYMI(J+1)
        IF(LS.EQ.33)RATIO=0.0
        AS(I,J)=0.0
        IF(N.EQ.2)U(I,J-1)=U(I,J)-RATIO*(U(I,J+1)-U(I,J))
        IF(N.EQ.3)V(I,J-1)=V(I,J)-RATIO*(V(I,J+1)-V(I,J))
        IF(N.EQ.4)W(I,J-1)=W(I,J)-RATIO*(W(I,J+1)-W(I,J))
        IF(N.EQ.5)E(I,J-1)=E(I,J)-RATIO*(E(I,J+1)-E(I,J))
        IF(N.EQ.6)D(I,J-1)=D(I,J)-RATIO*(D(I,J+1)-D(I,J))
        IF(N.EQ.7)T(I,J-1)=T(I,J)-RATIO*(T(I,J+1)-T(I,J))
        ENDIF
C  WALL
        IF(LS.EQ.34.OR.LS.EQ.36) THEN
        AS(I,J)=0.0
        DELTA=Y(J)-YC(J)
        AREA=RC(J)*DXP(I)
        UWAL=U(I,J-1)
        VWAL=V(I,J-1)
        WWAL=W(I,J-1)
        VISWAL=VIS(I,J-1)
        ANG=ATAN(W(I,J)/(U(I,J)+SMALL))
        VT=(U(I,J)-UWAL)*COS(ANG)+(W(I,J)-WWAL)*SIN(ANG)
```

```
          VTTAU=CMU**0.25*SQRT(ABS(E(I,J)))
          YPLUS=VTTAU*DELTA*RHOP/VISWAL
          EYPLUS=ELOG*YPLUS
          TMULT=VISWAL*AREA/DELTA
          TAUW(I,J-1)=VISWAL*VT/DELTA
          IF(TURBUL)TAUW(I,J-1)=RHO(I,J-1)*VTTAU**2
          IF(YPLUS.GT.11.6)TMULT=RHOP*CAPPA*VTTAU*AREA/ALOG(EYPLUS)
          IF(N.EQ.2) THEN
          SU(I,J)=TMULT*UWAL+SU(I,J)
          SP(I,J)=TMULT+SP(I,J)
          ELSE IF(N.EQ.3) THEN
          SU(I,J)=ASNOW*VWAL+SU(I,J)
          SP(I,J)=ASNOW+SP(I,J)
          ELSE IF(N.EQ.4) THEN
          SU(I,J)=TMULT*WWAL+SU(I,J)
          SP(I,J)=TMULT+SP(I,J)
          ELSE IF(N.EQ.5) THEN
          GENR=TMULT*VT/AREA*VT/DELTA
          TERM=RHOP**2*CMU*ABS(E(I,J))/VISWAL
          IF(YPLUS.GT.11.6)
     1    TERM=RHOP*VTTAU*ALOG(EYPLUS)*CMU**0.5/CAPPA/DELTA
          SU(I,J)=GENR*VOLP+SU(I,J)
          SP(I,J)=TERM*VOLP+SP(I,J)
          ELSE IF(N.EQ.6) THEN
          TERM=VTTAU**3/CAPPA/DELTA
          SU(I,J)=GREAT*TERM
          SP(I,J)=GREAT
          ELSE IF(N.EQ.7) THEN
          IF(TURBUL)THEN
          UPLUS=ABS(VT)/VTTAU
          STANTON=STAN(UPLUS,YPLUS,PR(7),PRT(7))
          TERM=RHOP*ABS(VT)*AREA*STANTON
          ELSE
          TERM=VISWAL/(PR(N)*DELTA)*AREA
          ENDIF
          IF(LS.EQ.34)THEN
          SU(I,J)=TERM*T(I,J-1)+SU(I,J)
          SP(I,J)=TERM+SP(I,J)
          QW(I,J-1)=TERM/AREA*(T(I,J-1)-T(I,J))*SPH(I,J-1)
          ELSE IF(LS.EQ.36)THEN
          SU(I,J)=QW(I,J-1)*AREA/SPH(I,J-1)+SU(I,J)
          T(I,J-1)=QW(I,J-1)/TERM*AREA/SPH(I,J-1)+T(I,J)
```

```
            ENDIF
            ENDIF
            ENDIF
C *** NORTH BOUNDARY
300     LN=NTAGN(I,J)
        IF(LN.EQ.0)GO TO 1
        ANNOW=AN(I,J)
C  INLET
        IF(LN.EQ.41.OR.LN.EQ.47)THEN
        AN(I,J)=0.0
        IF(N.EQ.2)SU(I,J)=ANNOW*U(I,J+1)+SU(I,J)
        IF(N.EQ.3)SU(I,J)=ANNOW*V(I,J+1)+SU(I,J)
        IF(N.EQ.4)SU(I,J)=ANNOW*W(I,J+1)+SU(I,J)
        IF(N.EQ.5)SU(I,J)=ANNOW*E(I,J+1)+SU(I,J)
        IF(N.EQ.6)SU(I,J)=ANNOW*D(I,J+1)+SU(I,J)
        IF(N.EQ.7)SU(I,J)=ANNOW*T(I,J+1)+SU(I,J)
        SP(I,J)=ANNOW+SP(I,J)
        ENDIF
C  SYMMETRY
        IF(LN.EQ.42)THEN
        IF(N.EQ.3)SP(I,J)=ANNOW+SP(I,J)
        AN(I,J)=0.0
        IF(N.EQ.3)V(I,J+1)=0.0
        IF(N.EQ.2)U(I,J+1)=U(I,J)
        IF(N.EQ.4)W(I,J+1)=W(I,J)
        IF(N.EQ.5)E(I,J+1)=E(I,J)
        IF(N.EQ.6)D(I,J+1)=D(I,J)
        IF(N.EQ.7)T(I,J+1)=T(I,J)
        ENDIF
C  EXIT
        IF(LN.EQ.43.OR.LN.EQ.45) THEN
        AN(I,J)=0.0
        RATIO=(YC(J+1)-Y(J))/DYMI(J)
        IF(LN.EQ.43)RATIO=0.0
        IF(N.EQ.2)U(I,J+1)=U(I,J)+RATIO*(U(I,J)-U(I,J-1))
        IF(N.EQ.3)V(I,J+1)=V(I,J)+RATIO*(V(I,J)-V(I,J-1))
        IF(N.EQ.4)W(I,J+1)=W(I,J)+RATIO*(W(I,J)-W(I,J-1))
        IF(N.EQ.5)E(I,J+1)=E(I,J)+RATIO*(E(I,J)-E(I,J-1))
        IF(N.EQ.6)D(I,J+1)=D(I,J)+RATIO*(D(I,J)-D(I,J-1))
        IF(N.EQ.7)T(I,J+1)=T(I,J)+RATIO*(T(I,J)-T(I,J-1))
        ENDIF
C  WALL
```

```
      IF(LN.EQ.44.OR.LN.EQ.46) THEN
      AN(I,J)=0.0
      DELTA=YC(J+1)-Y(J)
      AREA=RC(J+1)*DXP(I)
      UWAL=U(I,J+1)
      VWAL=V(I,J+1)
      WWAL=W(I,J+1)
      VISWAL=VIS(I,J+1)
      ANG=ATAN(W(I,J)/(U(I,J)+SMALL))
      VT=(U(I,J)-UWAL)*COS(ANG)+(W(I,J)-WWAL)*SIN(ANG)
      VTTAU=CMU**0.25*SQRT(ABS(E(I,J)))
      YPLUS=VTTAU*DELTA*RHOP/VISWAL
      EYPLUS=ELOG*YPLUS
      TMULT=VISWAL*AREA/DELTA
      TAUW(I,J+1)=-VISWAL*VT/DELTA
      IF(TURBUL)TAUW(I,J+1)=RHO(I,J+1)*VTTAU**2
      IF(YPLUS.GT.11.6)TMULT=RHOP*CAPPA*VTTAU*AREA/ALOG(EYPLUS)
      IF(N.EQ.2) THEN
      SU(I,J)=TMULT*UWAL+SU(I,J)
      SP(I,J)=TMULT+SP(I,J)
      ELSE IF(N.EQ.3) THEN
      SU(I,J)=ANNOW*VWAL+SU(I,J)
      SP(I,J)=ANNOW+SP(I,J)
      ELSE IF(N.EQ.4) THEN
      SU(I,J)=TMULT*WWAL+SU(I,J)
      SP(I,J)=TMULT+SP(I,J)
      ELSE IF(N.EQ.5) THEN
      GENR=TMULT*VT/AREA*VT/DELTA
      TERM=RHOP**2*CMU*ABS(E(I,J))/VISWAL
      IF(YPLUS.GT.11.6)
    1 TERM=RHOP*VTTAU*ALOG(EYPLUS)*CMU**0.5/CAPPA/DELTA
      SU(I,J)=GENR*VOLP+SU(I,J)
      SP(I,J)=TERM*VOLP+SP(I,J)
      ELSE IF(N.EQ.6) THEN
      TERM=VTTAU**3/CAPPA/DELTA
      SU(I,J)=GREAT*TERM
      SP(I,J)=GREAT
      ELSE IF(N.EQ.7) THEN
      IF(TURBUL)THEN
      UPLUS=ABS(VT)/VTTAU
      STANTON=STAN(UPLUS,YPLUS,PR(7),PRT(7))
      TERM=RHOP*ABS(VT)*AREA*STANTON
```

```
            ELSE
            TERM=VISWAL/(PR(N)*DELTA)*AREA
            ENDIF
            IF(LN.EQ.44)THEN
            SU(I,J)=TERM*T(I,J+1)+SU(I,J)
            SP(I,J)=TERM+SP(I,J)
            QW(I,J+1)=TERM/AREA*(T(I,J+1)-T(I,J))*SPH(I,J+1)
            ELSE IF(LN.EQ.46)THEN
            SU(I,J)=QW(I,J+1)*AREA/SPH(I,J+1)+SU(I,J)
            T(I,J+1)=QW(I,J+1)/TERM*AREA/SPH(I,J+1)+T(I,J)
            ENDIF
            ENDIF
            ENDIF
1           CONTINUE
            RETURN
            END
C ********************************************
            SUBROUTINE GRID
            INCLUDE 'COM2D.FOR'
C ********************************************
C CALCULATE CELL-FACE COORDINATES
            IF(GRNODE)THEN
            XC(2)=X(1)
            YC(2)=Y(1)
            XC(1)=XC(2)
            YC(1)=YC(2)
            DO 1 I=3,INM
1           XC(I)=0.5*(X(I)+X(I-1))
            XC(IN)=X(IN)
            DO 2 J=3,JNM
2           YC(J)=0.5*(Y(J)+Y(J-1))
            YC(JN)=Y(JN)
            ENDIF
C CALCULATE NODE COORDINATES
            IF(GRCELL)THEN
            X(1)=XC(2)
            Y(1)=YC(2)
            DO 11 I=2,INM
11          X(I)=0.5*(XC(I)+XC(I+1))
            DO 12 J=2,JNM
12          Y(J)=0.5*(YC(J)+YC(J+1))
            Y(JN)=YC(JN)
```

```
              X(IN)=XC(IN)
              ENDIF
C   *** CALCULATE INTERPOLATION FACTORS
              DXMI(1)=0.0
              DO 4 I=2,IN
4             DXMI(I)=X(I)-X(I-1)
              DYMI(1)=0.0
              DO 5 J=2,JN
5             DYMI(J)=Y(J)-Y(J-1)
              DO 6 J=1,JN
              R(J)=1.0
              RC(J)=1.0
              IF(AXISYMM)THEN
              R(J)=Y(J)
              RC(J)=YC(J)
              ENDIF
6             CONTINUE
C *** CALCULATE CELL VOLUME
              SUMVOL=0.0
              DO 7 J=2,JNM
              DO 7 I=2,INM
              VOL(I,J)=R(J)*(XC(I+1)-XC(I))*(YC(J+1)-YC(J))
              SUMVOL=SUMVOL+VOL(I,J)
7             CONTINUE
              WRITE(6,*)' DOMAIN VOLUME = ',SUMVOL
C *** CALCULATE AREAS
              DO 9 I=1,INM
              DXP(I)=(XC(I+1)-XC(I))
9             DXP(IN)=0.0
              DO 10 J=1,JNM
              DYP(J)=(YC(J+1)-YC(J))
10            DYP(JN)=0.0
              RETURN
              END
C *******************************************
              SUBROUTINE COEF(NN,PRN,PRTN)
              INCLUDE 'COM2D.FOR'
C *******************************************
              N=NN
              IF(N.EQ.1) GO TO 1000
C COEFFICIENTS OF TRANSPORT EQUATIONS
              PRINV=1./PRN
```

```
      PRTINV=1.0/PRTN
      DO 1 J=2,JNM
      DO 1 I=2,INM
      SU(I,J)=0.0
      SP(I,J)=0.0
C *** DIFFUSION COEFFICIENTS AND INTERPOLATED VALUES
      LW=NTAGW(I,J)/10
      LE=NTAGE(I,J)/20
      LS=NTAGS(I,J)/30
      LN=NTAGN(I,J)/40
C  **** LAMINAR VISCOSITY
      TERMW=(X(I)-XC(I))/VIS(I,J)+(XC(I)-X(I-1))/VIS(I-1,J)
      VISW=DXMI(I)/TERMW*(1-LW)+LW*VIS(I-1,J)
      TERME=(XC(I+1)-X(I))/VIS(I,J)+(X(I+1)-XC(I+1))/VIS(I+1,J)
      VISE=DXMI(I+1)/TERME*(1-LE)+LE*VIS(I+1,J)
      TERMS=(Y(J)-YC(J))/VIS(I,J)+(YC(J)-Y(J-1))/VIS(I,J-1)
      VISS=DYMI(J)/(TERMS+SMALL)*(1-LS)+LS*VIS(I,J-1)
      TERMN=(YC(J+1)-Y(J))/VIS(I,J)+(Y(J+1)-YC(J+1))/VIS(I,J+1)
      VISN=DYMI(J+1)/TERMN*(1-LN)+LN*VIS(I,J+1)
C  **** TURBULENT VISCOSITY
      IF(TURBUL)THEN
      TERMW=(X(I)-XC(I))/VIST(I,J)+(XC(I)-X(I-1))/VIST(I-1,J)
      VISTW=DXMI(I)/TERMW*(1-LW)+LW*VIST(I-1,J)
      TERME=(XC(I+1)-X(I))/VIST(I,J)+(X(I+1)-XC(I+1))/VIST(I+1,J)
      VISTE=DXMI(I+1)/TERME*(1-LE)+LE*VIST(I+1,J)
      TERMS=(Y(J)-YC(J))/VIST(I,J)+(YC(J)-Y(J-1))/VIST(I,J-1)
      VISTS=DYMI(J)/TERMS*(1-LS)+LS*VIST(I,J-1)
      TERMN=(YC(J+1)-Y(J))/VIST(I,J)+(Y(J+1)-YC(J+1))/VIST(I,J+1)
      VISTN=DYMI(J+1)/TERMN*(1-LN)+LN*VIST(I,J+1)
      ENDIF
C ***  CONVECTION COEFFICIENTS
      CW=FINTW(RHO,I,J)*FINTW(U,I,J)*R(J)*DYP(J)
      CE=FINTE(RHO,I,J)*FINTE(U,I,J)*R(J)*DYP(J)
      CS=FINTS(RHO,I,J)*FINTS(V,I,J)*RC(J)*DXP(I)
      CN=FINTN(RHO,I,J)*FINTN(V,I,J)*RC(J+1)*DXP(I)
C **** DIFFUSION COEFFICIENTS (ALLOWANCE FOR BLOCKED REGIONS )
      TERM=(1-LW)/DXMI(I)+LW/(X(I)-XC(I))
      DW=(VISTW*PRTINV+VISW*PRINV)*R(J)*DYP(J)*TERM
      TERM=(1-LE)/DXMI(I+1)+LE/(XC(I+1)-X(I))
      DE=(VISTE*PRTINV+VISE*PRINV)*R(J)*DYP(J)*TERM
      TERM=(1-LS)/DYMI(J)+LS/(Y(J)-YC(J))
      DS=(VISTS*PRTINV+VISS*PRINV)*RC(J)*DXP(I)*TERM
```

```
           TERM=(1-LN)/DYMI(J+1)+LN/(YC(J+1)-Y(J))
           DN=(VISTN*PRTINV+VISN*PRINV)*RC(J+1)*DXP(I)*TERM
C *** CALCULATE CELL-PECLET NUMBERS
           PECLW=CW/(DW+SMALL)
           PECLE=CE/(DE+SMALL)
           PECLS=CS/(DS+SMALL)
           PECLN=CN/(DN+SMALL)
C *** CONVECTION SCHEMES
           IF(UPWIND)THEN
           AAW=1.0
           AAE=1.0
           AAS=1.0
           AAN=1.0
           ELSE IF(HYBRID)THEN
           AAW=AMAX1(0.0,1.-0.5*ABS(PECLW))
           AAE=AMAX1(0.0,1.-0.5*ABS(PECLE))
           AAS=AMAX1(0.0,1.-0.5*ABS(PECLS))
           AAN=AMAX1(0.0,1.-0.5*ABS(PECLN))
           ELSE IF(POWER)THEN
           AAW=AMAX1(0.0,(1.-0.1*ABS(PECLW))**5)
           AAE=AMAX1(0.0,(1.-0.1*ABS(PECLE))**5)
           AAS=AMAX1(0.0,(1.-0.1*ABS(PECLS))**5)
           AAN=AMAX1(0.0,(1.-0.1*ABS(PECLN))**5)
           ENDIF
C *** TOTAL COEFFICIENTS
           AW(I,J)=DW*(AAW+AMAX1(PECLW,0.0))
           AE(I,J)=DE*(AAE+AMAX1(-PECLE,0.0))
           AS(I,J)=DS*(AAS+AMAX1(PECLS,0.0))
           AN(I,J)=DN*(AAN+AMAX1(-PECLN,0.0))
1          CONTINUE
           GO TO 2000
C COEFFICIENTS OF PRESSURE CORRECTION EQUATION
1000       DO 2 J=2,JNM
           DO 2 I=2,INM
           PP(I,J)=0.0
           SP(I,J)=0.0
           SU(I,J)=0.0
           LW=NTAGW(I,J)/10
           LE=NTAGE(I,J)/20
           LS=NTAGS(I,J)/30
           LN=NTAGN(I,J)/40
           LB=1-NTAG(I,J)
```

```
          DXW=X(IN)-X(INM)
          DXE=X(2)-X(1)
          DYS=Y(JN)-Y(JNM)
          DYN=Y(2)-Y(1)
C WEST
          SUMW=FINTW(APU,I,J)
          AW(I,J)=FINTW(RHO,I,J)*(R(J)*DYP(J))**2/SUMW*(1-LW)*LB
          IF(NTAGW(I,J).EQ.17)THEN
          JJ=J
          IF(IPERIOD.EQ.1)JJ=JN-J+1
          RHOW=(DXW*RHO(2,J)+DXE*RHO(INM,JJ))/(DXE+DXW)
          SUMW=(DXW*APU(2,J)+DXE*APU(INM,JJ))/(DXE+DXW)
          AW(I,J)=RHOW*(R(J)*DYP(J))**2/SUMW*2*LB
          ENDIF
C EAST
          SUME=FINTE(APU,I,J)
          AE(I,J)=FINTE(RHO,I,J)*(R(J)*DYP(J))**2/SUME*(1-LE)*LB
          IF(NTAGE(I,J).EQ.27)THEN
          JJ=J
          IF(IPERIOD.EQ.1)JJ=JN-J+1
          RHOE=(DXW*RHO(2,JJ)+DXE*RHO(INM,J))/(DXE+DXW)
          SUME=(DXW*APU(2,JJ)+DXE*APU(INM,J))/(DXE+DXW)
          AE(I,J)=RHOE*(R(J)*DYP(J))**2/SUME*2.0*LB
          ENDIF
C SOUTH
          SUMS=FINTS(APV,I,J)
          AS(I,J)=FINTS(RHO,I,J)*(RC(J)*DXP(I))**2/SUMS*(1-LS)*LB
          IF(NTAGS(I,J).EQ.37)THEN
          II=I
          IF(JPERIOD.EQ.1)II=IN-I+1
          RHOS=(DYS*RHO(I,2)+DYN*RHO(II,JNM))/(DYN+DYS)
          SUMS=(DYS*APV(I,2)+DYN*APV(II,JNM))/(DYN+DYS)
          AS(I,J)=RHOS*(RC(J)*DXP(I))**2/SUMS*2.0*LB
          ENDIF
C NORTH
          SUMN=FINTN(APV,I,J)
          AN(I,J)=FINTN(RHO,I,J)*(RC(J+1)*DXP(I))**2/SUMN*(1-LN)*LB
          IF(NTAGN(I,J).EQ.47)THEN
          II=I
          IF(JPERIOD.EQ.1)II=IN-I+1
          RHON=(DYS*RHO(II,2)+DYN*RHO(I,JNM))/(DYN+DYS)
          SUMN=(DYS*APV(II,2)+DYN*APV(I,JNM))/(DYN+DYS)
```

```
              AN(I,J)=RHON*(RC(J+1)*DXP(I))**2/SUMN*2.0*LB
              ENDIF
2             CONTINUE
2000          CONTINUE
              RETURN
              END
C *********************************************
              SUBROUTINE SORCE(NNV)
              INCLUDE 'COM2D.FOR'
C *********************************************
              DIMENSION PROD(IT,JT)
              N=NNV
              GO TO (10,20,30,40,50,60,70),N
C *** FOR PRESSURE CORRECTION
10            DO 11 J=2,JNM
              DO 11 I=2,INM
              CW=FINTW(RHO,I,J)*FINTW(U,I,J)*R(J)*DYP(J)
              CE=FINTE(RHO,I,J)*FINTE(U,I,J)*R(J)*DYP(J)
              CS=FINTS(RHO,I,J)*FINTS(V,I,J)*RC(J)*DXP(I)
              CN=FINTN(RHO,I,J)*FINTN(V,I,J)*RC(J+1)*DXP(I)
              SM=CE-CW+CN-CS
              IF(UNSTDY)SM=SM+(RHO(I,J)-RHOO(I,J))/DELT*VOL(I,J)
              SU(I,J)=SU(I,J)-SM*(1-NTAG(I,J))
11            CONTINUE
              GO TO 1000
C *** FOR U-VELOCITY
20            DO 21 J=2,JNM
              DO 21 I=2,INM
              DPDX=(FINTE(P,I,J)-FINTW(P,I,J))/DXP(I)
              SU(I,J)=SU(I,J)-DPDX*VOL(I,J)*(1-NTAG(I,J))
21            CONTINUE
              GO TO 1000
C *** FOR V-VELOCITY
30            DO 31 J=2,JNM
              DO 31 I=2,INM
              DPDY=(FINTN(P,I,J)-FINTS(P,I,J))/DYP(J)
              SU(I,J)=SU(I,J)-DPDY*VOL(I,J)*(1-NTAG(I,J))
              VISP=VIS(I,J)+VIST(I,J)
              IF(AXISYMM)SP(I,J)=SP(I,J)+VISP/R(J)**2*VOL(I,J)
31            CONTINUE
              GO TO 1000
C *** FOR W-VELOCITY
```

```
40         DO 41 J=2,JNM
           DO 41 I=2,INM
           SU(I,J)=SU(I,J)+0.0
41         CONTINUE
           GO TO 1000
C *** FOR KINETIC ENERGY
50         DO 51 J=2,JNM
           DO 51 I=2,INM
           IF(NTAG(I,J).EQ.1)GO TO 51
           LW=NTAGW(I,J)
           LE=NTAGE(I,J)
           LS=NTAGS(I,J)
           LN=NTAGN(I,J)
C EXCLUDE NEAR-WALL NODE
           IF(LW.EQ.14.OR.LW.EQ.16)GO TO 51
           IF(LE.EQ.24.OR.LE.EQ.26)GO TO 51
           IF(LS.EQ.34.OR.LS.EQ.36)GO TO 51
           IF(LN.EQ.44.OR.LN.EQ.46)GO TO 51
C PRODUCTION TERMS
           DUDX=(FINTE(U,I,J)-FINTW(U,I,J))/DXP(I)
           DVDX=(FINTE(V,I,J)-FINTW(V,I,J))/DXP(I)
           DUDY=(FININ(U,I,J)-FINTS(U,I,J))/DYP(J)
           DVDY=(FINTN(V,I,J)-FINTS(V,I,J))/DYP(J)
           TERM=2.0*(DUDX**2+DVDY**2)+(DUDY+DVDX)**2
           IF(AXISYMM)TERM=TERM+2*(V(I,J)/R(J))**2
           PROD(I,J)=VIST(I,J)*TERM
           ENP=AMAX1(E(I,J),0.0)
           SU(I,J)=PROD(I,J)*VOL(I,J) +SU(I,J)
           SP(I,J)=RHO(I,J)**2*CMU*ENP/(VIST(I,J)+SMALL)*VOL(I,J)+SP(I,J)
51         CONTINUE
           GO TO 1000
C *** FOR DISSIPATION
60         DO 61 J=2,JNM
           DO 61 I=2,INM
           IF(NTAG(I,J).EQ.1)GO TO 61
           RHOP=RHO(I,J)
           VOLP=VOL(I,J)
           DPEP=ABS(D(I,J)/(E(I,J)+SMALL))
           SU(I,J)=CD1*RHOP*DPEP*PROD(I,J)*VOLP +SU(I,J)
           SP(I,J)=CD2*RHOP*DPEP*VOLP+SP(I,J)
61         CONTINUE
           GO TO 1000
```

```
C *** FOR TEMPERATURE
70      DO 71 J=2,JNM
        DO 71 I=2,INM
        SU(I,J)=SU(I,J)+0.0
71      CONTINUE
 1000   CONTINUE
        RETURN
        END
C *******************************************
        SUBROUTINE APCOF(NN)
        INCLUDE 'COM2D.FOR'
C *******************************************
        N=NN
        RPINV=1./RP(N)
        DO 1 J=2,JNM
        DO 1 I=2,INM
        SUM=AW(I,J)+AE(I,J)+AS(I,J)+AN(I,J)
        IF(N.EQ.1)AP1(I,J)=(SUM+SP(I,J))*RPINV
        IF(N.GT.1)AP(I,J)=(SUM+SP(I,J))*RPINV
        IF(N.EQ.2)APU(I,J)=AP(I,J)
        IF(N.EQ.3)APV(I,J)=AP(I,J)
1       CONTINUE
        RETURN
        END
C *******************************************
        SUBROUTINE PROPS
        INCLUDE 'COM2D.FOR'
C *******************************************
        DO 1 J=1,JN
        DO 1 I=1,IN
        SPH(I,J)=SPHEAT
        RHO(I,J)=DENSIT
        VIS(I,J)=VISCOS
        IF(TURBUL)THEN
        VISO=VIST(I,J)
        VIST(I,J)=(CMU*RHO(I,J)*E(I,J)**2/(D(I,J)+SMALL)*RP(8)
     1          +(1.-RP(8))*VISO)
        IF(VIST(I,J).LE.0.0)VIST(I,J)=SMALL
        ENDIF
1       CONTINUE
        IF(BSOR(8))CALL ADSORB(8)
        RETURN
```

```
      END
C *******************************************
      SUBROUTINE UNST(NN)
      INCLUDE 'COM2D.FOR'
C *******************************************
      N=NN
      DO 1 J=2,JNM
      DO 1 I=2,INM
      SUU=SU(I,J)
      TERM=RHO(I,J)*VOL(I,J)/DELT*(1-NTAG(I,J))
      IF(UNSTDY)THEN
      TERM=TERM*RHOO(I,J)/RHO(I,J)
      IF(N.EQ.2)SU(I,J)=TERM*UO(I,J)+SUU
      IF(N.EQ.3)SU(I,J)=TERM*VO(I,J)+SUU
      IF(N.EQ.4)SU(I,J)=TERM*WO(I,J)+SUU
      IF(N.EQ.5)SU(I,J)=TERM*EO(I,J)+SUU
      IF(N.EQ.6)SU(I,J)=TERM*DO(I,J)+SUU
      IF(N.EQ.7)SU(I,J)=TERM*TO(I,J)+SUU
      ELSE IF(FTRAN)THEN
      IF(N.EQ.2)SU(I,J)=TERM*U(I,J)+SUU
      IF(N.EQ.3)SU(I,J)=TERM*V(I,J)+SUU
      IF(N.EQ.4)SU(I,J)=TERM*W(I,J)+SUU
      IF(N.EQ.5)SU(I,J)=TERM*E(I,J)+SUU
      IF(N.EQ.6)SU(I,J)=TERM*D(I,J)+SUU
      IF(N.EQ.7)SU(I,J)=TERM*T(I,J)+SUU
      ENDIF
      SP(I,J)=TERM +SP(I,J)
1     CONTINUE
      RETURN
      END
C *******************************************
      SUBROUTINE UPDATE
      INCLUDE 'COM2D.FOR'
C *******************************************
      DO 1 J=1,JN
      DO 1 I=1,IN
      RHOO(I,J)=RHO(I,J)
      PO(I,J)=P(I,J)
      UO(I,J)=U(I,J)
      VO(I,J)=V(I,J)
      WO(I,J)=W(I,J)
      EO(I,J)=E(I,J)
```

```
        DO(I,J)=D(I,J)
        TO(I,J)=T(I,J)
1       CONTINUE
        RETURN
        END
C ******************************************
        SUBROUTINE INFLUX
        INCLUDE 'COM2D.FOR'
C ******************************************
        DO 1 N=1,7
1       RNORM(N)=0.0
        DO 2 J=2,JNM
        DO 2 I=2,INM
        IF(NTAGW(I,J).EQ.11) THEN
        CW=ABS(RHO(I-1,J)*U(I-1,J)*DYP(J)*R(J))
        VT=SQRT(U(I-1,J)**2+V(I-1,J)**2+W(I-1,J)**2)
        RNORM(1)=RNORM(1)+CW
        RNORM(2)=RNORM(2)+CW*VT
        RNORM(3)=RNORM(2)
        RNORM(4)=RNORM(2)
        RNORM(5)=RNORM(5)+CW*ABS(E(I-1,J))
        RNORM(6)=RNORM(6)+CW*ABS(D(I-1,J))
        RNORM(7)=RNORM(7)+CW*ABS(T(I-1,J))
        ELSE IF(NTAGE(I,J).EQ.21) THEN
        CE=ABS(RHO(I+1,J)*U(I+1,J)*DYP(J)*R(J))
        VT=SQRT(U(I+1,J)**2+V(I+1,J)**2+W(I+1,J)**2)
        RNORM(1)=RNORM(1)+CE
        RNORM(2)=RNORM(2)+CE*VT
        RNORM(3)=RNORM(2)
        RNORM(4)=RNORM(2)
        RNORM(5)=RNORM(5)+CE*ABS(E(I+1,J))
        RNORM(6)=RNORM(6)+CE*ABS(D(I+1,J))
        RNORM(7)=RNORM(7)+CE*ABS(T(I+1,J))
        ELSE IF(NTAGS(I,J).EQ.31) THEN
        CS=ABS(RHO(I,J-1)*V(I,J-1)*DXP(I)*RC(J))
        VT=SQRT(U(I,J-1)**2+V(I,J-1)**2+W(I,J-1)**2)
        RNORM(1)=RNORM(1)+CS
        RNORM(2)=RNORM(2)+CS*VT
        RNORM(3)=RNORM(2)
        RNORM(4)=RNORM(2)
        RNORM(5)=RNORM(5)+CS*ABS(E(I,J-1))
        RNORM(6)=RNORM(6)+CS*ABS(D(I,J-1))
```

```
      RNORM(7)=RNORM(7)+CS*ABS(T(I,J-1))
      ELSE IF(NTAGN(I,J).EQ.41) THEN
      CN=ABS(RHO(I,J+1)*V(I,J+1)*DXP(I)*RC(J+1))
      VT=SQRT(U(I,J+1)**2+V(I,J+1)**2+W(I,J+1)**2)
      RNORM(1)=RNORM(1)+CN
      RNORM(2)=RNORM(2)+CN*VT
      RNORM(3)=RNORM(2)
      RNORM(4)=RNORM(2)
      RNORM(5)=RNORM(5)+CN*ABS(E(I,J+1))
      RNORM(6)=RNORM(6)+CN*ABS(D(I,J+1))
      RNORM(7)=RNORM(7)+CN*ABS(T(I,J+1))
      ENDIF
2     CONTINUE
      DO 3 N=1,7
      TERM=ABS(RNORM(N))
3     IF(TERM.LT.10.*SMALL)RNORM(N)=1.0
      WRITE(6,*)' RNORM VALUES'
      WRITE(6,*)(RNORM(N),N=1,7)
      RETURN
      END
C *****************************************
      SUBROUTINE MASBAL
      INCLUDE 'COM2D.FOR'
C *****************************************
      SUMFW=0.0
      SUMFE=0.0
      SUMFS=0.0
      SUMFN=0.0
      DO 2 J=2,JNM
      DO 2 I=2,INM
      IF(NTAGW(I,J).EQ.13.OR.NTAGW(I,J).EQ.15) THEN
      CW=RHO(I-1,J)*U(I-1,J)*DYP(J)*R(J)
      SUMFW=SUMFW+CW
      ELSE IF(NTAGE(I,J).EQ.23.OR.NTAGE(I,J).EQ.25) THEN
      CE=RHO(I+1,J)*U(I+1,J)*DYP(J)*R(J)
      SUMFE=SUMFE+CE
      ELSE IF(NTAGS(I,J).EQ.33.OR.NTAGS(I,J).EQ.35) THEN
      CS=RHO(I,J-1)*V(I,J-1)*DXP(I)*RC(J)
      SUMFS=SUMFS+CS
      ELSE IF(NTAGN(I,J).EQ.43.OR.NTAGN(I,J).EQ.45) THEN
      CN=RHO(I,J+1)*V(I,J+1)*DXP(I)*RC(J+1)
      SUMFN=SUMFN+CN
```

```
              ENDIF
2             CONTINUE
              SUMF=ABS(SUMFW)+ABS(SUMFE)+ABS(SUMFS)+ABS(SUMFN)
              FACTOR=RNORM(1)/(SUMF+SMALL)
              WRITE(6,8787)FACTOR
              WRITE(*,8787)FACTOR
C APPLY MASS CONSERVATION AT EXIT
              IF(CONMAS)THEN
              DO 3 J=2,JNM
              DO 3 I=2,INM
              IF(NTAGW(I,J).EQ.13.OR.NTAGW(I,J).EQ.15) THEN
              U(I-1,J)=U(I-1,J)*FACTOR
              V(I-1,J)=V(I-1,J)*FACTOR
              W(I-1,J)=W(I-1,J)*FACTOR
              ELSE IF(NTAGE(I,J).EQ.23.OR.NTAGE(I,J).EQ.25) THEN
              U(I+1,J)=U(I+1,J)*FACTOR
              V(I+1,J)=V(I+1,J)*FACTOR
              W(I+1,J)=W(I+1,J)*FACTOR
              ELSE IF(NTAGS(I,J).EQ.33.OR.NTAGS(I,J).EQ.35) THEN
              V(I,J-1)=V(I,J-1)*FACTOR
              U(I,J-1)=U(I,J-1)*FACTOR
              W(I,J-1)=W(I,J-1)*FACTOR
              ELSE IF(NTAGN(I,J).EQ.43.OR.NTAGN(I,J).EQ.45) THEN
              V(I,J+1)=V(I,J+1)*FACTOR
              U(I,J+1)=U(I,J+1)*FACTOR
              W(I,J+1)=W(I,J+1)*FACTOR
              ENDIF
3             CONTINUE
              ENDIF
8787          FORMAT(50X,F10.4,F10.4)
              RETURN
              END
C *********************************************
              SUBROUTINE PVCOR
              INCLUDE 'COM2D.FOR'
C *********************************************
C **** APPLY SMOOTHING PRESSURE CORRECTION
              DO 4 J=2,JNM
              DO 4 I=2,INM
              PMX=(DXMI(I)*P(I+1,J)+DXMI(I+1)*P(I-1,J))/(DXMI(I)+DXMI(I+1))
              PMY=(DYMI(J)*P(I,J+1)+DYMI(J+1)*P(I,J-1))/(DYMI(J)+DYMI(J+1))
              PSM(I,J)=(P(I,J)-(PMX+PMY)/2.0)*GAMMA
```

```
          PP(I,J)=(PP(I,J)-PSM(I,J))*(1-NTAG(I,J))
4         CONTINUE
C *** APPLY MASS-CONSERVING PRESSURE CORRECTION
          PREF=0
          RSP=0.0
          DO 6 J=2,JNM
          DO 6 I=2,INM
          P(I,J)=P(I,J)+(PP(I,J)-PREF)*RP(9)*(1-NTAG(I,J))
          IF(ABS(PP(I,J)).GT.RSP)RSP=ABS(PP(I,J))
6         CONTINUE
          FDIF(1)=RSP
          CALL BOUNDP
C *** CORRECT VELOCITIES
          RSU=0.0
          RSV=0.0
          DO 1 J=2,JNM
          DO 1 I=2,INM
          PSMW=FINTW(PP,I,J)
          PSME=FINTE(PP,I,J)
          PSMS=FINTS(PP,I,J)
          PSMN=FINTN(PP,I,J)
C CORRECT U-VELOCITY
          IF(SLVE(2))THEN
          DPDX=(PSME-PSMW)/DXP(I)
          UDASH=-DPDX*VOL(I,J)/APU(I,J)*(1-NTAG(I,J))
          IF(ABS(UDASH).GT.RSU)RSU=ABS(UDASH)
          U(I,J)=U(I,J)+UDASH
          ENDIF
C CORRECT V-VELOCITY
          IF(SLVE(3))THEN
          DPDY=(PSMN-PSMS)/DYP(J)
          VDASH=-DPDY*VOL(I,J)/APV(I,J)*(1-NTAG(I,J))
          IF(ABS(VDASH).GT.RSV)RSV=ABS(VDASH)
          V(I,J)=V(I,J)+VDASH
          ENDIF
1         CONTINUE
          FDIF(2)=RSU
          FDIF(3)=RSV
C CHECK MASS RESIDUAL
          SUM=0.0
          DO 9 J=2,JNM
          DO 9 I=2,INM
```

```
          TERM=AE(I,J)*PP(I+1,J)+AW(I,J)*PP(I-1,J)
    1         +AN(I,J)*PP(I,J+1)+AS(I,J)*PP(I,J-1)-AP1(I,J)*PP(I,J)
          IF(TERM.GT.GREAT*0.01)TERM=0.0
          SUM=SUM+TERM**2*(1-NTAG(I,J))
    9     CONTINUE
          RSDU(1)=SQRT(SUM)/RNORM(1)
          RETURN
          END
C *******************************************
          SUBROUTINE BOUNDP
          INCLUDE 'COM2D.FOR'
C *******************************************
          DO 2 J=2,JNM
          DO 2 I=2,INM
          IF (NTAG(I,J).EQ.1) GO TO 2
          LW=NTAGW(I,J)/10
          LE=NTAGE(I,J)/20
          LS=NTAGS(I,J)/30
          LN=NTAGN(I,J)/40
          DXW=X(IN)-X(INM)
          DXE=X(2)-X(1)
          DYS=Y(JN)-Y(JNM)
          DYN=Y(2)-Y(1)
C EAST-WEST PERIODICITY
          IF(NTAGW(I,J).EQ.17.OR.NTAGE(I,J).EQ.27)THEN
          JJ=J
          IF(IPERIOD.EQ.1)JJ=JN-J+1
          ENDIF
          IF(LW.EQ.1) THEN
          RATIO=(X(I)-XC(I))/DXMI(I+1)
          P(I-1,J)=P(I,J)-RATIO*(P(I+1,J)-P(I,J))
          PP(I-1,J)=PP(I,J)-RATIO*(PP(I+1,J)-PP(I,J))
          IF(NTAGW(I,J).EQ.17)THEN
          PMEAN=(DXW*P(2,J)+DXE*P(INM,JJ))/(DXE+DXW)
          P(1,J)=PMEAN+DP1/2.0
          PPMEAN=(DXW*PP(2,J)+DXE*PP(INM,JJ))/(DXE+DXW)
          PP(1,J)=PPMEAN
          ENDIF
          ENDIF
          IF(LE.EQ.1) THEN
          RATIO=(XC(I+1)-X(I))/DXMI(I)
          P(I+1,J)=P(I,J)+RATIO*(P(I,J)-P(I-1,J))
```

```
        PP(I+1,J)=PP(I,J)+RATIO*(PP(I,J)-PP(I-1,J))
        IF(NTAGE(I,J).EQ.27)THEN
        PMEAN=(DXW*P(2,JJ)+DXE*P(INM,J))/(DXE+DXW)
        P(IN,J)=PMEAN-DP1/2.0
        PPMEAN=(DXW*PP(2,JJ)+DXE*PP(INM,J))/(DXE+DXW)
        ENDIF
        ENDIF
C NORTH-SOUTH PERIODICITY
        IF(NTAGS(I,J).EQ.37.OR.NTAGN(I,J).EQ.47)THEN
        II=I
        IF(JPERIOD.EQ.1)II=IN-I+1
        ENDIF
        IF(LS.EQ.1) THEN
        RATIO=(Y(J)-YC(J))/DYMI(J+1)
        P(I,J-1)=P(I,J)-RATIO*(P(I,J+1)-P(I,J))
        PP(I,J-1)=PP(I,J)-RATIO*(PP(I,J+1)-PP(I,J))
        IF(NTAGS(I,J).EQ.37)THEN
        P(I,1)=(DYS*P(I,2)+DYN*P(II,JNM))/(DYN+DYS)
        P(I,1)=P(I,1)+DP2/2.0
        PP(I,1)=(DYS*PP(I,2)+DYN*PP(II,JNM))/(DYN+DYS)
        ENDIF
        ENDIF
        IF(LN.EQ.1) THEN
        RATIO=(YC(J+1)-Y(J))/DYMI(J)
        P(I,J+1)=P(I,J)+RATIO*(P(I,J)-P(I,J-1))
        PP(I,J+1)=PP(I,J)+RATIO*(PP(I,J)-PP(I,J-1))
        IF(NTAGN(I,J).EQ.47)THEN
        P(I,JN)=(DYS*P(II,2)+DYN*P(I,JNM))/(DYN+DYS)
        P(I,JN)=P(I,1)-DP2/2.0
        PP(I,JN)=(DYS*PP(II,2)+DYN*PP(I,JNM))/(DYN+DYS)
        ENDIF
        ENDIF
2       CONTINUE
        IF(BSOR(9))CALL ADSORB(9)
        RETURN
        END
C *****************************************
        SUBROUTINE INDATA
        INCLUDE 'COM2D.FOR'
C *****************************************
        WRITE(6,*)'*** THIS IS COLLOCATED GRID PROGRAM ***'
        WRITE(6,*)'***********************************'
```

```
      WRITE(6,*)'GRID INFORMATION'
      WRITE(6,*)' IN = ',IN,' JN = ',JN
      WRITE(6,*)'  X - COORDINATES '
      CALL PR1D(X,1,IN)
      WRITE(6,*)'  Y - COORDINATES '
      CALL PR1D(Y,1,JN)
      WRITE(6,*)' XC - COORDINATES '
      CALL PR1D(XC,1,IN)
      WRITE(6,*)' YC - COORDINATES '
      CALL PR1D(YC,1,JN)
      WRITE(6,*)' DXMI'
      CALL PR1D(DXMI,1,IN)
      WRITE(6,*)'  DYMI'
      CALL PR1D(DYMI,1,JN)
      WRITE(6,*)' PRESSURE REFERENCE POINT  IPREF = ',IPREF,
     1     ' JPREF = ',JPREF
      WRITE(6,*)'RELAXATION PARAMETERS ARE'
      WRITE(6,*)' RP(1) = ',RP(1),' RP(2) = ',RP(2),' RP(3) = ',RP(3),
     1         ' RP(4) = ',RP(4),' RP(5) = ',RP(5), ' RP(6) = ',RP(6),
     1         ' RP(7) = ',RP(7),' RP(8) = ',RP(8), ' RP(9) = ',RP(9)
      WRITE(6,*)'FLUID VISCOSITY = ',VISCOS
      WRITE(6,*)'FLUID DENSITY  = ',DENSIT
      WRITE(6,*)'FLUID PRANDTL NUMBERS ARE'
      WRITE(6,*)' PR(1) = ',PR(1),' PR(2) = ',PR(2),' PR(3) = ',PR(3),
     1         ' PR(4) = ',PR(4),' PR(5) = ',PR(5),' PR(6) = ',PR(6),
     1         ' PR(7) = ',PR(7)
      IF(TURBUL) THEN
      WRITE(6,*)'TURBULENT PRANDTL NUMBERS ARE'
      WRITE(6,*)
     1 ' PRT(1) = ',PRT(1),' PRT(2) = ',PRT(2),' PRT(3) = ',PRT(3),
     1 ' PRT(4) = ',PRT(4),' PRT(5) = ',PRT(5), ' PRT(6) = ',PRT(6),
     1 ' PRT(7) = ',PRT(7)
      ENDIF
      IF(STEADY)WRITE(6,*)' STEADY FLOW CALCULATIONS'
      IF(UNSTDY)WRITE(6,*)' UNSTEADY FLOW CALCULATIONS'
      IF(FTRAN)WRITE(6,*)' FALSE TRANSIENT DELT = ',DELT
      IF(CONMAS)WRITE(6,*)' MASS BALANCE IS IMPOSED'
      IF(UPWIND)WRITE(6,*)' CONVECTION SCHEME = UPWIND  '
      IF(HYBRID)WRITE(6,*)' CONVECTION SCHEME = HYBRID '
      IF(POWER)WRITE(6,*)' CONVECTION SCHEME = POWER LAW'
      WRITE(6,*)' THE FOLLOWING EQUATIONS ARE SOLVED'
      IF(SLVE(1))WRITE(6,*)' PRESSURE CORRECTION EQUN.'
```

```
      IF(SLVE(2))WRITE(6,*)' U-VELOCITY '
      IF(SLVE(3))WRITE(6,*)' V-VELOCITY '
      IF(SLVE(4))WRITE(6,*)' W-VELOCITY '
      IF(SLVE(5))WRITE(6,*)' T. KINETIC ENERGY'
      IF(SLVE(6))WRITE(6,*)' DISSIPATION '
      IF(SLVE(7))WRITE(6,*)' TEMPERATURE'
      WRITE(6,*)'*****************************************'
      RETURN
      END
C ********************************************
      SUBROUTINE SOLVE(F,RPP,RSUM)
      INCLUDE 'COM2D.FOR'
C ********************************************
      DIMENSION  F(IT,JT)
      DIMENSION SA(MXGR),SB(MXGR),SS(MXGR),PSI(MXGR)
C *** CALCULATION OF RESIDUALS
      RS=0.0
      DO 10 J=2,JNM
      DO 10 I=2,INM
      TERM=AW(I,J)*F(I-1,J)+AE(I,J)*F(I+1,J)
     1     +AS(I,J)*F(I,J-1)+AN(I,J)*F(I,J+1)
      TERM=TERM+SU(I,J)-F(I,J)*AP(I,J)*RPP
      FACTOR=1.0
      IF(SP(I,J).GT.GREAT*1.0E-10)FACTOR=0.0
      TERM=TERM*FACTOR
10    RS=RS+TERM*TERM
      RSUM=SQRT(RS)
C*** J-DIRECTION SWEEP
      DO 51 J=2,JNM
      DO 52 I=2,INM
      SOR=SU(I,J)
      DEN=1.0/(AP(I,J)+SMALL)
      SOR=SOR+(1.-RPP)/(DEN+SMALL)*F(I,J)
      SA(I)=AE(I,J)*DEN
      SB(I)=AW(I,J)*DEN
      SS(I)=(AS(I,J)*F(I,J-1)+AN(I,J)*F(I,J+1)+SOR)*DEN
52    CONTINUE
      PSI1=F(1,J)
      PSIN=F(IN,J)
      CALL TDMA(2,INM,PSI1,PSIN,SA,SB,SS,PSI)
      DO 53 I=2,INM
      LP=NTAG(I,J)
```

```
53        F(I,J)=PSI(I)*(1-LP)+LP*F(I,J)
51        CONTINUE
C*** I-DIRECTION SWEEP
          DO 54 I=2,INM
          DO 55 J=2,JNM
          SOR=SU(I,J)
          DEN=1.0/(AP(I,J)+SMALL)
          SOR=SOR+(1.-RPP)/DEN*F(I,J)
          SA(J)=AN(I,J)*DEN
          SB(J)=AS(I,J)*DEN
          SS(J)=(AW(I,J)*F(I-1,J)+AE(I,J)*F(I+1,J)+SOR)*DEN
55        CONTINUE
          PSI1=F(I,1)
          PSIN=F(I,JN)
          CALL TDMA(2,JNM,PSI1,PSIN,SA,SB,SS,PSI)
          DO 56 J=2,JNM
          LP=NTAG(I,J)
56        F(I,J)=PSI(J)*(1-LP)+LP*F(I,J)
54        CONTINUE
          RETURN
          END
C ******************************************
          SUBROUTINE SOLP
          INCLUDE 'COM2D.FOR'
C ******************************************
          DIMENSION SA(MXGR),SB(MXGR),SS(MXGR),PSI(MXGR)
          DO 100 L=1,NSWEEP(1)
C*** J-DIRECTION SWEEP
          DO 51 J=2,JNM
          DO 52 I=2,INM
          SOR=SU(I,J)
          DEN=1.0/(AP1(I,J)+SMALL)
          SOR=SOR+(1.-RP(1))/DEN*PP(I,J)
          SA(I)=AE(I,J)*DEN
          SB(I)=AW(I,J)*DEN
          SS(I)=(AS(I,J)*PP(I,J-1)+AN(I,J)*PP(I,J+1)+SOR)*DEN
52        CONTINUE
          PSI1=PP(1,J)
          PSIN=PP(IN,J)
          CALL TDMA(2,INM,PSI1,PSIN,SA,SB,SS,PSI)
          DO 53 I=2,INM
          LP=NTAG(I,J)
```

```
53        PP(I,J)=PSI(I)*(1-LP)+LP*PP(I,J)
51        CONTINUE
C*** I-DIRECTION SWEEP
          DO 54 I=2,INM
          DO 55 J=2,JNM
          SOR=SU(I,J)
          DEN=1.0/(AP1(I,J)+SMALL)
          SOR=SOR+(1.-RP(1))/DEN*PP(I,J)
          SA(J)=AN(I,J)*DEN
          SB(J)=AS(I,J)*DEN
          SS(J)=(AW(I,J)*PP(I-1,J)+AE(I,J)*PP(I+1,J)+SOR)*DEN
55        CONTINUE
          PSI1=PP(I,1)
          PSIN=PP(I,JN)
          CALL TDMA(2,JNM,PSI1,PSIN,SA,SB,SS,PSI)
          DO 56 J=2,JNM
          LP=NTAG(I,J)
56        PP(I,J)=PSI(J)*(1-LP)+LP*PP(I,J)
54        CONTINUE
          IF(NPERIOD.EQ.1)CALL BOUNDP
100       CONTINUE
          RETURN
          END
C ******************************************
          SUBROUTINE EQN
          INCLUDE 'COM2D.FOR'
C ******************************************
          MWRITE=NITER+MFREQ
          IF(NITER.EQ.0)NITER=1
          NADD=MXIT
5555      NBEGIN=NITER
          MXIT=NITER+NADD
          DO 2000 NTIME=1,MXSTEP
          TTIME=STIME+NTIME*DELT
          DO 1000 NITER=NBEGIN,MXIT
C **** U-VELOCITY
          IF(SLVE(2))THEN
          CALL COEF(2,PR(2),PRT(2))
          CALL SORCE(2)
          IF(UNSTDY.OR.FTRAN)CALL UNST(2)
          CALL BOUND(2)
          IF(BSOR(2))CALL ADSORB(2)
```

```
            CALL APCOF(2)
            CALL SOLVE(U,RP(2),RSU)
            RSDU(2)=RSU/(RNORM(2)+SMALL)
            CALL BOUND(2)
            ENDIF
C **** V-VELOCITY
            IF(SLVE(3))THEN
            CALL COEF(3,PR(3),PRT(3))
            CALL SORCE(3)
            IF(UNSTDY.OR.FTRAN)CALL UNST(3)
            CALL BOUND(3)
            IF(BSOR(3))CALL ADSORB(3)
            CALL APCOF(3)
            CALL SOLVE(V,RP(3),RSU)
            RSDU(3)=RSU/(RNORM(3)+SMALL)
            CALL BOUND(3)
            ENDIF
C **** PRESSURE CORRECION
            IF(SLVE(1))THEN
            CALL MASBAL
            CALL COEF(1,PR(1),PRT(1))
            CALL SORCE(1)
            IF(BSOR(1))CALL ADSORB(1)
            CALL APCOF(1)
            CALL SOLP
            CALL PVCOR
            ENDIF
C **** W-VELOCITY
            IF(SLVE(4))THEN
            CALL COEF(4,PR(4),PRT(4))
            CALL SORCE(4)
            IF(UNSTDY.OR.FTRAN)CALL UNST(4)
            CALL BOUND(4)
            IF(BSOR(4))CALL ADSORB(4)
            CALL APCOF(4)
            CALL SOLVE(W,RP(4),RSU)
            RSDU(4)=RSU/(RNORM(4)+SMALL)
            CALL BOUND(4)
            ENDIF
C **** KINETIC ENERGY
            IF(TURBUL)THEN
            IF(SLVE(5))THEN
```

```
      CALL COEF(5,PR(5),PRT(5))
      CALL SORCE(5)
      IF(UNSTDY.OR.FTRAN)CALL UNST(5)
      CALL BOUND(5)
      IF(BSOR(5))CALL ADSORB(5)
      CALL APCOF(5)
      CALL SOLVE(E,RP(5),RSU)
      RSDU(5)=RSU/(RNORM(5)+SMALL)
      CALL BOUND(5)
C **** DISSIPATION
      CALL COEF(6,PR(6),PRT(6))
      CALL SORCE(6)
      IF(UNSTDY.OR.FTRAN)CALL UNST(6)
      CALL BOUND(6)
      IF(BSOR(6))CALL ADSORB(6)
      CALL APCOF(6)
      CALL SOLVE(D,RP(6),RSU)
      RSDU(6)=RSU/(RNORM(6)+SMALL)
      CALL BOUND(6)
      ENDIF
      ENDIF
C **** TEMPERATURE
      IF(SLVE(7))THEN
      CALL COEF(7,PR(7),PRT(7))
      CALL SORCE(7)
      IF(UNSTDY.OR.FTRAN)CALL UNST(7)
      CALL BOUND(7)
      IF(BSOR(7))CALL ADSORB(7)
      CALL APCOF(7)
      CALL SOLVE(T,RP(7),RSU)
      RSDU(7)=RSU/(RNORM(7)+SMALL)
      CALL BOUND(7)
      ENDIF
C **** SPECIES AND ENTHALPY
      CALL OMEGA
C **** PROPERTIES
      CALL PROPS
C **** CHECK MAX RESIDUALS
      RSTOP=AMAX1(RSDU(1),RSDU(2),RSDU(3),RSDU(4),RSDU(5)
     1       ,RSDU(6),RSDU(7))
C STORE RESIDUALS FOR PLOTTING
C         RESIU(NITER)=RSDU(2)
```

```
C          RESIV(NITER)=RSDU(3)
C          RESIM(NITER)=RSDU(1)
          IF(STEADY)WRITE(6,1919)NITER,(RSDU(N),N=1,7)
          IF(STEADY)WRITE(6,1919)NITER,(FDIF(N),N=1,4)
          IF(STEADY)WRITE(*,1919)NITER,(FDIF(N),N=1,4)
          IF(STEADY)WRITE(*,1919)NITER,(RSDU(N),N=1,7)
1919      FORMAT(1X,I6,7(E10.3))
          IF(RSTOP.LT.CC) GO TO 1100
C INTERMEDIATE WRITE-OUT
          IF(MWRITE.EQ.NITER)THEN
          MWRITE=NITER+MFREQ
          CALL OPT
          WRITE(*,*)'OUTPUT IS WRITTEN AT NITER =  ',NITER
          WRITE(6,*)'OUTPUT IS WRITTEN AT NITER =  ',NITER
          ENDIF
1000      CONTINUE
1100      IF(STEADY)RETURN
          CALL UPDATE
          WRITE(6,*)'NTIME = ',NTIME,' TTIME = ',TTIME
          WRITE(6,*)(RSDU(N),N=1,7)
          WRITE(6,*)(FDIF(N),N=1,7)
2000      CONTINUE
          RETURN
          END
C ********************************************
          SUBROUTINE TDMA(IB,IL,Y1,YN,BA,BB,BS,YY)
          INCLUDE 'COM2D.FOR'
C ********************************************
          DIMENSION BA(MXGR),BB(MXGR),BS(MXGR),YY(MXGR),A(MXGR),B(MXGR)
          A(IB)=BA(IB)
          B(IB)=BB(IB)*Y1+BS(IB)
          DO 1 I=IB+1,IL
          TERM=1.0-BB(I)*A(I-1)
          A(I)=BA(I)/(TERM +SMALL)
1         B(I)=(BB(I)*B(I-1)+BS(I))/(TERM+SMALL)
          YY(IL)=B(IL)+A(IL)*YN
          DO 2 I=IL-1,IB,-1
2         YY(I)=A(I)*YY(I+1)+B(I)
          RETURN
          END
```

```
C **********************************************
      FUNCTION STAN(UPLUS,YPLUS,PR,PRT)
C **********************************************
C LAMINAR SUB LAYER
      IF(YPLUS.LT.11.6)THEN
      PF=(PR/PRT-1)*UPLUS
      ELSE
C TURBULENT LAYER
      PF=9.24*((PR/PRT)**0.75 -1.0)*(1+0.28*EXP(-0.007*PR/PRT))
      ENDIF
      STAN=1./(PRT*UPLUS*(PF+UPLUS))
      RETURN
      END
C **********************************************
      FUNCTION FINTW(F,II,JJ)
      INCLUDE 'COM2D.FOR'
C **********************************************
      DIMENSION F(IT,JT)
      I=II
      J=JJ
      LW=NTAGW(I,J)/10
      TW=((X(I) XC(I))*F(I-1,J)+(XC(I)-X(I-1))*F(I,J))/DXMI(I)
      FINTW=TW*(1-LW)+LW*F(I-1,J)
      RETURN
      END
C **********************************************
      FUNCTION FINTE(F,II,JJ)
      INCLUDE 'COM2D.FOR'
C **********************************************
      DIMENSION F(IT,JT)
      I=II
      J=JJ
      LE=NTAGE(I,J)/20
      TE=((X(I+1)-XC(I+1))*F(I,J)+(XC(I+1)-X(I))*F(I+1,J))/DXMI(I+1)
      FINTE=TE*(1-LE)+LE*F(I+1,J)
      RETURN
      END
C **********************************************
      FUNCTION FINTS(F,II,JJ)
      INCLUDE 'COM2D.FOR'
```

```
C *********************************************
      DIMENSION F(IT,JT)
      I=II
      J=JJ
      LS=NTAGS(I,J)/30
      TS=((Y(J)-YC(J))*F(I,J-1)+(YC(J)-Y(J-1))*F(I,J))/DYMI(J)
      FINTS=TS*(1-LS)+LS*F(I,J-1)
      RETURN
      END
C *********************************************
      FUNCTION FINTN(F,II,JJ)
      INCLUDE 'COM2D.FOR'
C *********************************************
      DIMENSION F(IT,JT)
      I=II
      J=JJ
      LN=NTAGN(I,J)/40
      TN=((Y(J+1)-YC(J+1))*F(I,J)+(YC(J+1)-Y(J))*F(I,J+1))/DYMI(J+1)
      FINTN=TN*(1-LN)+LN*F(I,J+1)
      RETURN
      END
C *********************************************
      SUBROUTINE OPT
      INCLUDE 'COM2D.FOR'
C *********************************************
      OPEN(12,FILE='NSOUT',FORM='UNFORMATTED')
      WRITE(12)NITER,TTIME
      DO 1 J=1,JN
      DO 1 I=1,IN
      WRITE(12)P(I,J),U(I,J),V(I,J),W(I,J),E(I,J),D(I,J),T(I,J)
      WRITE(12)VIS(I,J),VIST(I,J),RHO(I,J),SPH(I,J)
      WRITE(12)QW(I,J),AMW(I,J),TAUW(I,J),O(I,J),HH(I,J)
1     CONTINUE
      CLOSE(12)
      RETURN
      END
C *********************************************
      SUBROUTINE IPT
      INCLUDE 'COM2D.FOR'
C *********************************************
      OPEN(13,FILE='NSIN',FORM='UNFORMATTED')
      READ(13)NITER,STIME
```

```
       DO 1 J=1,JN
       DO 1 I=1,IN
       READ(13)P(I,J),U(I,J),V(I,J),W(I,J),E(I,J),D(I,J),T(I,J)
       READ(13)VIS(I,J),VIST(I,J),RHO(I,J),SPH(I,J)
       READ(13)QW(I,J),AMW(I,J),TAUW(I,J),O(I,J),HH(I,J)
1      CONTINUE
       CLOSE(13)
       RETURN
       END
C ****************************************
       SUBROUTINE PR1D(F,IB,IL)
       INCLUDE 'COM2D.FOR'
C ****************************************
       DIMENSION F(MXGR)
       I1=IB
       IE=I1+10
       IF(IE.GT.IN)IE=IL
100    CONTINUE
       WRITE(6,500) (F(I),I=I1,IE)
       WRITE(6,600) (I,I=I1,IE)
       IF(IE.LT.IL) THEN
       I1=IE+1
       IE=I1+10
       IF(IE.GT.IL)IE=IL
       GO TO 100
       ENDIF
500    FORMAT(11F10.4)
600    FORMAT(1X,6H      ,I3,11I10)
        RETURN
        END
C ****************************************
       SUBROUTINE PRINTK(F,IB,IL,JB,JL,HEADER,JSTEP)
       INCLUDE 'COM2D.FOR'
C ****************************************
       CHARACTER*10 HEADER
       DIMENSION F(IT,JT)
       WRITE(6,*)'***********************************'
       WRITE(6,*)' DISTRIBUTION OF  F(I,J) ',HEADER
       WRITE(6,*)'***********************************'
       I1=IB
       IE=I1+11
       IF(IE.GT.IL) IE=IL
```

```
100       CONTINUE
          DO 1 J=JL,JB,JSTEP
1         WRITE(6,500)J,(F(I,J),I=I1,IE)
          WRITE(6,600) (I,I=I1,IE)
          IF(IE.LT.IL) THEN
          I1=IE+1
          IE=I1+11
          IF(IE.GT.IL) IE=IL
          GO TO 100
          ENDIF
500       FORMAT(I3,12E10.3)
600       FORMAT(1X,6H      ,I3,11I10)
          RETURN
          END
```

Bibliography

[1] Abu-Ghannam, B. J. and Shaw, R., Natural Transition of Boundary Layers – The Effects of Turbulence, Pressure Gradient and Flow History, J. Mech. Eng. Sci., Vol. 22, No. 5, pp. 213–238 (1980).

[2] Antonopoulos, K. A., Heat Transfer in Tube Banks under Conditions of Turbulent Inclined Flow, Int. J. Heat Mass Transfer, Vol. 28, pp. 1645–1656 (1985).

[3] Barrett, R., Berry, M., Chan, T., Demmel, J., Donato, J., Dongarra, J., Eijkhout, V., Pozo, R., Romine, C., and van der Vorst, H., *Templates for the Solution of Linear Systems: Building Blocks for Iterative Methods*, SIAM, Philadelphia (1994).

[4] Cebeci, T. and Bradshaw, P., *Physical and Computational Aspects of Convective Heat Transfer*, Springer Verlag, New York (1984).

[5] Chai, J., Zhang, M., Moder, J., and Patankar, S. V., Conduction Heat Transfer Calculations Using Structured and Unstructured Grids, Proc. 2nd ICHMT Conference on Advances in Computational Heat Transfer, Palm Cove, Australia, pp. 519–526, May 2001.

[6] Chiu, C. K. and Caldwell, J., Application of Broyden's Method to the Enthalpy Method for Phase-Change Problems, Numerical Heat Transfer A, Vol. 30, pp. 575–587 (1996).

[7] Caldwell, J. and Chan, C. C., Spherical Solidification by Enthalpy Method and the Heat Balance Integral Method, Appl. Math. Modelling, Vol. 24, pp. 45–53 (2000).

[8] Courant, R., Isacson, E., and Rees, M., On the Solution of Non-Linear Hyperbolic Differential Equations by Finite Differences, Comm. Pure Appl. Math., Vol. 5, p. 243 (1952).

[9] Craft, T. J., Launder, B. E., and Suga, K., Development and Application of a Cubic Eddy-Viscosity Model of Turbulence, Int J. Heat Fluid Flow, Vol. 17, No. 12, p. 108 (1996).

[10] Crawford, M. E. and Kays, W. M., STAN5 – A Program for Numerical Computation of Two-Dimensional Internal/External Boundary Layer Flows, NASA CR-2742 (1976).

[11] Date, A. W., A Strong Enthalpy Formulation of the Stefan Problem, Int. J. Heat Mass Transfer, Vol. 34, pp. 2231–2235 (1991).

[12] Date, A. W., Novel Strongly Implicit Enthalpy Formulation for Multidimensional Stefan Problems, Numerical Heat Transfer B, Vol. 21, pp. 231–251 (1992).

[13] Date, A. W., A Novel Enthaply Formulation for Multidimensional Solidification and Melting of a Pure Substance, SADHANA, Academy Proc. in Engineering Sciences, Indian Academy of Sciences, Vol. 19, Part 5, pp. 833–850 (1994).

[14] Date, A. W., Complete Pressure Correction Algorithm for Solution of Incompressible Navier–Stokes Equations on a Nonstaggered Grid, Numerical Heat Transfer B, Vol. 29, p. 441 (1996).

[15] Date, A. W., Solution of Navier–Stokes Equations on Non-Staggered Grids at All Speeds, Numerical Heat Transfer B, Vol. 33, p. 451 (1998).

[16] Date, A. W., Smoothing Pressure Correction on Colocated Grids with Structured and Unstructured Meshes, 12th Int. Heat Transfer Conference, Greenoble, France (2002).

[17] Date, A. W., Fluid Dynamic View of Pressure Checkerboarding Problem and Smoothing Pressure Correction on Meshes with Colocated Variables, Int. J. Heat Mass Transfer, Vol. 46, pp. 4885–4898 (2003).

[18] Date, A. W., Solution of Transport Equations on Unstructured Meshes with Cell-Centered Collocated Variables. Part I, Discretisation, Int. J. Heat Mass Transfer, Vol. 48, pp. 1117–1127 (2005).

[19] Davidson, L., A Pressure Correction Method for Unstructured Meshes with Arbitrary Control Volumes, Int. J. Numerical Methods Fluids, Vol. 22, pp. 265–281 (1996).

[20] Demirdzic, I. and Muzaferija, S., Numerical Method for Coupled Fluid Flow, Heat Transfer and Stress Analysis Using Unstructured Moving Meshes with Cells of Arbitrary Topology, Comput. Methods Appl. Mech. Eng., Vol. 125, pp. 235–255 (1995).

[21] Deshpande, M. D. and Giddens, D. P., Direct Solution of Two Linear Systems of Equations Forming a Coupled Tridiagonal Type Matrices, Int. J. Numerical Method, Eng., Vol. 11, pp. 1049–1052 (1977).

[22] Dippery, D. F. and Sebersky, R. H., Heat and Momentum Transfer in Smooth and Rough Tubes, Int. J. Heat Mass Transfer, Vol. 6, pp. 328–356 (1963).

[23] Eckert, E. R. G. and Drake, R. M., Jr., *Analysis of Heat and Mass Transfer*, Int. Student Ed., McGraw-Hill–Kogakusha, Tokyo (1972).

[24] Gosman, A. D., Lockwood, F. C., and Saluja, A. P., The Prediction of Cylindrical Furnaces Gaseous Fueled with Premixed and Diffusion Burners, 17th Int. Symposium on Combustion. The Combustion Institute, Pittsburgh, pp. 747–760 (1979).

[25] Grimison, E. D., Correlation and Utilization of New Data on Flow Resistance and Heat Transfer for Cross Flow of Gases over Tube Banks, Trans. ASME, Vol. 59, p. 583 (1937).

[26] Gupta, V. and Srinivasan, J., *Heat and Mass Transfer*, Tata McGraw-Hill, New Delhi (1978).

[27] Ho-Le, K., Finite Element Mesh Generation Methods: A Review and Classification, Comput. Aided Design J., Vol. 20, pp. 27–38 (1988).

[28] Holman, J. P., *Thermodynamics*, 3rd Ed., McGraw-Hill–Kogakusha, Tokyo (1980).

[29] Jesshope, C. R., SIPSOL – A Suite of Subprograms for the Solution of the Linear Equations Arising from Elliptic Partial Differential Equations, Comput. Phys. Commun., Vol. 17, pp. 383–389 (1979).

[30] Jones, W. P. and Launder, B. E., The Prediction of Laminarization with a Two-Equation Model of Turbulence, Int. J. Heat Mass Transfer, Vol. 15, p. 301 (1972).

[31] Karki, K. C. and Patankar, S. V., A Pressure Based Calculation Procedure for Viscous Flows at All Speeds in Arbitrary Configurations, AIAA, 26th Aerospace Science Meeting, Paper No. AIAA-88-0058, Nevada, USA (1988).

[32] Karki, K. C. and Patankar, S. V., Solution of Two-Dimensional Incompressible Flow Problems Using a Curvilinear Coordinate System Based Calculation Procedure, Numerical Heat Transfer, Vol. 14, pp. 309–321 (1988).

[33] Kays, W. M. and Crawford, M. E., *Convective Heat and Mass Transfer*, 3rd Ed., McGraw-Hill Int. Editions, New York (1993).

[34] Kreith, F. and Bohn, M. S., *Principles of Heat Transfer*, 6th Ed., Asian Books, New Delhi (2001).

[35] Keller, H. B. and Cebeci, T., Accurate Numerical Methods for Boundary-Layer Flows II: Two-Dimensional Turbulent Flows, AIAA J., Vol. 10, No. 9, p. 1193 (1972).

[36] Krall, K. M. and Sparrow, E. M., Turbulent Heat Transfer in the Separated, Reattached and Redevelopment Regions of a Circular Tube, ASME J. Heat Transfer, Vol. 88, pp. 131–136 (1966).

[37] Kuehn, T. H. and Goldstein, R. J., An Experimental Study of Natural Convection Heat Transfer in Concentric and Eccentric Horizontal Cylindrical Annuli, ASME J. Heat Transfer, Vol. 100, pp. 635–640 (1978).

[38] Kuo, K. K., *Principles of Combustion*, Wiley-Interscience, New York (1986).

[39] Launder, B. E. and Spalding, D. B., *Mathematical Models of Turbulence*, Academic Press, New York (1972).

[40] Launder, B. E. and Spalding, D. B., The Numerical Computation of Turbulent Flows, Comput. Methods Appl. Mech. Eng., Vol. 3, p. 269 (1974).

[41] Launder, B. E., On the Computation of Convective Heat Transfer in Complex Turbulent Flows, Trans. ASME, J. Heat Transfer, Vol. 110, p. 1113 (1988).

[42] Leonard, B. P., Simple, High-Accuracy Resolution Program for Convective Modeling of Discontinuities, Int. J. Numerical Methods Fluids, Vol. 81, pp. 1291–1381 (1988).

[43] Lin, C. H. and Lin, C. A., Simple High-Order Bounded Convection Scheme to Model Discontinuities, AIAA J., Vol. 35, No. 3, pp. 563–565 (1997).

[44] Magnussen, B. F. and Hjertager, B. W., On Mathematical Modeling of Turbulent Combustion with Special Emphasis on Soot Formation and Combustion, 16th Int. Symposium on Combustion, The Combustion Institute, Pittsburgh, pp. 719–729 (1976).

[45] Mason, M. L., Putnam, L. E., and Re, R. J., The Effect of Throat Contouring on Two-Dimensional Converging-Diverging Nozzles at Static Conditions, NASA Technical Paper 1704 (1980).

[46] Mathur, S. R. and Murthy, J. Y., A Pressure-Based Method for Unstructured Meshes, Numerical Heat Transfer B, Vol. 31, pp. 195–215 (1997).

[47] McBain, G. D., Suehrecke, H., and Harris, J., Evaporation from an Open Cylinder, Int. J. Heat Mass Transfer, Vol. 43, pp. 2117–2128 (2000).

[48] Modest, M. F., *Radiative Heat Transfer*, McGraw-Hill, New York (1993).

[49] Patankar, S. V., *Numerical Fluid Flow and Heat Transfer*, Hemisphere, New York (1981).

[50] Patankar, S. V. and Spalding, D. B., A Finite-Difference Procedure for Solving the Equations of the Two-Dimensional Boundary Layer, Int. J. Heat Mass transfer, Vol. 10, p. 1389 (1967); also see *Heat and Mass Transfer in Boundary Layers*, 2nd Ed., International Textbook, London (1970).

[51] Patankar, S. V. and Spalding, D. B., A Calculation Procedure for Heat Mass and Momentum Transfer in Three-Dimensional Parabolic Flows, Int. J. Heat Mass Transfer, Vol. 15, p. 1787 (1971).

[52] Patankar, S. V., Parabolic Systems: Finite-Difference Method – I, in *Handbook of Numerical Heat Transfer*, Minkowycz, W. J., Sparrow, E. M., Schnieder, G. E., and Pletcher, R. H. (eds.), Wiley, New York (1988).

[53] Patankar, S. V., *Computation of Conduction and Duct Flow Heat Transfer*, Innovative Research, Maple Grove, Minnesota (1991).

[54] Patankar, S. V. and Prakash, C., An Analysis of the Effect of Plate Thickness on Laminar Flow and Heat Transfer in Interrupted-Plate Passages, Int. J. Heat Mass Transfer, Vol. 24, pp. 1801–1810 (1981).

[55] Perry, R. H. and Chilton, C. H., *Chemical Engineers' Handbook*, 5th Ed., McGraw-Hill–Kogakusha, Tokyo (1973).

[56] Prakash, C. and Patankar, S. V., Combined Free and Forced Convection in Vertical Tubes with Radial Internal Fins, J. Heat Transfer, Vol. 103, pp. 566–472 (1981).

[57] Raithby, G. D., Skew Upstream Difference Schemes for Problems Involving Fluid Flow, Comput. Methods Appl. Mech. Eng., Vol. 9, pp. 153–164 (1976).

[58] Ray, S. and Date, A. W., A Calculation Procedure for solution of Incompressible Navier–Stokes Equations on Curvilinear Non-Staggered Grid, Numerical Heat Transfer B, Vol. 38, pp. 93–111 (2000).

[59] Rhie, C. M. and Chow, W. L., A Numerical Study of the Turbulent Flow Past an Isolated Airfoil with Trailing Edge Separation, AIAA J., Vol. 21, p. 1525 (1983).

[60] Runchal, A. K. and Wolfshtein, M., Numerical Integration Procedure for the Steady-State Navier–Stokes Equations, J. Mech. Eng. Sci., Vol. 11, p. 445 (1969).

[61] Runchal, A. K., CONDIF: A Modified Central Difference Scheme for Convective Flows, Int. J. Numerical Methods Eng., Vol. 24, pp. 1593–1608 (1987).

[62] Runchal, A. K., Mass Transfer Investigation in Turbulent Flow Downstream of Sudden Enlargement of a Circular Pipe for Very High Schmidt Numbers, Int. J. Heat Mass Transfer, Vol. 14, pp. 781–792 (1971).

[63] Sastry, S. S., *Introductory Methods of Numerical Analysis*, Prentice-Hall of India, New Delhi (1981).

[64] Scarborough, J. B., *Numerical Mathematical Analysis*, 6th Ed., Oxford and IBH, New Delhi (1966).

[65] Schlichting, H., *Boundary-Layer Theory*, 6th Ed., English translation by Kestin J., McGraw-Hill, New York (1968).

[66] Shah, R. K. and London, A. L., *Laminar Forced Convection in Ducts*, Advances in Heat Transfer, Vol. 15, Academic Press, New York (1978).

[67] Sloan, S. W. and Houlsby, G. T., An Implementation of Watson's Algorithm for Computing 2-Dimensional Delaunay Triangulations, Adv. Eng. Software, Vol. 6, pp. 192–197 (1984).

[68] Smoot, L. D. and Pratt, D. T., *Pulverised-Coal Combustion and Gasification*, Plenum Press, New York (1979).

[69] Sofialidis, D. and Prinos, P., Fluid Flow and Heat Transfer in a Pipe with Suction, Int. J. Heat Mass Transfer, Vol. 15, pp. 3627–3640 (1997).

[70] Sokolnikoff, I. S. and Redheffer, R. M., *Mathematics of Physics and Modern Engineering*, McGraw-Hill–Kogakusha Int. Student Ed., New York (1958).

[71] Sorenson, R. L., A Computer Program to Generate Two-Dimensional Grids about Airfoils and Other Shapes by the Use of Poisson's Equation, NASA Tech Memo. 81198 (1980).

[72] Spalding, D. B., A Standard Formulation of Steady Convective Mass Transfer Problem, Int. J. Heat Mass Transfer, Vol. 1, pp. 192–207 (1960).

[73] Spalding, D. B., *Convective Mass Transfer*, McGraw-Hill, New York (1963).

[74] Spalding, D. B., Mixing and Chemical Reaction in Steady, Confined Turbulent Flames, 13th Int. Symposium on Combustion, The Combustion Institute, Pittsburgh, pp. 649–657 (1971).

[75] Spalding, D. B., A Novel Finite-Difference Formulation for Differential Expressions Involving Both First and Second Derivatives, Int. J. Numerical Methods Eng., Vol. 4, p. 551 (1972).

[76] Spalding, D. B., Basic Equations of Fluid Flow and Heat and Mass Transfer and Procedures for their Solution, Rep HTS/76/6, Mech Eng. Dept., Imperial College, London (1976).

[77] Spalding, D. B., *Genmix: A General Computer Program for Two-Dimensional Parabolic Phenomena*, Pergamon Press, Oxford (1977).

[78] Spalding, D. B., *Combustion and Mass Transfer*, Pergamon Press, Oxford (1979).

[79] Stone, H. L., Iterative Solution of Implicit Approximations of Multidimensional Partial Differential Equations, SIAM J. Numerical Anal., Vol. 5, pp. 530–558 (1968).

[80] Sukhatme, S. P., *A Textbook on Heat Transfer*, 3rd Ed., Orient Longman, Bombay (1989).

[81] Thompson, J. F., Warsi, Z. U. A., and Mastin, C. W., *Numerical Grid Generation, Foundations and Applications*, North-Holland, New York (1985).

[82] Turns, S. R., *An Introduction to Combustion*, McGraw-Hill, New York (1996).

[83] Venkatakrishnan, V., Perspectives on Unstructured Grid Flow Solvers, AIAA J., Vol. 34, pp. 533–547 (1996).

[84] Voller, V. R., Development and Application of a Heat Balance Integral Method for Analysis of Metallurgical Solidification, Appl. Math. Modelling, Vol. 13, pp. 3–11 (1989).

[85] Voller, V. R. and Swaminathan, C. R., General Source-Based Method for Solidification Phase-Change, Numerical Heat Transfer B, Vol. 19, pp. 175–189 (1991).

[86] Warsi, Z. U. A., *Fluid Dynamics – Theoretical and Computational Approaches*, CRC Press, London (1993).

[87] Watson, D. T., Computing n-Dimensional Delaunay Triangulation with Application to Voronoi Polytopes, Comput. J., Vol. 24, pp. 167–173 (1981).

[88] White, F. M., *Fluid Mechanics*, 2nd Ed., McGraw-Hill, New York (1986).

[89] Wolfshtein, M., Numerical Smearing in One-Sided Difference Approximations to Equations of Non-Viscous Flow, Rep No. EF/TN/A/3, Heat Transfer Section, Mech. Eng. Dept, Imperial College, London (1968).

[90] Zhu, J., A Low Diffusive and Oscillation-Free Convection Scheme, Comm. Appl. Numerical Methods, Vol. 7, pp. 225–232 (1991).

[91] Zhukauskas, A., Heat Transfer from Tubes in Cross Flow, in Hartnett, J. P., and Irvine T. F. (eds.), Advances in Heat Transfer, Vol. 8, Academic Press, New York (1972).

Index